Outstanding Contributions to Logic

Volume 9

More information about this series at http://www.springer.com/series/10033

Geoffrey Hellman · Roy T. Cook
Editors

Hilary Putnam on Logic and Mathematics

Springer

Editors
Geoffrey Hellman
Department of Philosophy
University of Minnesota Twin Cities
Minneapolis, MN, USA

Roy T. Cook
Department of Philosophy
University of Minnesota Twin Cities
Minneapolis, MN, USA

ISSN 2211-2758 ISSN 2211-2766 (electronic)
Outstanding Contributions to Logic
ISBN 978-3-319-96273-3 ISBN 978-3-319-96274-0 (eBook)
https://doi.org/10.1007/978-3-319-96274-0

Library of Congress Control Number: 2018950963

This Springer imprint is published by the registered company Springer Nature Switzerland AG
The registered company address is: Gewerbestrasse 11, 6330 Cham, Switzerland

Preface

At its inception, the plan for this volume was to follow the guidelines for volumes in this series honoring the contributions of distinguished logicians, which specify including replies to the essays by the honored logician, a pattern also followed by the Library of Living Philosophers series. In the present instance, Hilary Putnam, though already in his mid-eighties, was still very active, in fact having just completed his replies to all the essays in the LLP volume dedicated to him, which appeared two years ago. Unfortunately, however, Putnam's health then declined, and this led to his passing before he had the chance to write his replies to the essays of the present volume in all but one case, viz. to Tim McCarthy's essay, "Normativity and Mechanism," which reply is included here following McCarthy's paper. Thus, in the spirit of a memorial volume, we solicited and received reminiscences reflecting on the authors' and others' associations with Hilary; we are publishing these here, following the essays.

As brought out in a moving obituary by Martha Nussbaum (published in the Huffington Post, 3-14-'16), one has to go back to Aristotle to find a philosopher who has contributed so significantly to so many areas of philosophy as has Hilary Putnam, ranging from technical logic and mathematics, through the philosophy of those vast subjects, to philosophy of physics, especially quantum mechanics, philosophy of language, philosophy of mind and psychology, metaphysics, epistemology, ethics, and philosophy of literature. We believe that the essays in the present volume speak to the breadth and depth of Putnam's work in logic, mathematics proper, philosophy of logic, and philosophy of mathematics.

In addition, it is fitting that we honor Hilary for his greatness as a teacher and advisor (as experienced directly by one of us). His courses in philosophy of science, logic, and set theory were high points for many of us. He had a remarkable gift for conveying the essentials of complex, technical materials, in the classes he taught as well as in his publications, with the effect of strongly encouraging and motivating his students. As an adviser, he set the bar high but within reach, and he showed genuine appreciation of students' efforts, treating them more like colleagues engaged in joint inquiry than students at the seat of the master. And, in an era when graduate students were expected to address faculty as "Professor X," Hilary insisted that we call him "Hilary." He was not only our esteemed teacher; he became our true friend.

Minneapolis, USA

Geoffrey Hellman
Roy T. Cook

Contents

Editors and Contributors

About the Editors

Geoffrey Hellman received his AB and Ph.D. (1973) from Harvard. Having published widely in analytic philosophy and philosophy of science, he has, since the 1980s, concentrated on philosophy of quantum mechanics and philosophy and foundations of mathematics, where he has, following the lead of his adviser, Hilary Putnam, developed modal-structural interpretations of mathematical theories, including number theory, analysis, and set theory. He has also worked on predicative foundations of arithmetic (with Solomon Feferman) and pluralism in mathematics (with J. L. Bell). In 2007, he was elected as a Fellow of the American Academy of Arts and Sciences.

Roy T. Cook is CLA Scholar of the College and John M. Dolan Professor of Philosophy. He is the author of *Key Concepts in Philosophy*: *Paradoxes* (2013) and *The Yablo Paradox: An Essay on Circularity*, as well as numerous articles and essays in the philosophy of mathematics, the philosophy of logic, and the aesthetics of popular art.

Contributors

John P. Burgess Department of Philosophy, Princeton University, Princeton, NJ, USA

Roy T. Cook University of Minnesota, Minneapolis, MN, USA

Martin Davis Courant Institute of Mathematical Sciences, New York University, New York, NY, USA; Berkeley, CA, USA

Michael Detlefsen University of Notre Dame, Notre Dame, IN, USA

Harvey M. Friedman The Ohio State University, Columbus, OH, USA

Warren Goldfarb Department of Philosophy, Harvard University, Cambridge, MA, USA

Geoffrey Hellman University of Minnesota, Minneapolis, MN, USA

Kate Hodesdon Department of Philosophy, University of Bristol, Bristol, UK

Akihiro Kanamori Boston University, Boston, MA, USA

Øystein Linnebo University of Oslo, Oslo, Norway

Timothy McCarthy Departments of Philosophy, University of Illinois, Urbana, IL, US

Stewart Shapiro The Ohio State University, Columbus, OH, USA

Chapter 1
Memories of Hilary Putnam

Roy T. Cook and Geoffrey Hellman

Hilary Whitehall Putnam was many things. He was one of the most important American philosophers of the twentieth century, certainly; a lifelong socialist (and occasional Marxist); a convert to Judaism[1]; a native speaker of French[2]; a childhood "friend" of Samuel Beckett, Ford Maddox Ford, and Luigi Pirandello[3]; and much more. But most important, perhaps, is the fact that Putnam was, according to all that knew him, first-and-foremost four things: an excellent teacher, a great colleague, an unparalled scholar in both mathematics and philosophy, and, perhaps most importantly, a good friend to those who were lucky enough to count themselves amongst that select group.

It is traditional in volumes such as this – especially volumes appearing relatively soon after the death of their subject – to include a lengthy, and often rather dry and boring, intellectual biography. In the case of Putnam, such an essay would not, in fact, be boring (although, given the richness of Putnam's life, such an essay might have to be rather lengthy!) But we wanted to do something a little different, and at any rate we knew we couldn't do a better job at biography than what is already in the gripping intellectual autobiography Putnam wrote for his installment in the Library of Living Philosophers series (Auxier et al. 2015).[4] Instead, we thought we would take a slightly different approach, focusing on the roles mentioned above.

[1]Although Putnam's mother was Jewish, his childhood was secular and he only rediscovered Judaism when his son requested a Bar Mizvah.
[2]Although he was born in America in 1926, his family lived in and around Paris during his early childhood and he only learned English upon their return to the United States in 1933.
[3]Putnam was a childhood friend of these literary luminaries in the sense that, as a young child, he often sat on their laps or interacted with them in other ways when they were visiting his father Samuel Putnam, a translator and writer!
[4]Seriously – if you are interested in a detailed account of both Putnam's intellectual work and his extremely interesting life, you really should read the autobiographical essay. It's amazing!

R. T. Cook (✉) · G. Hellman
University of Minnesota, Minneapolis, MN, USA
e-mail: roycookparadox@gmail.com

G. Hellman
e-mail: hellm001@umn.edu

G. Hellman and R. T. Cook (eds.), *Hilary Putnam on Logic and Mathematics*,
Outstanding Contributions to Logic 9, https://doi.org/10.1007/978-3-319-96274-0_1

Actually, we will focus here on Putnam as a teacher, colleague, and friend. No one reading this volume is likely to have doubts regarding Putnam's monumental importance as a mathematician and philosopher. And for the odd bird that does have such doubts, the essays that make up the bulk of this volume will quickly set them straight. But we thought it was worth emphasizing the other, more personal aspects of the positive impact Putnam had on philosophy during his six-plus decade career. And there is no better way to appreciate such positive effects than in the words of those positively affected. Thus, instead of attempting an exhaustive accounting of Putnam's life, we have instead collected a handful of anecdotes that, we think, paint an illuminating portrait of Putnam, the man.

Although there will obviously be overlap in these stories (after all, Putnam, like anyone else, could be more than one of friend, teacher, and colleague at once!), we'll start with some that paint a picture of Putnam the teacher. Warren Goldfarb, who was a student of Putnam's but years later became one of his colleagues at Harvard, notes that although Putnam was far from an easy teacher, he was definitely a teacher who could get students (at least, those who could keep up) excited about the material being taught[5]:

> I first encountered Hilary Putnam in the Fall of 1966, when, as a sophomore math concentrator, I decided to take Philosophy 140, the introduction to logic offered at Harvard. That course was Quine's baby. By then he had taught it for 25 years and would continue to teach it for another 10, and he had written *Methods of Logic* to be its textbook. But Quine was at Oxford that year, and so Hilary filled in. It was his second year at Harvard. Now as Quine taught it, the course was considered pretty tough, particularly since it was a required course for the philosophy concentrators, and many of them had no background in mathematical reasoning and were not comfortable with it. Early on in the course, Hilary announced that the teaching of the actual methods of logic in the book *Methods of Logic* – for example, the paraphrase of ordinary language into logical notation, various routines for the assessment of truth-functional and monadic schemata, and the use of natural deduction for quantification theory, particularly skill at constructing deductions – all those were straightforward enough to be handled simply in the weekly sections by the teaching fellows, and he would not lecture on them. In his lectures, he went on, he would discuss other things. And so, starting the third week of the semester in this introductory logic course, Hilary introduced the semantic paradoxes, discussed Tarski's solution of them, and speculated as to whether a non-hierarchical solution was possible and what it might look like (this was ten years before Kripke's work). Hilary then returned to *Methods* to expound the completeness proof, which is usually the very endpoint of this course. But not for Hilary. For then, in the ensuing weeks, he introduced the idea of effectively decidability, using Emil Post's canonical systems rather than Turing machines, and gave a rigorous proof of the undecidability of the halting problem. That's quite an introductory course, in a philosophy department, for a largely non mathematical audience! The bulk of the students, I have to report, did not fare very happily. But, for me, coming from mathematics, it was a great ride. I decided then and there that logic, in both its technical and philosophical dimensions, was what I wanted to pursue, and before the end of the semester I changed my concentration from math to philosophy. Although I learned a great deal from Hilary through the thirty-five years we were Departmental colleagues, in the end Hilary's greatest influence on me was this: he got me interested.

---

[5] Warren Goldfarb is Walter Beverly Pearson Professor of Modern Mathematics and Mathematical Logic at Harvard University.

Of course, Putnam's impact as a teacher was not limited to those students sitting in his classroom at Harvard – instead, he went out of his way to help students from all over the world and all walks of life. Bahram Assadian illustrates the lengths to which Putnam was willing to go in order to help budding young philosophers[6]:

> In 2006, when I was a philosophy undergraduate at the University of Tehran, I was utterly fascinated by Putnam's philosophy of mathematics. I was sympathetic with mathematical structuralism and also struggling, separately, with questions about modality. Putnam's delightful synthesis of structuralism with modality has been one of the most important moments in my philosophical life. Perhaps, the only thing I could do, and I did, to express my excitement was to translate "Mathematics without Foundations" into Persian. His prose style, humour, and framings were sitting very well with Persian, as if a perfect philosophy of mathematics, in every possible sense, has been introduced into it!
>
> With not much hope of receiving a reply, I emailed Putnam to ask some of my questions about his paper. A couple of days later, he replied with his own personal email. He had answered the questions, informed me that he has returned to the philosophy of mathematics, and expressed his hope of writing a paper in the next six months, so that he may have something new to send me. I wrote him again after six months, and he sent me a draft of his "Indispensability Arguments in the Philosophy of Mathematics", which he had read at the 40th Chapel Hill Colloquium in Philosophy in October 2006.
>
> Six months later, I was checking my inbox and came across an email from Putnam. Initially, I felt that a friend has fooled me, but the truth was that Hilary Putnam had sent me his new paper, "Set Theory: Realism, Replacement and Modality", which he had read for the Paul Benacerraf retirement conference in Princeton. I was most honoured and flattered that I have been in Putnam's brain!
>
> Both papers were later published in 2012, in *Philosophy in an Age of Science*. Although when I started my postgraduate studies, it was overshadowed by neo-Fregeanism, Putnam's modal interpretation of mathematics is still very alive in me, and I owe a pleasant debt to his responsibility and generosity for discussing and sharing his philosophy over email with an undergraduate student.

Putnam's role as a teacher and communicator wasn't limited to teaching or talking about philosophy and mathematics, however. On the contrary, he could hold an audience enthralled speaking on a wide variety of topics (no doubt reflecting the wide range of his own experiences). As Michael Lynch notes, such moments not only inspired interest in the topic under discussion, but also inspired the listener to try to live up to Putnam's example[7]:

> I remember having dinner with Hilary Putnam when he came to Syracuse University around 1994 to give some lectures. I was a graduate student working on a dissertation on realism and truth. Putnam was my philosophical hero and he was incredibly generous with his time during the week he was there, meeting with me and offering me advice and philosophical wisdom. During this particular dinner he told me about traveling to Mexico when he was 18 and going to Diego Rivera's house and getting invited in for dinner by Frida Kahlo. He told me about his discussions with Einstein. Near the end of the dinner he turned to me and remarked on how lucky he was to have met such great minds, and how grateful he was to

[6]A the time of writing this essay Bahram Assadian had just completed his PhD in philosophy at Birkbeck, University of London.

[7]Michael Lynch is Professor of Philosophy and Director of the Humanities Institute at the University of Connecticut - Storrs.

> them for their willingness to talk to a young person. I remember thinking then, as I do now, that I was having the exact same experience talking to Hilary. How lucky I was, and am, for having met him. He was one of the greats.

Putnam's interactions with colleagues were marked by a similar kind of generosity and openness, but they were also marked by Putnam's sense of humor and mischievousness, and sometimes by his radical views both with regard to academic philosophy and with regard to other, more pedestrian matters. For example, Martin Davis, with whom Putnam (along with Julia Robinson and Yuri Matiyasevich) solved Hilbert's Tenth Problem, relates the following pair of anecdotes, the first of which involves Matiyasevich[8]:

> After the work Julia Robinson, Hilary, and I had done on Hilbert's 10th problem, the 22 year-old Yuri Matiyasevich in Leningrad provided the final piece of the puzzle in January 1970. I met Yuri in Nice that summer, and he met Julia in Bucharest the following year. Learning that Hilary was also at the Congress in Bucharest, Yuri wanted to meet him. Yuri relates: "I was told, 'Don't meet Hilary Putnam; Hilary Putnam is a Maoist." '
>
> When a carton of my newly published *Computability & Unsolvability* arrived smelling of printer's ink during the summer of 1958 when Hilary and I were working together, I proudly showed them to him. He offered to find an error on any page. Taking up the challenge, I showed him the reverse side of the title page which was almost blank. Hilary pointed to the word "permission": it was misspelled, missing its second "i".

Aki Kanamori tells another story about the same conference at which Matiyasevich was warned to avoid Putnam – one that also involves Putnam's idiosyncratic political ideology[9]:

> In the summer of 1971, while I was a research student at Cambridge University, I attended the grand sounding Fourth International Congress for Logic, Methodology and the Philosophy of Science at Bucharest. There, I saw Tarski lecturing in a small amphitheatre and musing about set theory that "old soldiers never die, they just fade away", and Kreisel in a panel discussion suddenly roll up some paper and whack a fat man snoring in the front row on his head. What was just as memorable, however, was a sudden remark by Hilary. Sauntering around the city in the fading afternoon, I chanced into a pub on a side street, set into a century-old building of gray splintering wood. There was a small group from the conference having some white beer, and, connecting names with people, I soon realized that I was chatting with Hilary Putnam about mathematical logic. Suddenly, the doors swung open, and a crew of workers came marching in. They were big, ruddy, and every one of them wore denim overalls, broad and seemingly starched. They went up to the bar and started ordering beers, and the sedate atmosphere became one of commotion and all-around talk. Taking all this in, Hilary exclaimed, "It is so wonderful to see workers coming back from work!" Later I would be told that he was coming out of his Marxist activist phase, but at the time I was impressed less by the naïveté and more by a kind of sophistication of the remark, a sudden clarity of thought and purpose amid the details of mathematical logic. Yes, Hilary would turn from one way of thinking to another, from one subject to the next, but there was always enthusiasm and sophistication on the other side of apparent naïveté.

---

[8]Martin Davis is Emeritus Professor at the Courant Institute of Mathematical Sciences at New York University.

[9]Akihiro Kanamori is Professor of Mathematics at Boston University.

Tim McCarthy, who, like Davis, Goldfarb and Kanamori, also contributed an essay to this volume, shared some recollections that continue in this theme of generosity to colleagues, students, and even random audience members, but also clearly illustrate the care and consideration he extended to his friends (and their food)[10]:

> I knew Hilary Putnam in one way or another for almost 40 years; we became good friends in the last 25 years of his life. I first saw him in action at a conference on the occasion of Hempel's retirement from Princeton University, held at Princeton in November, 1975. At this point, I was a third-year graduate student. Many of the luminaries in my own area were present: Quine, Putnam, Davidson, Kripke, and the Princeton contingent including Benacerraf, Lewis, Harman, Rorty and Hempel himself. Even in that company, I thought that weekend, Putnam stood out. He tended to get the better of any argument he embarked on, and was capable of cutting worthy opponents down. But Hilary was also capable of exercising amused restraint in dealing with difficult people, especially in public contexts. As a first-year instructor at Michigan, I helped organize the Tanner Symposium for that year, which included a talk by Putnam to which the public was invited. In the Q&A an Ann Arbor resident asked an utterly unintelligible question that went on for at least three minutes. Hilary stood at the lectern, patiently waiting for the question to come to an end. "Excellent! Excellent!", he replied. "That's the perfect preamble to the question Professor Sklar was going to ask!" Fortunately, Larry Sklar did have a question to ask.
>
> In later years my encounters with Hilary were both philosophical and personal, and frequently included Ruth Anna and my wife Noreen. They also as frequently involved food as a central theme. In the Fall of 1995, Peter Winch and I organized a conference on Wittgenstein in Urbana. Noreen, Peter and I produced several excellent meals for this meeting – a somewhat peculiar emphasis for a conference on a figure who said that he didn't care what he ate as long as it was always the same, and who is reported to have exclaimed "Hot ziggety!" when a peanut butter sandwich was put before him. In any case, one of the culinary high points of the conference came at the conclusion of a dinner at which Noreen produced her signature Tiramisu. Hilary had extracted from her a solemn promise that he was to receive the first piece. When the moment for serving desert arrived, people actually began to pile up at the kitchen door. A woman of her word, my wife cut a sizable serving of cake and presented it to Hilary, much to the consternation of Winch and Stanley Cavell, who were waiting in line. "Why does he get the first slice?", Stanley wanted to know. Noreen replied, "Because he's special!"
>
> Hilary and Ruth Anna were regular visitors to our home in Urbana. One year they were able to join us for Thanksgiving. The two of them spent hours that week playing Upwords with our daughter Johanna. Our turkey preparation that year called for 130 cloves of garlic, peeled – an onerous chore. The day before Thanksgiving, Noreen and I left Hilary and Ruth Anna to run an errand. Returning about an hour later, we found 130 cloves of garlic, peeled and nicely placed into several small dishes, a remarkable occurrence that has ever after in our household been known as the Miracle of the Cloves and the Dishes.

[10]Timothy McCarthy is Professor of Linguistics and of Philosophy at the University of Illinois at Urbana Champaign.

Finally, we'll conclude with one last collection of anecdotes that we believe nicely ties all these themes together, but also illustrate Putnam's lifelong connection to France. These brief stories are recounted by Karine Chemla and Bruno Belhoste[11]:

> Hilary spent the first years of his life in France, first in a small village and then in Paris, where his father, Samuel, was part of a group that counted many key figures of the artistic scene at the time. Hilary liked to evoke his father's depiction of his life in Paris during these years, in *Paris Was Our Mistress: Memoirs of a Lost & Found Generation* (New York: The Viking Press, 1947). Hilary also liked to recall how, as a kid, he had played on Samuel Beckett's knees, and how the journals and the book that his father had co-edited while in Paris (the literary journals like *This Quarter*, and then *The New Review*, and the anthology of European poetry in English translation titled *The European Caravan: An Anthology of the New Spirit in European Literature*, 1931) had published Beckett's first poems. In relation to this, during one of his trips to Paris in the 2000s, we introduced Hilary to Barbara Bray, who had been one of the closest friends of Beckett for decades, and they exchanged souvenirs about him.
>
> We think that this background explains why Paris occupied a very special place in Hilary's mind and heart, and why he and his wife Ruth Anna made sure to bring each of their grandchildren to Paris when they reached the age of 10. Hilary liked to speak and write French, which in fact had been his first spoken language. He also loved to come with Ruth Anna to Paris, where for years they stayed with us in a most relaxed fashion, often on the way between Israel (where they flew to spend warmer winters) and Boston. Hilary and Ruth Anna were thus regularly with us when Spring broke out, and it was a delight (and a lesson) to see them marveling at the new buds in the chestnut tree in front of our building. However simple the accommodation and the meals we could offer might have been (and in times of renovation of the house, comfort verged towards a minimum), Hilary and Ruth Anna would not mind, and shared what we could offer. This does not mean that Hilary did not know what good food was, and did not like it: he devoured croissants and everything that French cuisine could offer, as he devoured art museums, where he taught us a singular and quite amazing movement of the hand to look at paintings, which he had learnt from his childhood.
>
> Hilary read and thought broadly. His interests had no limit that we could identify. He read about the world at large, and he read publications in any kind of discipline. This included anthropology, where he once enthusiastically brought to our attention a reference to an article of Karine's in Mary Douglas's book *Leviticus as Literature* (Oxford UP, 1999) of which we had not been aware. This was typical of Hilary's generosity and his consideration for others.
>
> Hilary also read a great deal of literature. It was with enthusiasm and a manifest emotion that he visited with us the house of Aunt Leonie in Combray, evoked in the first pages of Marcel Proust's *Remembrance of Things Past*. Later on, he would remember this beautiful sunny day and talk about it for years, like on the last occasion when Karine ever saw him, during a visit she paid Hilary and Ruth Anna in Arlington. Perhaps Hilary's father had been to Combray, and perhaps also Proust's novels had had an important place in his family's readings, when Hilary was a boy.
>
> Over the years, we also witnessed Hilary's concern to make the world a better place. This was evident in his concern to improve "human and social diversity" (we prefer this expression for what is commonly called "cultural diversity") in philosophy – to increase diversity both in the topical concerns of philosophers and in the scholars working in philosophy departments. Hilary was also always ready to lend support to scholars from any part of the world when they were facing a hostile audience, and he was equally friendly with and interested in anybody, independently of the interlocutor's status. Karine experienced this when she met Hilary at the

[11]Karine Chemla is a Director of Research at the Centre National de la Recherche Scientifique (CNRS), and Bruno Belhoste is Professor and Director of the Institut d'Histoire Moderne et Contemporaine (IHMC) at Université Paris 1 Panthéon-Sorbonne (Paris 1).

> Wissenschaftskolleg in Berlin in 1994, and this was the starting point of the friendship they formed at the time. When Hilary met Bruno a couple of years later, they also quickly realized the values they shared in common and which became the basis of their close friendship. It was easy to love Hilary, though. In addition to his sense of humor and wit, which maintained a sparkling atmosphere around him, his unforgettable smile (even when he may have wanted to criticize) bespoke his true gentleness and fellow-feeling, which naturally drew people to him.
>
> The last year that Hilary and Ruth Anna visited Paris, Ruth Anna was bothered by the effects of her Parkinson's disease, even though Hilary looked after her every minute. We managed a stalemate while we visited the Orsay museum, with Ruth Anna in a wheelchair. When we took them to the airport to leave Paris, we read in Hilary's slightly curved and loaded back in his deep blue rain coat, when he left us, a sense that he might never return. Unfortunately, he was right.

During the planning stages for this volume, Hilary Putnam had generously agreed to write short essays responding to each of the papers, but unfortunately he passed away on March 13, 2016, before such plans could be carried out. This is, of course, a great philosophical loss for the volume you hold in your hands which, however excellent, would have been all the more insightful with Putnam's incisive comments. But it is a much greater *personal* loss for the editors of this volume (one of whom was advised by Putnam in graduate school; the other, although he had admired Putnam from afar, only got to know him personally during the early stages of this project); for the distinguished scholars who contributed essays for this volume on or connected to to Putnam's work in mathematics and logic; for the generous friends and colleagues who contributed reminiscences to this essay; and for all the other people whose life was changed for the better through knowing Putnam. He will be deeply missed.

Reference

Auxier, R., Anderson, D., & Hahn, L. (2015). *The philosophy of Hilary Putnam: library of living philosophers* (Vol. 34). Chicago: Open Court.

Roy T. Cook is CLA Scholar of the College and John M. Dolan Professor of Philosophy. He is the author of *Key Concepts in Philosophy: Paradoxes* (2013) and *The Yablo Paradox: An Essay on Circularity*, as well as numerous articles and essays in the philosophy of mathematics, the philosophy of logic, and the aesthetics of popular art.

Geoffrey Hellman received his AB and Ph.D. (1973) from Harvard. Having published widely in analytic philosophy and philosophy of science, he has, since the 1980s, concentrated on philosophy of quantum mechanics and philosophy and foundations of mathematics, where he has, following the lead of his adviser, Hilary Putnam, developed modal-structural interpretations of mathematical theories, including number theory, analysis, and set theory. He has also worked on predicative foundations of arithmetic (with Solomon Feferman) and pluralism in mathematics (with J. L. Bell). In 2007, he was elected as a fellow of the American Academy of Arts and Sciences.

Chapter 2
Bibliography of Hilary Putnam's Writings in Logic and Mathematics

This bibliography records Hilary Putnam's contributions to mathematics, logic, their philosophy, and related matters. Needless to say, the border between those works that, strictly speaking, address logic and mathematics, and those that address other areas of philosophy, is fuzzy when dealing with a philosopher as systematic and as wide-ranging as Putnam. Thus, we have had to make some decisions – decisions that might not be the same decisions that the reader would have made if given the same task. For example, we have included many (but not all) of the more philosophical papers on truth, but have only included papers on confirmation when they directly address mathematical or logical, rather than scientific, issues and concerns. Nevertheless, we hope that this bibliography will be useful to readers interested in Putnam's extensive and important work in mathematics and logic.

We have given original publication information, and (with the exception of Putnam's PhD dissertation) noted reprints only when they occur in collections of Putnam's essays also included on this list (i.e. those that contain a significant number of papers on mathematics and logic).

(1951) *The Meaning of the Concept of Probability in Application to Finite Sequences*, PhD Dissertation, University of California – Los Angeles, 1951. Reprinted New York: Garland, 1991; London, Routledge: 2011.

(1956) "Mathematics and the Existence of Abstract Entities", *Philosophical Studies* 7(6): 81–88.

(1957a) "Arithmetic Models for Consistent Formulae of Quantification Theory", *Journal of Symbolic Logic* 22(1): 110–111.

(1957b) "Decidability and Essential Undecidability", *Journal of Symbolic Logic* 22(1): 39–54.

(1957c) "Eine Unableitbarkeitsbeweismethode für den Intuitionistischen Aussagenkalkül" (w/ George Kreisel), *Archiv für Mathematische Logik und Grundlagenforschung* 3(1–2): 74–78.

G. Hellman and R. T. Cook (eds.), *Hilary Putnam on Logic and Mathematics*, Outstanding Contributions to Logic 9, https://doi.org/10.1007/978-3-319-96274-0_2

(1957d) "Review of Hughes Leblac", *An Introduction to Deductive Logic*, *Philosophical Review* 66(4): 551–554.

(1957e) "Three-Valued Logic", *Philosophical Studies* 8(5): 73–80. Reprinted in (1975a): 166–173.

(1958a) "Elementary Logic and Foundations of Set Theory", in *Philosophy in the Mid-Century* (Raymond Klibansky, ed.), Florence: La Nuova Italia Editrice: 56–61.

(1958b) "Feasible Computational Methods in the Propositional Calculus" (w/ Martin Davis), Troy, NY: Rensselaer Polytechnical Institute, Research Division.

(1958c) "Formalization of the Concept 'About'", *Philosophy of Science* 25(2): 125–130.

(1958d) "Reduction of Hilbert's Tenth Problem", (w/ Martin Davis), *Journal of Symbolic Logic* 23(2): 183–187.

(1959) "Review of Raphael Robinson, "Arithmetical Representation of Recursively Enumerable Sets"", *Journal of Symbolic Logic* 24(2): 170–171.

(1960a) "A Computing Procedure for Quantification Theory" (w/ Martin Davis), *Journal of the Association of Computing Machinery* 7(3): 201–215.

(1960b) "An Unsolvable Problem in Number Theory", *Journal of Symbolic Logic* 25(3): 220–232.

(1960c) "Exact Separation of Recursively Enumerable Sets Within Theories" (w/ Raymond Smullyan), *Proceedings of the American Mathematical Society* 11(4): 574–577.

(1960d) "Minds and Machines", in *Dimensions of Mind*, (Sydney Hook, ed.), New York: New York University Press: 138–164. Reprinted in (1975b): 362–385.

(1960e) "Review of Ernest Nagel and James E. Newman, *Gödel's Proof*", *Philosophy of Science* 27(2): 205–207.

(1961a) "The Decision Problem for Exponential Diophantine Equations" (w/ Martin Davis & Julia Robinson), *Annals of Mathematics* 74(3): 425–436.

(1961b) "Some Issues in the Theory of Grammar", in *The Structure of Language and Its Mathematical Aspects: Proceedings of a Symposium in Applied Mathematics* (Roman Jakobson ed.), Providence, Rhode Island: American Mathematical Society: 25–42. Reprinted in (1975b): 85–106.

(1961c) "Uniqueness Ordinals in Higher Constructive Number Classes", in *Essays on the Foundations of Mathematics Dedicated to A. A. Fraenkel on his Seventieth Anniversary* (Yoshua Bar-Hillel et alia, eds.) Jerusalem: The Hebrew University Magness Press: 190–206.

(1962a) "Dreaming and 'Depth Grammar'", in *Analytical Philosophy, First Series* (R. J. Butler, ed.), Oxford: Basil Blackwell: 211–235. Reprinted in (1975b): 304–324.

(1962b) "On Families of Sets Represented in Theories", *Archiv für Mathematische Logik und Grundlagenforschung* 6(1–2): 66–70.

(1962c) "Review of Hakan Törnebohm 'On Two Logical Systems Proposed in the Philosophy of Quantum-mechanics'", *Journal of Symbolic Logic* 27(1): 115.

(1963a) "A Note on Constructible Sets of Integers", *Notre Dame Journal of Formal Logic* 4(4): 270–273.

(1963b) "An Examination of Grünbaum's Philosophy of Geometry", in *Philosophy of Science: The Delaware Seminar Volume 2, 1962–1963*, (Bernard Baumrin, ed.), New York: John Wiley: 205–255.

(1963c) "'Degree of Confirmation' and Inductive Logic", in *The Philosophy of Rudolph Carnap* (Paul A. Schilpp, ed.): 761–783. Reprinted in (1975a): 270–292.

(1963d) "Diophantine Sets Over Polynomial Rings", (w/ Martin Davis), *Illinios Journal of Mathematics* 7(2): 251–256.

(1963e) "Probability and Confirmation", *The Voice of America Forum Lectures: Philosophy of Science Series* Number 10: 1–11. Reprinted in (1975a): 293–304.

(1963f) "Review of Georg Henrik von Wright, *Logical Studies*", *Philosophical Review* 72(2): 242–249.

(1964a) *Philosophy of Mathematics: Selected Readings* (ed., w/ Paul Benacerraf), Englewood Cliffs, New Jersey: Prentice Hall. 2nd Edition 1983, Cambridge UK: Cambridge University Press.

(1964b) "On Hierarchies and Systems of Notations", *Proceedings of the American Mathematical Society* 15(1): 44–50.

(1965a) "Craig's Theorem", *Journal of Philosophy* 62(10): 251–260. Reprinted in (1975a): 228–236.

(1965b) "On Minimal and Almost-Minimal Systems of Notation" (w/ David Luckham), *Transactions of the American Mathematical Society* 119(1): 86–100.

(1965c) "On the Notational Independence of Various Hierarchies of Degrees of Unsolvability" (w/ Gustav Hensel), *Journal of Symbolic Logic* 30(1): 69–86.

(1965d) "Recursively Enumerable Classes and Their Application to Recursive Sequences of Formal Theories", (w/ Marian Boykan Pour-el), *Archiv für Mathematische Logik und Grundlagenforschung* 8(3–4): 104–121.

(1965e) "Trial and Error Predicates and the Solution to a Problem of Mostowski", *Journal of Symbolic Logic* 30(1): 49–57.

(1967a) "The Craig Interpolation Lemma" (w/ Burton Dreben), *Notre Dame Journal of Formal Logic* 8(3): 229–233.

(1967b) "Mathematics without Foundations", *Journal of Philosophy* 64(1): 5–22. Reprinted in (1975a): 43–59. Reprinted in (1983b): 295–313.

(1967c) "The Thesis that Mathematics is Logic", in *Bertrand Russell: Philosopher of the Century* (Ralph Schoenman ed.), London: Allen & Unwin: 273–303. Reprinted in (1975a): 12–42.

(1967d) "Time and Physical Geometry", *Journal of Philosophy* 64(8): 240–247. Reprinted in (1975a): 198–205.

(1968a) "Degrees of Unsolvability of Constructible Sets of Integers", (w/ George Boolos), *Journal of Symbolic Logic* 33(4): 497–513.

(1968b) "Is Logic Empirical?", in *Boston Studies in the Philosophy of Science* Volume 5 (Robert Cohen & Marx Wartofsky eds.), Dordrecht: D. Reidel: 216–241. Reprinted as "The Logic of Quantum Mechanics" in (1975a): 174–197.

(1969a) "Normal Models and the Field of Σ_1" (w/ Gustav Hensel) *Fundamentae Mathematicae* 64(2): 231–240.

(1969b) "A Recursion-theoretic Characterization of the Ramified Analytic Hierarchy", (w/ Gustav Hensel & Richard Boyd), *Transactions of the American Mathematical Society* 141: 37–62.

(1970) "A Note on the Hyperarithmetical Hierarchy", (w/ Herbert Enderton), *Journal of Symbolic Logic* 35(3): 429–430.

(1971a) *Philosophy of Logic*, New York: Harper and Row. Reprinted in (1975a), 2nd edition: 323–357.

(1971b) "An Intrinsic Characterization of the Hierarchy of Constructible Sets of Integers" (w/ Stephen Leeds), in *Logic Colloquium '69* (Robin Grandy & Charles Yates eds.), Amsterdam: North Holland: 311–350.

(1973) "Recursive Functions and Hierarchies", *American Mathematical Monthly: Supplement: Papers in the Foundations of Mathematics* 80(6): 68–86.

(1974a) "How to Think Quantum-Logically", *Synthese* 29(1–4): 55–61.

(1974b) "Solution to a Problem of Gandy's" (w/ Stephen Leeds), *Fundamentae Mathematica* 81(2): 99–106.

(1974c) "Systems of Notations and the Ramified Analytic Hierarchy" (w/ Joan Lukas), *Journal of Symbolic Logic* 39(2): 243–253.

(1975a) *Mathematics, Matter, and Method: Philosophical Papers, Volume 1*, Cambridge MA: Cambridge University Press. 2nd Edition 1979, Cambridge UK: Cambridge University Press.

(1975a) *Mind, Language, and Reality: Philosophical Papers, Volume 2*, Cambridge MA: Cambridge University Press.

(1975c) "What is Mathematical Truth?", *Historia Mathematica* 2(4): 529–533. Reprinted in (1975a): 60–78.

(1978) "Quantum Logic, Conditional Probability, and Interference", (w/ Michael Friedman), *Dialectica* 32(3–4): 305–315.

(1979) "Philosophy of Mathematics: A Report", *Current Research in Philosophy of Science: Proceedings of the P.S.A. Critical Research Problems Conference* (Peter Asquith & Henry Kyburg eds.), East Lansing MI: Philosophy of Science Association: 386–398. Reprinted as "Philosophy of Mathematics: Why Nothing Works" in (1994a): 499–512.

(1980) "Models and Reality", *Journal of Symbolic Logic* 45(3): 464–482. Reprinted in (1983a): 1–25. Reprinted in (1983b): 421–445.

(1982) "Peirce the Logician", *Historia Mathematica* 9(3): 290–301. Reprinted in (1990): 252–260.

(1983) "Vagueness and Alternative Logic", *Erkenntnis* 19(1–3): 297–314. Reprinted in (1983a): 271–285.

(1984) "Proof and Experience", *Proceedings of the American Philosophical Society* 128(1): 31–34.

(1983a) *Realism and Reason: Philosophical Papers, Volume 3*, Cambridge MA: Cambridge University Press.

(1983b) *Philosophy of Mathematics: Selected Readings* (ed., w/ Paul Benacerraf), 2nd Edition 1983, Cambridge UK: Cambridge University Press.

(1989) "Model Theory and the 'Factuality' of Semantics", in *Reflections on Chomsky* (Alex George ed.), Oxford: Basil Blackwell: 213–232. Reprinted in (1994a): 351–375.

(1990) *Realism with a Human Face*, James Conant (ed.), Cambridge MA: Cambridge University Press.

(1991) "Does the Disquotational Theory Really Solve All Philosophical Problems?", *Metaphilosophy* 22(1–2): 1–13. Reprinted as "Does the Disquotational Theory of Truth Really Solve All Philosophical Problems?" in (1994a): 264–278.

(1992a) "Comments on the Lectures", *Reasoning and the Logic of Things*, Charles Sanders Peirce (Kenneth Ketner ed.), Cambridge: MA: Harvard University Press: 1–54.

(1992b) "Introduction: The Consequence of Mathematics", *Reasoning and the Logic of Things*, Charles Sanders Peirce (Kenneth Ketner ed.), Cambridge: MA: Harvard University Press: 1–54.

(1994a) *Words and Life*, James Conant (ed.), Cambridge MA: Harvard University Press.

(1994b) "Afterthoughts on 'Models and Reality'", *Diálogos* 29(63): 7–39.

(1994c) "Logic and Psychology", in *The Logical Foundations of Cognition: Vancouver Studies in Cognitive Science* Volume 4 (John Macnamara & Gonzalo Reyes eds.), Oxford: Oxford University Press: 35–42.

(1994d) "Mathematical Necessity Reconsidered", in *On Quine: New Essays* (Paulo Leonardi & Marco Santambrogio eds.), Cambridge UK: Cambridge University Press: 267–282. Reprinted as "Rethinking Mathematical Necessity" in (1994a): 245–263.

(1995) "Peirce's Continuum", in *Peirce and Contemporary Thought: Philosophical Inquiries* (Kenneth Ketner ed.), New York: Fordham Press: 1–22.

(1996) "On Wittgenstein's Philosophy of Mathematics", *Proceedings of the Aristotelian Society* 70 Supplement: 243–264.

(1997a) "James' Theory of Truth", in *The Cambridge Companion to William James* (Ruth Putnam, ed.), Cambridge UK: Cambridge University Press: 166–185.

(2000a) "A Note on Wittgenstein's 'Notorious Paragraph' About the Gödel Theorem" (w/ Juliet Floyd) *Journal of Philosophy* 97(11): 624–632.

(2000b) "Nonstandard Models and Kripke's Proof of the Gödel Theorem", *Notre Dame Journal of Formal Logic* 41(1): 53–58. Reprinted in (2012a): 263–269.

(2000c) "Paradox Revisited I: Truth", in *Between Logic and Intuition: Essays in Honor of Charles Parsons* (Gila Sher & Richard Tieszen eds.), Cambridge UK: Cambridge University Press: 3–15. Reprinted as "Revisiting the Liar Paradox" in (2012a): 202–215.

(2000c) "Paradox Revisited II: Sets", in *Between Logic and Intuition: Essays in Honor of Charles Parsons* (Gila Sher & Richard Tieszen eds.), Cambridge UK: Cambridge University Press. 16–26.

(2001) "Was Wittgenstein *Really* an Anti-Realist about Mathematics?", *Philosophical Explorations* 4(1): 2–16. An Expanded version of (1996). Reprinted in (2012a): 495–513.

(2005) "James on Truth (Again)", in *William James and the Varieties of Religious Experience: A Centenary Celebration* (Jeremy Carrette ed.), London & New York: Routledge: 172–182.

(2006a) "After Gödel", *Logic Journal of the IGPL* 14(5): 745–759. Reprinted in (2012a): 256–262.

(2006b) "Bays, Steiner, and Wittgenstein's 'Notorious' Paragraph about the Gödel Theorem" (w/ Juliet Floyd), *Journal of Philosophy* 103(2): 101–110.

(2007) "Wittgenstein and the Real Numbers", in *Wittgenstein and the Moral Life* (Alice Crary ed.), Cambridge MA: MIT Press: 235–250. Reprinted in (2012a): 442–457.

(2008) "A Note on Steiner on Wittgenstein, Gödel, and Tarski", *Proceedings and Addresses of the American Philosophical Association* 82(2): 101–115.

(2010) "Between Dolev and Dummett: Some Comments on Antirealism, Presentism, and Bivalence", *International Journal of Philosophical Studies* 18(1): 91–96.

(2011) "The Gödel Theorem and Human Nature", in *Kurt Gödel and the Foundations of Mathematics: Horizons of Truth* (Matthias Baaz, ed.) Cambridge UK: Cambridge University Press. Reprinted in (2012a): 239–255.

(2012a) *Philosophy in an Age of Science: Physics, Mathematics, and Skepticism*, (Mario De Caro & David MacArthur eds.), Cambridge MA: Cambridge University Press.

(2012b) "On Mathematics, Realism, and Ethics", *Harvard Review of Philosophy* 18(1): 143–160.

Part I
Logic and the Philosophy of Logic

Chapter 3
Logic, Counterexamples, and Translation

Roy T. Cook

Abstract In "Is Logic Empirical" (Putnam 1968), Putnam formulates an empirical argument against classical logic—in particular, an apparent counterexample to the distributivity laws. He argues further that this argument is also an argument in favor of quantum logic. Here we challenge this second conclusion, arguing instead that counterexamples in logic are counterexamples not to particular inferences, but to logics as a whole. The key insight underlying this argument is that what counts as a legitimate translation from natural language to formal language is dependent on the background logic being assumed. Hence, in the face of a counterexample, one can move to a logic that fails to validate the inference seemingly counter-instanced, or one can move to a logic where the best translation of the natural language claims involved in the counterexample are no longer best translated as an instance of the inference in question.

3.1 Introduction

In "Is Logic Empirical?" (Putnam 1968), Hilary Putnam formulates a now-famous, *empirical* argument against classical logic.[1] The argument hinges on the fact that quantum mechanics seems to provide examples where at least one of the standard distributivity laws:

$$A \wedge (B_1 \vee B_2) \dashv\vdash_{\mathsf{C}} (A \wedge B_1) \vee (A \wedge B_2)$$
$$A \vee (B_1 \wedge B_2) \dashv\vdash_{\mathsf{C}} (A \vee B_1) \wedge (A \vee B_2)$$

[1]Although arguments for the in-principle possibility of purely empirical challenges to the correctness of a logic – classical or otherwise – go back to Quine (1951), Putnam was amongst the first to provide a serious putative example of such an empirical challenge.

R. T. Cook (✉)
University of Minnesota, Minneapolis, MN, USA
e-mail: roycookparadox@gmail.com

G. Hellman and R. T. Cook (eds.), *Hilary Putnam on Logic and Mathematics*, Outstanding Contributions to Logic 9, https://doi.org/10.1007/978-3-319-96274-0_3

(in particular, the left-to-right direction of the first equivalence) fails. As a result, Putnam suggests we ought to abandon classical logic, and adopt quantum logic as the correct account of logical consequence.

The purpose of the present essay is not to challenge the correctness of Putnam's conclusion – that is, that quantum logic is the one, true, correct account of logical consequence.[2] Nothing in the sections to follow implies that Putnam got that bit *wrong*.[3] Instead, the worry that I will develop here concerns the methodology by which that conclusion was reached. In short, Putnam has shown, at best, that quantum logic is one amongst a number of logics that we might adopt in the face of the recalcitrant evidence (seemingly) provided by quantum mechanics.

Thus, I am going to grant without argument or critical examination many of the claims that Putnam spends a great deal of ink discussing – claims that have also played central roles in much of the discussion of "Is Logic Empirical?". In particular, I will assume without argument that:

1. Our account of the correct logic(s) should be sensitive to the empirical evidence provided by our best science.
2. Quantum mechanics provides sufficient evidence for abandoning classical logic – that is, it provides a counterexample to classical logic.

It is perhaps worth noting, as a matter of autobiographical detail, that I am extremely sympathetic to the first claim – and to the more general idea that logical theorizing should be attentive to all sorts of considerations in addition to those associated with the armchair. The second claim strikes me as somewhat more questionable, however.[4] But, since we have other logical fish to methodologically fry, I shall assume both claims throughout what follows.

---

[2]Of course, I personally don't believe that there is *one* correct logic (i.e. I am a logical pluralist of some sort), nor do even I believe that quantum logic is a plausible candidate for being one of the multitude of 'correct' or 'best' logics – see Cook (2014). The point is that nothing in the present paper depends on these further views.

[3]There are, of course, a number of extant criticisms of his argument along these lines. There are four primary themes running through such criticisms. First, there are objections to the role that realism plays in Putnam's argument – for a prominent example of this sort, see Dummett (1976). Second, there are objections to Putnam's presentation of the physics, the logic, or the connections between the two – for a recent example of this sort, see Maudlin (2005). Third, there are objections to the claim that formal quantum logic – that is, the propositional/first-order theory obtained via constructing a semantics in terms of the lattice of 'quantum propositions' – blocks the problematic inferences anyway. Gardner (1971) and Gibbons (1987) are notable examples of this approach. Finally, there is the claim – forcefully argued for in Hellman (1980) – that the language and logic within which quantum mechanics is formulated is fully classical, and hence shifting to a different logic once these puzzles arise (regardless of whether, *pace* Gardner (1971) and Gibbons (1987), such a logic actually blocks the problematic inferences) amounts to ignoring the problem rather than addressing it. The present essay will address none of these specific concerns.

[4]Of course, a fully sufficient examination of whether or not quantum mechanics really does provide counterexamples to classical logic would require a significantly deeper understanding of the relevant science than I possess. Fortunately, as I have already emphasized, the point of this paper, which really concerns the methodology of logic rather than that of science, does not depend on answering this distinct question.

Accepting, even merely for the sake of argument, that quantum mechanics provides a counterexample to classical logic is one thing, however. Understanding what such a counterexample might amount to, however, and in particular, what such a counterexample tells us about the identity of the correct logic (or logics), is something else. The main goal of this essay is to provide an examination of the structure and methodology of Putnam's argument for logical revision that is a good bit more careful and more complete than previous treatments. The previous sentence might seem a bit overstated if one does not attend to the wording carefully: as already emphasized, the point here is not to determine whether or not Putnam is right about quantum logic being the one true correct logic – a point that *has* been debated extensively in the literature. Instead, the focus here will be on the structure of Putnam's argument, and arguments like it, that take apparent counterexamples to classical logic as premises and arrive at revisions to classical logic as conclusion. In short, the topic at issue here involves the *general strategy* underlying Putnam's argument, and arguments like it. I am only interested (in the present essay, at least) on examining the methodology of logical revision, and not on the particular logic one might arrive at by correctly applying such methodology.[5]

Now that we are clear about what we are going to grant to Putnam (and, more generally, to philosophers, mathematicians, and logicians engaged in relevantly similar arguments for logical revision based on apparent counterexamples – empirical or not – to inferences previously accepted as valid), it is time to clearly identify where, exactly, we are going to disagree with Putnam. As already noted, we are going to grant for the sake of argument that quantum mechanics provides a counterexample to classical logic. What will be denied, however, is the following third claim:

3. Quantum mechanics provides sufficient evidence for abandoning the laws of distributivity – that is, it provides a counterexample to one of the laws of distributivity.

At first glance this third claim might seem nearly synonymous, in the present context, to the second claim above. After all, isn't the point of Putnam's argument to show that classical logic is not the correct logic (at least, for reasoning about the quantum realm) by showing that one of the distributivity laws fails when reasoning about quantum mechanics?

In fact, Putnam has not provided a counterexample to the distributivity laws. The reason, as we shall see, is simple (although teasing out the subtleties involved will take some work, of course): it is, in fact, *impossible* to provide a counterexample to any logical law *tout court*. Rather, counterexamples such as Putnam's quantum mechanical example do not show that individual inferences are invalid – rather, they show that individual inferences *understood from the perspective of a particular background logic* are invalid. As a result, it is not the individual inference that is impugned by a counterexample, but the background logic against which this inference is judged.

[5]Of course, it goes without saying that any novel conclusions we arrive at with respect to the correct methodology for dealing with purported counterexamples to our favored logics will have very real consequences for what candidate logics we might take seriously as correct or (if one has pluralist leanings) legitimate. The point is merely that we are not focusing here on answering the latter question.

As a result, when confronted with a purported counterexample C to our favored logic $\mathcal{L}$ (where Putnam's quantum mechanical example is a paradigm instance of such a C) we find ourselves in a situation where:

- Counterexample C shows that logic $\mathcal{L}$ is not the correct logic.
- Counterexample C does not show that any *particular* inference in $\mathcal{L}$ is incorrect.

In fact, as we shall see, this odd pair of claims is not merely a correct description of the situation in Putnam's quantum mechanical counterexample, but is the right way to view any counterexample (empirical or not) to any logic that is a candidate for correctness.

This essay proceeds as follows: In Sect. 3.2 we shall briefly set up a version of the famous double-slit experiment which provides a version of Putnam's celebrated counterexample, and then rehearse a simplified version of Putnam's argument against classical logic, and for quantum logic. Then, in Sect. 3.3, we shall look a bit more closely at the structure of Putnam's argument, and identify an un- (or at least under-) appreciated flexibility in arguments of this sort for logical revision. In Sect. 3.4 we will then show how to take advantage of this flexibility in order to formulate an alternative solution to Putnam's quantum puzzle. Finally, in Sect. 3.5 we shall tie up some loose ends, and show how the work of the previous sections shows us that, briefly put, counterexamples are never counterexamples to particular inference patterns, but are instead always counterexamples to a logic as a whole. Finally, in Sect. 3.6, we shall draw some more general conclusions about what, exactly, is involved in choosing one formal logic over another as 'correct', emphasizing that such a choice involves much more than merely settling on a particular set of rules delineating which conclusions follow from which premises.

3.2 Logic and Quantum Mechanics

The simple example with which we shall begin – which I am merely using as a rough illustration, and thus will describe only briefly and somewhat simplistically – is the famous double-slit experiment. In this experiment, photons are projected so that they pass through a plate with two slits cut into it and then collide with a detection screen. When the photons are projected through the plate without any observation regarding the slit through which they passed, the resulting pattern of impacts on the detection screen displays an *interference pattern* associated with wavelike behavior, and seemingly incompatible with each photon having traveled particle-like through exactly one or the other of the slits.

Given this (admittedly rather informal) description of the double-slit experiment, assume that we fire some photons, one-at-a-time, through the apparatus and we observe the expected interference pattern. Then, letting p be any one of the photons, the following seems to be a true claim:

> p impacted the detection screen at location λ, and either p passed through the first slit, or p passed through the second slit.

The first conjunct is true by assumption, and we can argue for the second conjunct as follows: Assume that p did not pass through either slit – that is, p did not pass through the first slit, and p did not pass through the second slit. Then p could not have reached the detection screen. But it did. Contradiction, so by *reductio* it is not the case that p did not pass through either slit, and by *DeMorgan's Law* this implies the disjunction in question.

Adopting the following translation manual:

$$A =_{df} p \text{ impacted the detection screen at location } \lambda.$$
$$B_1 =_{df} p \text{ passed through the first slit.}$$
$$B_2 =_{df} p \text{ passed through the second slit.}$$

we can formalize this as:

$$A \wedge (B_1 \vee B_2)$$

In the same situation, however, the following is not true.

> Either p impacted the detection screen at location λ and p passed through the first slit, or p impacted the detection screen at location λ and p passed through the second slit.

After all, if this claim were true, then either the first disjunct is true, or the second disjunct is true. Without loss of generality, assume it is the first disjunct that is true. Then p impacted the detection plate at location λ and p (definitely) passed through the first slit. But if p (definitely) passed through the first slit, then it (definitely?) did not pass through the second slit. But then (since if the reasoning held for p, it should generalize to all of the particles that pass through the plate and impact the screen) no interference pattern should be present.

We can formalize this second claim, using the same translation manual, as:

$$(A \wedge B_1) \vee (A \wedge B_2)$$

It is worth emphasizing that the formulation of the argument just given appears to be an argument that this claim is *false*. Of course, in the present context – classical logic with classical semantics – there is no distinction between a claim being false and it merely failing to be true. Once we move from the classical context to the quantum logical or intuitionistic context, however, there is room to distinguish between a claim merely failing to be true, and its being false. Thus, we shall return to examine the status of the offset disjunction above in more detail below (and the rather conspicuously marked "definitely?" will be important in that discussion).

Returning to our observations regarding the double-slit experiment, the upshot becomes clear when we observe that the relevant instance of the distributivity law is classically valid – that is:

$$A \wedge (B_1 \vee B_2) \vdash_{\mathsf{C}} (A \wedge B_1) \vee (A \wedge B_2)$$

Thus, if quantum mechanics does indeed tell us that the premise of this argument is true, and the conclusion is false (or at the very least, fails to be true), then we are forced to abandon classical logic.

Note that this much is exactly what was highlighted at the beginning of this essay as the, for-present-purposes and for-the-sake-of-argument, assumptions that will be granted to Putnam: physics (or any science), in general, can provide data relevant to the correctness of a particular logic, and quantum mechanics, in particular, provides a counterexample to classical logic. Thus, I shall not challenge any of the story sketched above, although we shall return to examine bits of the story in more detail later on.[6]

Putnam's argument, loosely put, then proceeds as follows (his discussion of the physics, as already noted, is of course much more sophisticated than the discussion above): All of this is okay, since we have at the ready an alternative logic – quantum logic – that does not validate the problematic instance of distributivity. In short:

$$A \wedge (B_1 \vee B_2) \nvdash_{\mathsf{Q}} (A \wedge B_1) \vee (A \wedge B_2)$$

Furthermore, quantum logic is independently well-motivated, built as it is 'on top' of the mathematical structures of quantum mechanics (loosely put, we substitute quantum structures for classical truth tables or models). As a result, we ought to abandon classical logic in favor of quantum logic.

But should we? As per the assumptions granted here, we certainly need to abandon classical logic in favor of *some* other logic. One might think that since it seems to be the distributivity law that is causing all the trouble, it must follow that we need to move to a logic that fails to validate the relevant instance of distributivity. Since quantum logic lacks exactly this law (and is motivated by the science at issue), this would seem to provide strong (albeit defeasible) evidence that we should adopt quantum logic.

It turns out, however, that we need not move to a new logic that fails to validate a particular logical law, just because *from the context of our current logic* we have a counterexample to that law. It turns out that an empirical situation that provides a counterexample to a particular logical law from within the context of one logical framework might not provide a counterexample to that very same law from within the context of a different logical framework. In order to show this, we need to attend a bit more closely to the structure of Putnam's argument (and arguments like it) for logical revision – a task to which we now turn.

[6]The fact that we are granting all of this for the sake of argument also, it is hoped, goes some ways towards excusing the looseness of the science in this section!

3.3 The Argument for Empirical Logic Revision

In general terms, we can sum up the general pattern of reasoning found in Putnam's argument for quantum logic as an instance of the following schema, which we shall call the *Flawed Argument for Revising Logic* (or FARL):

The (Flawed) Argument for Revising Logic:

(*Prem*$_1$) We have evidence in favor of accepting natural language claim $\Phi_{\mathcal{NL}}$.
(*Prem*$_2$) We have evidence in favor of rejecting natural language claim $\Psi_{\mathcal{NL}}$.

(*Prem*$_3$) Within the context of our current formal logic $\mathcal{FL}_1$, $\Phi_{\mathcal{NL}}$ is best formalized as $\Phi_{\mathcal{FL}_1}$.
(*Prem*$_4$) Within the context of our current formal logic $\mathcal{FL}_1$, $\Psi_{\mathcal{NL}}$ is best formalized as $\Psi_{\mathcal{FL}_1}$.
(*Prem*$_5$) The argument from $\Phi_{\mathcal{FL}_1}$ to $\Psi_{\mathcal{FL}_1}$ is valid in our current formal logic $\mathcal{FL}_1$, that is:

$$\Phi_{\mathcal{FL}_1} \vdash_{\mathcal{FL}_1} \Psi_{\mathcal{FL}_1}$$

(*Conc*) We should abandon formal logic $\mathcal{FL}_1$ in favor of a weaker (or at least different) logic $\mathcal{FL}_2$ where:

$$\Phi_{\mathcal{FL}_1} \nvdash_{\mathcal{FL}_2} \Psi_{\mathcal{FL}_1}$$

If we fill in some of the schematic variables in FARL, however, a point of potential resistance becomes obvious. In particular, the conclusion becomes:

We should abandon classical logic (C) in favor of quantum logic (Q) where:

$$\Phi_{\mathsf{C}} \nvdash_{\mathcal{Q}} \Psi_{\mathsf{C}}$$

This, however, seems to get things exactly wrong: Surely, when we are evaluating Q as a replacement for C in terms of its formalizing the problematic inference from the natural language claim $\Phi_{\mathcal{NL}}$ to the natural language claim $\Psi_{\mathcal{NL}}$, we should be using the best formalization of these claims with respect to Q, not with respect to C. In short, the conclusion of the argument should read.

(*Conc**) We should abandon formal logic $\mathcal{FL}_1$ in favor of a weaker (or at least different) logic $\mathcal{FL}_2$ where:

$$\Phi_{\mathcal{FL}_2} \nvdash_{\mathcal{FL}_2} \Psi_{\mathcal{FL}_2}$$

(and where $\Phi_{\mathcal{FL}_2}$ and $\Psi_{\mathcal{FL}_2}$ are the best formalizations of $\Phi_{\mathcal{NL}}$ and $\Psi_{\mathcal{NL}}$, respectively, within the context of $\mathcal{FL}_2$.)

On this, improved version of the argument, which we shall henceforth call the *Corrected Argument for Revising Logic* (or CARL), the relevant instance of the conclusion becomes:

We should abandon classical logic (C) in favor of quantum logic (Q) where:

$$\Phi_{\mathsf{Q}} \nvdash_{\mathcal{Q}} \Psi_{\mathsf{Q}}$$

If this is right, then at this point the defender of quantum logic owes us an argument that the best translation of the natural language sentences ($\Phi_{\mathcal{NL}}$ and $\Psi_{\mathcal{NL}}$) involved in the empirical counterexample to the relevant distributive law – that is, Φ_{Q} and Ψ_{Q} – are indeed identical to the best formalizations of these claims in classical logic – that is, Φ_{C} and Ψ_{C}. Let us grant this point, however.[7]

The real issue is this: On the corrected version of the argument, we should not just *assume* that the formal argument involved in the counterexample to our original logic $\mathcal{FL}_1$ must turn out to be invalid in the new logic $\mathcal{FL}_2$ chosen to replace $\mathcal{FL}_1$, since we need not translate the relevant natural language sentences the same way in the two distinct logical contexts. Instead, all we can *assume* is that either (1) the formal argument involved in the counterexample to $\mathcal{FL}_1$ will be invalid in $\mathcal{FL}_2$, or (2) the best translation of the premise(s) and conclusion of the counterexample to $\mathcal{FL}_1$ will, in the context of $\mathcal{FL}_2$, be distinct from the best translations of these natural language sentences from the perspective of $\mathcal{FL}_1$.

Of course, the *simplest*, *easiest* way to apply CARL, and obtain a new logic $\mathcal{FL}_2$ that satisfies the constraints laid out in the conclusion, is to select a logic where the best formalization of the claims in question in $\mathcal{FL}_1$ and $\mathcal{FL}_2$ are identical, and the inference in question is invalid in $\mathcal{FL}_2$. But this is not the only option – and likely not always the best option – since it might turn out that the claims involved in the counterexample *ought* to be formalized differently. In short, when revising logics, simple and easy don't always equate to correct.

This is, really, the central point of the present essay: Logicians often contrast and comparatively evaluate different logics by taking a particular natural language argument and then asking whether *the* formalization of that argument is or is not valid in the various logics in question. This obscures the fact that a single natural language argument might not have a single, logically neutral, correct formalization – instead, what counts as the best formalization of that argument may vary from logical framework to logical framework. We will return to this general point in Sect. 3.6 below, but before doing so let's look at a simple example – one simpler, even, than Putnam's quantum mechanical example – in some detail. Doing so will provide us with basic insights into the general phenomenon at issue, and will also provide some formal tools that will be useful when we return to Putnam's example.

Our toy example will be familiar to readers who have taken or taught a basic course on formal logic. In such courses, students are often initially confused with regard to how we ought to translate the natural language expression "unless". One common strategy for providing students with some basic insights regarding this translational conundrum is to point out (typically via clear examples) that "unless" seems to obey the following two rules of inference:

[7] There are two reasons for doing so: First, as already noted, the point is not to show that we cannot arrive at quantum logic via a correct application of CARL, but merely that quantum logic is not the *only* destination that we might arrive at via application of that pattern of argumentation. Second, it seems quite plausible to me that, from within the context of quantum logic, these are (up to logical equivalence within Q) the best translations of the claims in question!

$$\frac{\Phi \text{ unless } \Psi \quad \text{Not: } \Phi}{\Psi} \qquad\qquad \frac{\Phi \text{ unless } \Psi \quad \text{Not: } \Psi}{\Phi}$$

These facts suggest that "Φ unless Ψ" could be plausibly translated as "$\neg\Phi \rightarrow \Psi$", or perhaps "$\neg\Psi \rightarrow \Phi$" (or perhaps even "$(\neg\Phi \rightarrow \Psi) \wedge (\neg\Psi \rightarrow \Phi)$" or "$(\neg\Phi \rightarrow \Psi) \vee (\neg\Psi \rightarrow \Phi)$"). The instructor then typically points out that:

$$\neg\Phi \rightarrow \Psi \dashv\vdash_{\mathsf{C}} \Phi \vee \Psi$$
$$\neg\Psi \rightarrow \Phi \dashv\vdash_{\mathsf{C}} \Phi \vee \Psi$$

Hence, the proper translation of "Φ unless Ψ" is "$\Phi \vee \Psi$".

Note that all of this depends on the fact that introductory courses on formal logic are typically restricted to instruction on, and from the perspective of, classical logic. Imagine, however, that an intuitionistic logician teaches a course on basic logic (something that happens all the time) and further that she teaches her students intuitionistic logic (H) and teaches it from the perspective of an intuitionist (something that happens far less frequently).[8] Now, when discussing the proper translation of "unless" claims, the intuitionist can presumably point out, just as the classical logician did, that both of the argument patterns identified above seem valid. And, like her classical counterpart, she can use these facts to suggest that "$\neg\Phi \rightarrow \Psi$", or "$\neg\Psi \rightarrow \Phi$" (or, again, perhaps "$(\neg\Phi \rightarrow \Psi) \wedge (\neg\Psi \rightarrow \Phi)$" or "$(\neg\Phi \rightarrow \Psi) \vee (\neg\Psi \rightarrow \Phi)$") is a plausible first stab at a translation of "unless".

The problem comes in the last step. The classical logician uses the logical equivalence of these more complex formulations and "$\Phi \vee \Psi$" to argue that the latter is the preferred, simplest formalization of the natural language expression "unless". For the intuitionist, however, none of these formulas are equivalent. Of course, the intuitionist can use the following facts to narrow down the field of potential formalizations:

- The proper formalization of "Φ unless Ψ" should be at least as strong as $\neg\Phi \rightarrow \Psi$.
- The proper formalization of "Φ unless Ψ" should be at least as strong as $\neg\Psi \rightarrow \Phi$.
- The proper formalization of "Φ unless Ψ" should be symmetric.[9]

But, while this allows the intuitionist to narrow down the 'right' formalization of "unless" to two candidates, "$\Phi \vee \Psi$" and "$(\neg\Phi \rightarrow \Psi) \wedge (\neg\Psi \rightarrow \Phi)$", it is insufficient to decide between these.[10]

[8]And perhaps for good pedagogical reasons, irrespective of what one's final view is on the correct logic or logics!

[9]That is, "Φ unless Ψ" ought to be logically equivalent to "Ψ unless Φ". This is not implied by the validity of the two rules mentioned above, but seems like a plausible additional constraint on formalizations of "unless" – one that undergraduate logic students would likely agree to relatively easily.

[10]I am not suggesting that there might not be other criteria that would allow us to decide between these two distinct translations of the natural language expression "unless" in intuitionistic contexts.

The point is a general one: when comparing two formal logics $\mathcal{FL}_1$ and $\mathcal{FL}_2$ (in a shared formal language) where the first is strictly stronger that the second in the sense that:

For all Φ, Ψ:

$$\text{If: } \Phi \vdash_{\mathcal{FL}_2} \Psi \text{ then } \Phi \vdash_{\mathcal{FL}_1} \Psi$$

and there exist Φ, Ψ such that:

$$\Phi \vdash_{\mathcal{FL}_1} \Psi \text{ and } \Phi \nvdash_{\mathcal{FL}_2} \Psi$$

it will usually be the case that there are logical equivalences that hold in the stronger logic but that fail to hold in the weaker logic. In particular, if the logics in question obey the standard structural rules, and the standard rules for conjunction, then, if:

$$\Phi \vdash_{\mathcal{FL}_1} \Psi$$

and:

$$\Phi \nvdash_{\mathcal{FL}_2} \Psi$$

then:

$$\vdash_{\mathcal{FL}_1} (\Phi \wedge \Psi) \leftrightarrow \Phi$$

but:

$$\nvdash_{\mathcal{FL}_2} (\Phi \wedge \Psi) \leftrightarrow \Phi$$

And if two formulas are logically equivalent in one logic, but not in another, then when formalizing a natural language statement in the formal language in question we need to *choose* between the two formulas in question if working in a framework based on the latter logic, but we need not choose between them in the stronger logic.[11]

Now that we have a feel for how weaker logics, such as H, allow for a wider range of legitimate choices with respect to formalizing natural language claims, it turns out that we can use this fact to provide an example of how we might respond to the quantum mechanical challenge to classical logic without adopting a logic that fails to validate the inference in question, but instead insists on a different formalization of the natural language claims in question.

But I do not know what such considerations might look like, other than, perhaps, an empirical analysis of the actual inferential practices of intuitionistists with respect to "unless".

[11] This is not to say that there are *no* reasons to prefer one formalization over another, logically equivalent one. After all, one might be syntactically much simpler than the other. The point is that, insofar as the primary criterion for successful formalization is getting the truth conditions right – at least, in situations like the present one where the issue is whether one claim follows logically from another – we have no reason to prefer one formalization over another logically equivalent one.

3.4 Disjunction(s) and Distributivity

In Sect. 3.2 of this essay we considered following natural language claim describing the double-slit experiment:

> p impacted the detection screen at location λ, and either p passed through the first slit, or p passed through the second slit.

and argued that it was true, and hence its (classical) formalization:

$$A \wedge (B_1 \vee B_2)$$

was also true. Here is the argument again:

> The first conjunct is true by assumption, and we can argue for the second conjunct as follows: Assume that p did not pass through either slit – that is, p did not pass through the first slit, and p did not pass through the second slit. Then p could not have reached the detection screen. But it did. Contradiction, so by *reductio* it is not the case that p did not pass through either slit, and by *DeMorgan's Law this implies the disjunction in question.*

This argument is, of course, perfectly valid from the perspective of classical logic. But once we are using the argument to *challenge* classical logic, its status becomes much more problematic. In particular, if intuitionistic logic (or any of a host of other 'constructive' logics, including the continuum-many *intermediate* or *superintuitionistic* logics strictly between C and H) is in the running as a potential replacement for classical logic, then the argument looks much less benign. In fact, by intuitionistic lights the argument is just outright invalid. I have helpfully underlined the step in the argument that would and should be rejected by an intuitionstic logician, since the relevant instance of the DeMorgan's law used in the last step of the argument is not intuitionsitically valid – that is:

$$\neg(\neg\Phi \wedge \neg\Psi) \nvdash_{\mathsf{H}} \Phi \vee \Psi$$

Of course, the portion of the argument up to but not including the underlined text is intuitionistically valid, and thus the intuitionist presumably has just as much reason to accept this sub-argument as does his classical colleague, and hence accept:

> p impacted the detection screen at location λ, and it is not the case that: p did not pass through the first slit and p did not pass through the second slit.

This claim can be formalized rather naturally as:

$$A \wedge \neg(\neg B_1 \wedge \neg B_2)$$

We now merely need to note that the argument from *this* formula as premise, to the problematic conclusion, is not intuitionistically valid:

$$A \wedge \neg(\neg B_1 \wedge \neg B_2) \nvdash_{\mathsf{H}} (A \wedge B_1) \vee (A \wedge B_2)$$

Thus, when confronted with classical-logic challenging examples from quantum mechanics such as the double slit experiment, adopting quantum logic is not our only option: we can instead adopt intuitionistic logic and reconstrue what counts as the best formalization of the empirical phenomena.

Actually, the situation is not so simple. The fragment of the original argument that remains valid from an intuitionistic perspective doesn't actually show that "$A \wedge \neg(\neg B_1 \wedge \neg B_2)$" is the correct formalization of the empirical phenomena, but instead shows that whatever formalization we adopt, it ought to be at least as strong as this (in short: the reasoning shows this claim is true, but is also compatible with stronger formalizations being true).

In addition, if the intuitionistic framework is to be a viable option in the present context, then it had better be the case that "$A \wedge (B_1 \vee B_2)$" is not an acceptable formalization. This seems right, however: intuitionistic logic typically treats disjunction as determinate, in the sense that "$\Phi \vee \Psi$" is taken to be equivalent to something like:

> Φ is definitely the case, or Ψ is definitely the case.

or:

> Φ is the case, or Ψ is the case, and we can tell which.

and these sorts of strong claims with regard to it being determinately the case that p went through one (but not the other) of the two slits seem like exactly what we *don't* want to claim in the case at hand.[12]

So the proper formalization of the empirical phenomena should be at least as strong as "$A \wedge \neg(\neg B_1 \wedge \neg B_2)$", and weaker than "$A \wedge (B_1 \vee B_2)$". Within intuitionistic logic there are a multitude of distinct (that is, logically non-equivalent) formulas involving A, B_1, and B_2 that meet this criteria (in fact, infinitely many!) Thus, if we want to explore the idea of using intuitionistic logic as a viable alternative to quantum logic in responding to the logical challenge posed by the double-slit experiment, then we need to do a bit more work.

Restricting our attention to the disjunction-like right conjunct (that is, we are looking only at formalizations of the form "$A \wedge \star(B_1, B_2)$" for some intuitionistically definable propositional operator $\star$), and considering only those relatively simple disjunction-like formulas that we have already discussed, we get the following possibilities[13]:

[12]For more on how we might read the intuitionistic connectives as synonymous with the corresponding classical connectives while nevertheless tracking determinacy or knowability in a manner in which their classical counterparts do not, see Cook (2014).

[13]Note that we need not restrict our attention to this handful of simple translations. There are many other interesting, disjunction-like operators definable within intuitionistic logic. Interesting examples include *pseudo-disjunction*:

$$B_1 \dot{\vee} B_2 =_{df} ((B_1 \rightarrow B_2) \rightarrow B_2) \wedge ((B_2 \rightarrow B_1) \rightarrow B_1)$$

Church disjunction:

$$B_1 \dddot{\vee} B_2 =_{df} (B_1 \rightarrow B_2) \rightarrow ((B_2 \rightarrow B_1) \rightarrow B_1)$$

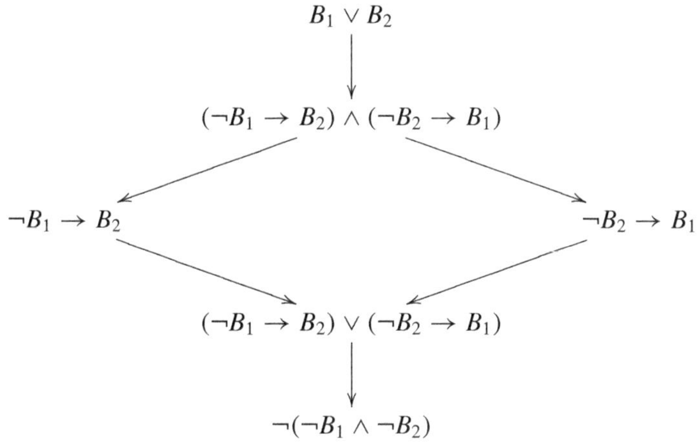

Each formula in the diagram is classical logically equivalent to all of the others, and no two are intuitionistically equivalent. Transitive closure of the arrows indicates entailment.

We can simplify further by noting that (much like the case of "unless" discussed in the previous section) we can restrict our attention to symmetric formalizations – after all, anything we can say with respect to p and the first slit is also something we can say with respect to p and the second slit (and vice versa). Thus, we can restrict our attention to the four disjunction-like translations running vertically through the center of the diagram.

Thus, the intuitionist has (at least) four distinct ways in which she might formalize the claim that p impacted the detection screen at λ and didn't not pass through either screen[14]:

$$
\begin{aligned}
P_1 &=_{df} A \wedge (B_1 \vee B_2)\\
P_2 &=_{df} A \wedge ((\neg B_1 \rightarrow B_2) \wedge (\neg B_2 \rightarrow B_1))\\
P_3 &=_{df} A \wedge ((\neg B_1 \rightarrow B_2) \vee (\neg B_2 \rightarrow B_1))\\
P_4 &=_{df} A \wedge \neg(\neg B_1 \wedge \neg B_2)
\end{aligned}
$$

At this point in our argument, any of these formalizations other than the already rejected strongest – "$A \wedge (B_1 \vee B_2)$" – will work, since none of them entail the problematic (since untrue) conclusion:

---

and *Cornish disjunction*:

$$B_1 \star B_2 =_{df} (((B_1 \rightarrow B_2) \rightarrow B_2) \rightarrow B_1) \rightarrow B_1$$

These are examined in detail in Humberstone (2011) – the absolutely definitive and authoritative study of propositional connectives in classical and non-classical logics – on pages 555, 235, and 235 respectively.

[14]This last bit is formulated a bit obscurely so as not to prejudice which formalization we, in the end, prefer.

$$A \wedge (B_1 \vee B_2) \vdash_{\mathsf{H}} (A \wedge B_1) \vee (A \wedge B_2)$$
$$A \wedge ((\neg B_1 \rightarrow B_2) \wedge (\neg B_2 \rightarrow B_1)) \nvdash_{\mathsf{H}} (A \wedge B_1) \vee (A \wedge B_2)$$
$$A \wedge ((\neg B_1 \rightarrow B_2) \vee (\neg B_2 \rightarrow B_1)) \nvdash_{\mathsf{H}} (A \wedge B_1) \vee (A \wedge B_2)$$
$$A \wedge \neg(\neg B_1 \wedge \neg B_2) \nvdash_{\mathsf{H}} (A \wedge B_1) \vee (A \wedge B_2)$$

So far, so good. But of course, if the intuitionist is motivated to formalize the claim describing the premise of the argument differently from the way in which the quantum logician (or the classical logician) formalizes this claim, then it is of course open to the defender of quantum logic to ask whether, from the perspective of the intuitionist, the correct translation of the conclusion should be modified as well. After all, there might be some alternative formulation, classically equivalent to:

$$(A \wedge B_1) \vee (A \wedge B_2)$$

and equally unacceptable in the present circumstances to classical, quantum, and intuitionistic logician alike, that *does* follow from the intuitionist's preferred formalization of the (true) premise.

Again, we restrict our 'reformulations' of disjunction to the four options given above. We then have four distinct ways in which we might formulate the unacceptable conclusion:

$$C_1 =_{df} (A \wedge B_1) \vee (A \wedge B_2)$$
$$C_2 =_{df} (\neg(A \wedge B_1) \rightarrow (A \wedge B_2)) \wedge (\neg(A \wedge B_2) \rightarrow (A \wedge B_1))$$
$$C_3 =_{df} (\neg(A \wedge B_1) \rightarrow (A \wedge B_2)) \vee (\neg(A \wedge B_2) \rightarrow (A \wedge B_1))$$
$$C_4 =_{df} \neg(\neg(A \wedge B_1) \wedge \neg(A \wedge B_2))$$

The following chart summarizes the logical situation (where "Yes" indicates that the relevant premise-formulation P_n intuitionistically entails the relevant conclusion-formulation C_m, and "No" indicates a failure of entailment):

Entails?	C_1	C_2	C_3	C_4
P_1	*Yes*	*Yes*	*Yes*	*Yes*
P_2	*No*	*Yes*	*Yes*	*Yes*
P_3	*No*	*No*	*Yes*	*Yes*
P_4	*No*	*No*	*Yes*	*Yes*

Thus, so long as the intuitionist can do one of the following:

1. Successfully argue that the proper formalization of the premise of the argument shown to be invalid by the twin slit experiment is no stronger than:

$$P_2 : A \wedge ((\neg B_1 \rightarrow B_2) \wedge (\neg B_2 \rightarrow B_1))$$

and the proper formalization of the conclusion of that argument is at least as strong as:

$$C_1 : (A \wedge B_1) \vee (A \wedge B_2)$$

2. Successfully argue that the proper formalization of the premise of the argument shown to be invalid by the twin slit experiment is no stronger than:

$$P_3 : A \wedge ((\neg B_1 \rightarrow B_2) \vee (\neg B_2 \rightarrow B_1))$$

and the proper formalization of the conclusion of that argument is at least as strong as

$$C_2 : (\neg(A \wedge B_1) \rightarrow (A \wedge B_2)) \wedge (\neg(A \wedge B_2) \rightarrow (A \wedge B_1))$$

then the intuitionistic response to the empirical counterexample in question can serve as a viable alternative to Putnam's quantum logical solution.

If we wanted to pursue a detailed defense of an intuitionistic account of the double-slit experiment, rather than merely use the example to illustrate the flexibility inherent in CARL, then the first option seems like the more promising one.[15] After all, even if, from the perspective of the intuitionist, we can reject formulating the natural language premise of Putnam's argument as P_1:

$$A \wedge (B_1 \vee B_2)$$

it *does* seem that the intuitionist should accept that, if the photon definitely did not pass through the first slit, then it passed through the second slit (and, symmetrically, if the photon definitely did not pass through the second slit, then it passed through the first slit). Hence, it seems like the intuitionist should accept P_2:

$$A \wedge ((\neg B_1 \rightarrow B_2) \wedge (\neg B_2 \rightarrow B_1))$$

as an acceptable formalization of the facts described by the natural language premise in question (and hence accept the truth of P_3 and P_4 as well). But there would seem to be no problem for the intuitionist. As we have already noted, she can agree with the classical logician that C_1:

[15] Of course, a full defense of an intuitionistic approach to the puzzle would also need to provide an intuitionistic account of probability theory – one that, in particular, allowed for classically equivalent claims to receive distinct probabilities. After all, a full and careful presentation of the puzzle in question – such as the one given by Putnam himself in Putnam (1968) – presents the issue in terms of probabilities and not in terms of truth and falsity. Since the purpose here is not to provide a full-fledged defense of the intuitionistic approach to this puzzle, but merely to demonstrate the existence and potential viability of methodological alternatives such as, but not limited to, the intuitionistic approach, I shall not pursue such an account here. For a recent account of intuitionistic probability that might do the job, however, the reader is encouraged to consult (Weatherson 2003).

$$(A \wedge B_1) \vee (A \wedge B_2)$$

fails to be true, but this is perfectly okay, since it does not follow from her strongest formalization of the premise (i.e. P_2), while C_2:

$$(\neg(A \wedge B_1) \rightarrow (A \wedge B_2)) \wedge (\neg(A \wedge B_2) \rightarrow (A \wedge B_1))$$

which does follow from P_2, seems like something that the intuitionist should accept as formalizing a true claim about the double-slit experiment. After all, this amounts, in natural language, to something like:

> If it is definitely not the case that the photon impacted the detection screen and passed through the first slit, then it is the case that the photon impacted the detection screen and passed through the second slit, and if it is definitely not the case that the photon impacted the detection screen and passed through the second slit, then it is the case that the photon impacted the detection screen and passed through the first slit.

and this seems plausible for the same reasons that P_2 seems plausible: if we were to have mounted detectors at each slit, and then fired the photon through the apparatus, a failure to detect the photon at the first slit would entail that the detector at the second slit detected the photon (and vice versa).

The upshot of all of this is that, even if we grant that the double-slit experiment does show that classical logic (**C**) is not the correct logic, it does not on its own show that the correct logic $\mathcal{L}$, whatever that is, does not validate the instance of distributivity in question – that is, it does not show that:

$$A \wedge (B_1 \vee B_2) \nvdash_{\mathcal{L}} (A \wedge B_1) \vee (A \wedge B_2)$$

where, again, $\mathcal{L}$ is whatever the correct logic (or logics) turn(s) out to be. After all, as we have seen, intuitionistic logic validates the relevant instance of distributivity yet allows for an adequate treatment of the apparent counterexample to classical logic supplied by the double-slit experiment. Or does it?

3.5 Falsity and Failure to Be True

Hopefully by this point the reader is at least convinced that a counterexample to classical logic does not require that we move to a logic where that inference is invalid – that is, a counterexample to classical logic need not be (and typically will not be) a counterexample to any particular inference validated by classical logic. Of course, moving to a logic that does not validate the inference challenged by the counterexample *in the classical setting* is one way that we might respond to such a counterexample. But another option, nicely illustrated by the intuitionistic approach to the double-slit experiment sketched above, is to move to a logic where the best formalization of the inference in question is not the same as the best formalization of that inference from the context of classical logic. This is all that was intended in the previous sections – to show that the general strategy is a viable one, even if in the end application of the alternative, intuitionionistic strategy turns out to be inadequate as a response to the quantum mechanical case. Nevertheless, the example we have focused on until now runs the risk of misleading in two ways.

First, the attention we have lavished on the double-slit experiment might give the impression that the flexibility of CARL is somehow specific to *empirical* counterexamples to one's favored logic. We can easily eliminate this potential source of misapprehension simply by providing an addition example of the phenomenon in question that relies, not on an empirical counterexample such as the double-slit experiment, but on a counterexample produced a priori, from the armchair, and craft a similar translating-shifting response to the puzzle. Fortunately, such an a priori counterexample already exists.

Cook (2014) argues that one natural way to formulate the core claim underlying various versions of *logical pluralism* defended in the literature (e.g. Beall and Restall 2006) is as follows:

$$\begin{aligned}\mathsf{LP}: (\exists \Rightarrow_1)(\exists \Rightarrow_2)[(\exists\Delta)(\exists\Phi)((\Delta \Rightarrow_1 \Phi) \wedge (\Delta \not\Rightarrow_2 \Phi))\wedge \\ (\forall\Delta)(\forall\Phi)((I(\Delta) \rhd I(\Phi)) \leftrightarrow (\Delta \Rightarrow_1 \Phi))\wedge \\ (\forall\Delta)(\forall\Phi)((I(\Delta) \rhd I(\Phi)) \leftrightarrow (\Delta \Rightarrow_2 \Phi))]\end{aligned}$$

where $\Rightarrow_1$ and $\Rightarrow_2$ are the validity relations (i.e. 'turnstiles') of distinct formal logics $\mathcal{L}_1$ and $\mathcal{L}_2$, $\rhd$ is the natural language consequence relation that we are trying to capture, and I is any translation function mapping formulas from the language of $\mathcal{L}_1$ and $\mathcal{L}_2$ to sentences in natural language.[16] In short, on this understanding of logical pluralism, the view amounts to claiming that there are two logics $\mathcal{L}_1$ and $\mathcal{L}_2$ (with a shared language) such that:

1. There is at least one inference that is valid in $\mathcal{L}_1$ but not valid in $\mathcal{L}_2$.
2. Every argument is valid in $\mathcal{L}_1$ if and only if all corresponding natural language arguments are instances of the natural language consequence relation (i.e. $\mathcal{L}_1$ *agrees exactly* with logical consequence in natural language).
3. Every argument is valid in $\mathcal{L}_2$ if and only if all corresponding natural language arguments are instances of the natural language consequence relation (i.e. $\mathcal{L}_2$ *agrees exactly* with logical consequence in natural language).

The problem, however, is that LP is inconsistent in classical logic:

$$\mathsf{LP} \vdash_{\mathsf{C}} \bot$$

[16]It is assumed that the logics $\mathcal{L}_1$ and $\mathcal{L}_2$ are built on the same language.

The logical pluralist, presumably, does not think that her view actually entails a contradiction.[17] Thus, at first glance, it appears as if she must reject classical logic in favor of a weaker logic where LP does not entail a contradiction.

To do so, however, would be to apply FARL without recognizing the flexibility that the superior methodology of CARL supplies. As pointed out in Cook (2014), although a pluralist of this sort must reject classical logic,[18] she need not move to a logic where LP does not entail a contradiction. Instead, she can reformulate her characterization of logical pluralism as *Weak Logical Pluralism*:

$$\begin{aligned}\mathsf{WLP} : (\exists \Rightarrow_1)(\exists \Rightarrow_2)[&\neg(\forall\Delta)(\forall\Phi)((\Delta \Rightarrow_1 \Phi) \leftrightarrow (\Delta \Rightarrow_2 \Phi))\wedge\\ &(\forall\Delta)(\forall\Phi)(\neg(I(\Delta) \rhd I(\Phi)) \leftrightarrow \neg(\Delta \Rightarrow_1 \Phi))\wedge\\ &(\forall\Delta)(\forall\Phi)(\neg(I(\Delta) \rhd I(\Phi)) \leftrightarrow \neg(\Delta \Rightarrow_2 \Phi))]\end{aligned}$$

In short, she could argue that the best formalization of the natural language claims that make up her philosophical account of, and defense of, the idea that there is more than one correct logic is not LP but WLP. WLP is classically, but not intuitionistically, equivalent to LP, and Cook (2014) proves that while LP is inconsistent in both classical and intuitionistic logic, WLP is consistent in intuitionistic logic.[19] To put all of this bluntly: the truth of logical pluralism (if it is true), where logical pluralism is best formalized by LP (in a classical setting), plus the classical inconsistency of LP, does not force us to move to a logic where LP is consistent. Rather, it forces us to reject any logic where LP is the best formalization of logical pluralism *and* LP is inconsistent. But we can achieve that by moving to intuitionistic logic and accepting that WLP is a better formalization of logical pluralism from within the new intuitionistic perspective.[20]

---

[17]For simplicity of discussion I set aside possible logical pluralists who accept a multitude of paraconsistent or dialethic logics, and as a result might accept the problematic inference in question.

[18]As a former teacher succinctly put the point when I expressed sympathy to something like the intuitonistic version of pluralism sketched here and developed in more detail in Cook (2014):

> So let me get this straight. Now you're a logical pluralist, but classical logic doesn't even get to be *one of* the correct logics?

[19]In fact, it is consistent in Gödel-Dummett logic, which results from supplementing intuitionistic logic with the linearity axiom:

$$(\Phi \rightarrow \Psi) \vee (\Psi \rightarrow \Phi)$$

[20]It should be emphasized that there is nothing in CARL that privileges or prefers intuitionistic logic as opposed to other non-classical logics. But I am most familiar with, and most sympathetic to, intuitionistic logic and similarly constructive intermediate logics between H and C. Hence I leave the construction of additional examples that use non-classical logics other than intuitionistic logic to the ambitious reader.

In addition, there is nothing in CARL that requires that there be a *single* logic and translation manual that is *correct* or *best*. Instead, in the face of an apparent counterexample to classical logic such as the puzzles examined here, there might be multiple, equally good ways of changing both one's logic and one's translations of the problematic statements.

Thus, the flexibility provided by recognizing that it is CARL, rather than FARL, that we should apply when faced with apparent counterexamples to our favored logic or logics applies to both empirical and a priori armchair counterexamples. But there is a second reason why the double-slit experiment is somewhat less than ideal for present purposes – the fact that the reasoning involved in the apparent counterexample is simple enough that it can and should be formalized in propositional logic. The reason why this is problematic in the present situation has to do with the very close connections between what is provable in classical logic and what is provable in intuitionistic logic – in particular, to the fact that, loosely put, any set of propositional (i.e. quantifier-free) formulas that is classically inconsistent is also intuitionistically inconsistent. The following fact, which is easily proven from familiar theorems (see, e.g., Troelstra and van Dalen 1988), nicely focuses the problem:

Proposition 3.5.1 *Given formulas* $\Phi_1, \Phi_2, \Phi_3, \Phi_4 \in \mathcal{L}$ *if:*

$$\Phi_1 \vdash_{\mathsf{C}} \Phi_2$$

and:

$$\Phi_1 \dashv\vdash_{\mathsf{C}} \Phi_3$$
$$\Phi_2 \dashv\vdash_{\mathsf{C}} \Phi_4$$

then:

$$\Phi_3 \vdash_{\mathsf{H}} \neg\neg\Phi_4$$

Given this fact, the strategy outlined above for reconstruing the argument based on the double-slit experiment within intuitionistic logic can only work if we think that the natural language premise formalized by classical logician as:

$$A \wedge (B_1 \vee B_2)$$

is true, and the natural language conclusion formalized by the classical logician as:

$$(A \wedge B_1) \vee (A \wedge B_2)$$

fails to be true, but also fails to be false (i.e. its negation fails to be true by intuitionistic lights). Assume that the latter claim is actually false, and that Φ_{H} and that Ψ_{H} are our imagined intuitionist's favored translations of the natural language premise and conclusion in question. Then we have:

$$A \wedge (B_1 \vee B_2) \vdash_{\mathsf{C}} (A \wedge B_1) \vee (A \wedge B_2)$$

and:

$$\Phi_{\mathsf{H}} \dashv\vdash_{\mathsf{C}} A \wedge (B_1 \vee B_2)$$
$$\Psi_{\mathsf{H}} \dashv\vdash_{\mathsf{C}} (A \wedge B_1) \vee (A \wedge B_2)$$

Hence, by the fact above, we obtain:

$$\Phi_{\mathsf{H}} \vdash_{\mathsf{H}} \neg\neg\Psi_{\mathsf{H}}$$

Loosely put, however, this just says that the truth of Φ_{H} entails the non-falsity of Ψ_{H}. Hence, if the natural language premise of the argument based on the double-slit experiment is true, but the natural language conclusion is false, then so long as the intuitionist, in formalizing these claims, chooses formulas that are classically (but not necessarily intuitionistically) equivalent to those chosen by the classical logician, then the intuitionist is in no better shape with respect to the apparent counterexample than the classical logician. This problem with our present example – the double-slit experiment – can be further divided into two subproblems.

First, it seems to make the methodological point, as applied to Putnam's example in particular, hostage to subtle issues regarding the interpretation of quantum mechanics. If we think that the conclusion of the original argument:

$$(A \wedge B_1) \vee (A \wedge B_2)$$

in question fails to be true, but could also fail to be false (in the same sense that intuitionists believe that some instances of exclude middle fail to be true, but also fail to be false), then it seems, contrary to what Putnam (1968) suggests, that we need not reject the instance of distributivity in question, but could instead adopt something like the intuitionistic strategy sketched in the previous section. If, however, we conclude that the conclusion of the argument in question is genuinely false – that is, if we accept:

$$\neg((A \wedge B_1) \vee (A \wedge B_2))$$

then intuitionistic logic is of no help. Instead, we would either have to reject classical logic in favor of a logic that (unlike H) failed to validate the relevant instance of distributivity, or find a logic that did validate distributivity but whose consequence relation is less intimately related, in a technical sense, to the consequence relation of classical logic (in particular, a logic for which **Proposition** 3.5.1 fails).

Second, and closely related to the first problem, is the fact that the intuitionistic approach to dealing with the double-slit experiment relies fundamentally on the distinction, within intuitionistic logic, between a sentence failing to be true and a sentence being false. This distinction is particularly tricky within propositional intuitionistic logic, and although the distinction is clear enough from 'outside', so to speak (e.g. from the perspective of studying intuitionistic formalisms from the perspective of a classical metatheory), it is not clear that intuitionists themselves, limited as they are to strictly intuitionistic linguistic and logical resources, are able to coherently describe the distinction in question in an internally coherent manner (see Cook 2014 for some general discussion of this, and Hellman 1980 for related observations specific to the quantum logical context). They certainly do not have the resources to do so from within standard intuitonistic propositional logic, since $\neg\Phi$ expresses that Φ is false, there is no expression $*$ such that $*\Phi$ expresses that Φ merely fails to be true.

Thus, Putnam's example, although important in helping us to recognize the in-principle flexibility involved in logical revision in the first place, is less than ideal with respect to providing a clear example of such flexibility given the manner in which subtle issues regarding both the interpretation of quantum mechanics and the workings of intuitionistic propositional logic affect the plausibility of the example. If the double slit experiment were the only possible example where the phenomenon arose, then this would present a significant challenge. But as both the discussion of "unless" and the discussion of the proper formulation of logical pluralism show, the phenomenon in question is not isolated to quantum physics.[21] Instead, the alternative

---

[21] If the reader would like another possible example, consider a situation where we have a notion of some series of objects being "constructed", which we can formalize as $\mathcal{C}(x)$ and understand along the lines of familiar accounts in the philosophy of mathematics that involve notions such as *potential infinity* or *constructivity*. In such cases we might want to say that that:

It is not the case (at any particular time) that every object is constructed.

which the classical logician might (quite correctly, from her perspective) formalize as:

$$\neg(\forall x)(\mathcal{C}(x))$$

yet we might also want to claim that:

No object fails (forever) to be constructed.

which the classical logician might (again, correctly) formalize as:

$$(\forall x)(\mathcal{C}(x))$$

Of course, these claims are inconsistent in classical logic, but acceptance of this pair of natural language claims this need not imply a rejection of:

$$\neg(\forall x)(\mathcal{C}(x)), (\forall x)(\mathcal{C}(x)) \vdash_{\mathcal{L}} \bot$$

(which is, after all, just a complex instance of the standard negation elimination rule). Instead, we might move to intuitionistic logic, but argue that the best translation of the second natural language claim is not the one given above, but is instead:

$$(\forall x)(\neg\neg\mathcal{C}(x))$$

Of course, the legitimacy of this move depends on the fact that:

$$\neg(\forall x)(\mathcal{C}(x)), (\forall x)(\neg\neg\mathcal{C}(x)) \nvdash_{\mathrm{H}} \bot$$

strategy sketched here – to move to a new logic where the proper translation of problematic sentences differs from their best translations in one's current logic – is a general strategy that can be applied, in principle at least, whenever one is confronted with an apparent counterexample to one's favored logic.[22]

3.6 Some Final Observations on Logical Form

To sum all of this up: An apparent counterexample to a logic is never a counterexample to a particular inference validated by that logic. Instead, as is demonstrated by the flexibility of CARL, counterexamples must be judged more holistically, as counterexamples to a logic as a whole without being counterexamples to any particular inference pattern within that logic. When faced with any such counterexample, and a particular inference or law apparently violated by the counterexample in question, it is always possible, at least in principle, to move to a logic where the proper or best translation of the natural language sentences describing the counterexample no longer corresponds to the formulas occurring in the inference pattern or law, rather than moving to a logic that no longer validates that inference or law.

Before concluding, it is worth saying a bit more of a general nature about the source of this flexibility in the methodology of logical revision. One means for helping us to see what is going on is to consider the idea that logic is *formal*. In his famous paper on logical consequence, Alfred Tarski described the enterprise at issue, and the role that formality plays in that enterprise, as follows:

> Consider any class Δ of sentences and a sentence Φ which follows from the sentences of this class. From an intuitive standpoint it can never happen that both the class Δ consists only of true sentences and the sentence Φ is false. Moreover, since we are concerned here with the concept of logical, i.e. *formal*, consequence, and thus with a relation which is to be uniquely determined by the form of the sentences between which it holds ...the consequence relation cannot be affected by replacing the designations of the objects referred to in these sentences by the designations of any other objects (Tarski 1983: 414–415).

In this crucially important passage, Tarski introduces two important ideas that have guided research on logic, and logical consequence, ever since. First, and most obvious, is the idea that logical consequence is a *modal* relation that holds between premises and conclusions – that is, it is *necessary* – a thought we can summarize as[23]:

> [Necessity]: If (natural language) statement Φ is a logical consequence of a set of (natural language) statements Δ then the simultaneous truth of every member of Δ guarantees the truth of Φ.

---

[22]The reader interested in seeing another attempt to formulate and defend an intuitionistic response to the puzzles of quantum mechanics – one that combines the insights and machinery of intuitionistic logic and quantum logic – is encouraged to consult (Caspers et al. 2009).

[23]If the reader is interested in learning more about what, exactly, the necessity of logical consequence amounts to, and how Tarski himself understood this notion, the reader will find no better place to begin than Etchemendy (1999).

The second, and for our purposes more important, of Tarski's ideas is that the notion of logical consequence depends on the notion of *logical form*:

> [Formality]: If (natural language) statement Φ is a logical consequence of a set of (natural language) statements Δ then this fact follows solely from the logical form of Φ and of the members of Δ.

There is a good bit of controversy over how, exactly, to understand the formality constraint. One simple (and partial) means of fleshing out the idea that logical consequence is formal is to note that it seems to require that the following *substitutivity* requirement holds of any logic $\mathcal{L}$ that might be a candidate for the correct (or one of the legitimate, if pluralist) logic(s)[24]:

> [Logical Substitutivity]: For any formula Φ, set of formulas Δ, primitive non-logical expression Ψ, and (possibly complex) expression Γ of the same logical type as Ψ:
>
> If:
>
> $$\Delta \vdash_{\mathcal{L}} \Phi$$
>
> is a logical truth, then:
>
> $$\Delta[\Psi/\Gamma] \vdash_{\mathcal{L}} \Phi[\Psi/\Gamma]$$
>
> is a logical truth.

Although this much seems correct – that is, any logic that is to meet Tarski's requirement of formality ought to satisfy the substitutivity requirement – there is an obvious mismatch between Tarski's original formulation of the notion and substitutivity: Tarski formulates formality as a criterion that applies directly to natural language sentences, while the substitutivity requirement applies directly to formal logics; it is internal to the formalism, so to speak.

We can rectify this by noting that, although logical consequence, and thus the determination of the correct logic or logics, depends solely on logical form, logical consequence does not distinguish between logically equivalent sentences, regardless of whether or not they have the same logical form. As a result, for the purposes of determining which logic is the correct logic, logical consequence is a congruence over logical equivalence. Thus, we can re-cast formal logics as relations that operate, not on formulas and sets of formulas, but on equivalence classes of formulas modulo logical equivalence. The relata of formal logics, on this understanding, are equivalence classes (or, for premises, sets of equivalence classes) of the form[25]:

---

[24] $\Phi[\Psi/\Gamma]$ is the formula that results from uniformly replacing every occurrence of the (primitive) expression Ψ in Φ with Γ. $\Delta[\Psi/\Gamma]$ is:

$$\{\Theta[\Phi/\Gamma] : \Theta \in \Delta\}$$

[25] It is no accident that algebraic semantics for logics usually involve either exactly this construction, or something closely akin to it. See, e.g., Dunn and Hardegree (2001) for a standard treatment of algebraic semantics for logics. Also note that the account developed here – and in particular this construction of equivalence classes of formulas – depends on the logic in question satisfying the standard structural rules. Similar, albeit much more mathematically sophisticated, constructions that will support similar arguments for substructural logics are possible, but are left to the reader interested in such things.

$$[\Phi]_{\mathcal{L}} = \{\Psi : \ \vdash_{\mathcal{L}} \Phi \leftrightarrow \Psi\}$$

If formal logics can be understood as relations between such equivalence classes, however, then translation from natural language sentences to formal languages must likewise be characterizable as a function from sentences of natural language to equivalence classes of formulas from our formal logic.[26] Since logical consequence is formal, and formality implies insensitivity to the difference between logically equivalent sentences, nothing is lost by making such a move. But if this is right, then it has profound consequences for how we ought to think about translation from natural language to formal languages, and the manner in which we think of the role that translation from natural language to formal languages plays in our logical theorizing.

If the natural language consequence relation is insensitive to the distinction between distinct but logically equivalent formulas, and translation is for all intents and purposes a function from natural language sentences to equivalence classes of logically equivalent formulas in our formal language, then, if we reject one logic in favor of another, then we also reject one collection of equivalence classes for another, and hence *we reject one translation function for another*.

Consider again our conundrum regarding how to translate natural language sentences involving "unless". In Sect. 3.3 we identified two potential translations of Φ unless Ψ":

$$\Phi \vee \Psi$$
$$(\neg\Phi \rightarrow \Psi) \wedge (\neg\Psi \rightarrow \Phi)$$

For the classical logician this does not pose a problem. Either translation is as good as the other, since the formulas in question are logically equivalent in classical logic. Further, on the equivalence class approach to translation, there is no choice to be made: the classical logician should translate "Φ unless Ψ" as:

$$[\Phi \vee \Psi]_{\mathsf{C}} (= [(\neg\Phi \rightarrow \Psi) \wedge (\neg\Psi \rightarrow \Phi)]_{\mathsf{C}}$$

[26] Of course, we could also replace the sentences of the natural language with equivalence classes of logically equivalent natural language sentences. The methodological problem with this approach, of course, is that our primary means by which to determine which pairs of natural language sentences are logically equivalent is to first determine which logic we will take to be the correct logic, and then project logical equivalence from the formal logic to natural language via the inverse of our translation function. This is the reason we study formal languages in the first place: they are more amenable to mathematical study and manipulation than their natural language counterparts, and thus it is usually easier to delineate philosophically relevant phenomena like the logical consequence relation in the formal sphere, and then project them to natural language, than it is to detect those same phenomena directly in natural language. That being said, if there were some means for identifying, in general, which pairs statements in natural language were logically equivalent to each other that was independent of the methods and tools of formal logic, then we could replace natural language sentences with their equivalence classes in the treatment above. Nothing significant would change with respect to the issues being examined here.

Things are different for the intuitionistic logician, however, since she *does* have a choice, since two options are not logically equivalent from her perspective. Hence, she must choose either $[\Phi \vee \Psi]_{\mathsf{H}}$ or $[(\neg\Phi \rightarrow \Psi) \wedge (\neg\Psi \rightarrow \Phi)]_{\mathsf{H}}$ and this choice matters, since different things follow from each.[27]

Let us now return to the more common approach where we treat translation as a function from individual sentences in our natural language to individual sentences in our formal languages, keeping in mind the insights from the previous few paragraphs. Given the insensitivity of logical consequence to the differences between logically equivalent sentences, we can now further flesh out the formality requirement in a manner that addresses the role played by natural language via the following translation requirement:

> [Translation] Let $\mathcal{T}_1$ and $\mathcal{T}_2$ be any translation functions from natural language sentences to formulas, where, for any natural language sentences $\Phi_{\mathcal{NL}}$:
>
> $$\vdash_{\mathcal{L}} \mathcal{T}_1(\Phi_{\mathcal{NL}}) \leftrightarrow \mathcal{T}_2(\Phi_{\mathcal{NL}})$$
>
> That is:
>
> $$[\mathcal{T}_1(\Phi_{\mathcal{NL}})]_{\mathcal{L}} = [\mathcal{T}_2(\Phi_{\mathcal{NL}})]_{\mathcal{L}}$$
>
> Then (insofar as we are interested in logical consequence) $\mathcal{T}_1$ is exactly as good a translation function as is $\mathcal{T}_2$.

The translation requirement, however, gets to the crux of the matter: what counts as a good, correct, or legitimate translation from natural language sentences to our formal language is not something that can be decided before we begin to argue about which inferences are or are not valid. Instead, translation from natural language to a formal language presupposes that we have accepted a particular formal logic that governs the acceptability conditions for such translation generally – hence the presence of subscripted $\mathcal{L}$s in the translation requirement. In short, we don't formalize natural language sentences and then decide which logic is to govern the logical relations between these formalizations, but instead adopt a logic and then formalize natural language expressions in light of the logic we have adopted.[28]

It is this more general insight – that translation from natural language to formal language presupposes acceptance of a particular logic as correct or legitimate – that lies at the heart of our development and defense of CARL above. If one mistakenly

[27]Note that the fact that the classical logician and the intuitionistic logician are faced with different possibilities with respect to translating "unless" does not automatically entail that they *mean* different things by "unless". Translation, in the context of constructing a formal logic that correctly codifies the natural language consequence relation, need only preserve logical form. But although synonymous statements presumably have the same logical form, the converse is not obviously true. Thus, the classical logician could treat both options are equally legitimate given that her purpose is to study logical consequence, but agree with the intuitionistic logician that these formulas mean different things. See Cook (2014) for more discussion.

[28]It is worth noting that standard pedagogical practice in introduction to logic classes violates this methodological observation: typically, in such courses, we introduce the formal language, then teach students how to translate between natural language and our formal language, and only then, after mastering the translation rules we have inculcated in them, do we ask them to consider which rules are and are not valid.

believes that translation is prior to arguments about the correctness or incorrectness of this or that inference – that is, if translation is prior to the acceptance of a logic or logics as correct or legitimate – then the inadequate **FARL** seems plausible. But once we realize that arguments about the correctness of inferences presuppose accepting a particular translation function, which in turn presupposes that the correct logic is amongst those compatible with that translation function, an apparent counterexample to a particular inference $\mathcal{I}$ in a particular logic $\mathcal{L}$ need not be construed as a counterexample to $\mathcal{I}$ generally, forcing us to move to a logic that does not validate $\mathcal{I}$. Instead, we can treat this as evidence of the inadequacy of the translation function that mapped the natural language sentences in question to the formulas involved in the relevant instance of $\mathcal{I}$, and so move to a logic that no longer supports the translation function in question (a logic that, to emphasize one last time, might still validate $\mathcal{I}$). This is the freedom provided by **CARL**, but not **FARL** – the freedom to recognize that counterexamples to a particular logic might be counterexamples to the kinds of translation supported by that logic, rather than being counterexamples to any particular inference validated by that logic.[29]

References

Beall, J., & Restall, G. (2006). *Logical pluralism*. Oxford: Oxford University Press.

Ben-Menahem, Y. (Ed.). (2005). *Hilary Putnam*. Cambridge: Cambridge University Press.

Caspers, M., Heunen, C., Landsman, N., & Spitters, B. (2009). Intuitionistic quantum logic of an *n*-level system. *Foundations of Physics*, *39*, 731–759.

Cook, R. (2014). Should anti-realists be anti-realists about anti-realism? *Erkenntnis*, *79*, 233–258.

Cohern, R., & Wartofsky, M. (Eds.). (1968). *Boston studies in the philsoophy of science* (Vol. 5). Dordrecht: D Reidel.

Dummett, M. (1976). Is logic empirical? In Lewis 1976 (pp. 45–68), reprinted in (Dummett 1978, (pp. 269–289)).

Dummett, M. (1978). *Truth and other enigmas*. London: Duckworth.

Dunn, J., & Hardegree, G. (2001). *Algebraic methods in philosophical logic*. Oxford: Oxford University Press.

Etchemendy, J. (1999). *The concept of logical consequence*. Stanford: CSLI.

Gardner, M. (1971). Is quantum logic really logic? *Philosophy of Science*, *38*(4), 508–529.

Gibbons, P. (1987). *Particles and paradoxes: The limits of quantum logic*. Cambridge: Cambridge University Press.

Hellman, G. (1980). Quantum logic and meaning. *Proceedings of the Philosophy of Science Association*, *2*, 493–511.

Humberstone, L. (2011). *The connectives*. Cambridge: MIT Press.

Lewis, H. (Ed.). (1976). *Contemporary British philosophy*, 4th series. London: Allen and Unwin.

Maudlin, T. (2005). The tale of quantum logic. In Ben-Menahem 2005 (pp. 156–187).

Putnam, H. (1968). Is logic empirical? In Cohen & Wartofsky 1968 (pp. 216–241), reprinted as *The logic of quantum mechanics*. (In Putnam 1975 (pp. 174–197)).

Putnam, H. (1975). *Mathematics, matter, and method: Philosophical papers* (Vol. I). Cambridge: Cambridge University Press.

---

[29]Thanks are owed to Geoffrey Hellman, Stewart Shapiro, and Jos Uffink for helpful conversations on matters related to this paper.

Quine, W. (1951). Two dogmas of empiricism. *The Philosophical Review*, *60*, 20–43. (reprinted in Quine 1953 (pp. 20–46)).
Quine, W. (1953). *From a logical point of view*. Camrbridge: Harvard University Press.
Tarski, A. (1983). *Logic, semantics, metamathematics* (2nd ed.). Indianapolis: Hackett.
Troelstra, A., & van Dalen, D. (1988). *Constructivism in mathematics* (Vol. 1). Amsterdam: Elsevier.
Weatherson, B. (2003). From classical to intuitionistic probability. *Notre Dame Journal of Formal Logic*, *44*(2), 111–123.

Roy T. Cook is CLA Scholar of the College and John M. Dolan Professor of Philosophy. He is the author of *Key Concepts in Philosophy: Paradoxes* (2013) and *The Yablo Paradox: An Essay on Circularity*, as well as numerous articles and essays in the philosophy of mathematics, the philosophy of logic, and the aesthetics of popular art.

Chapter 4
Putnam's Theorem on the Complexity of Models

Warren Goldfarb

Abstract A streamlined proof of a theorem of Putnam's: any satisfiable schema of predicate calculus has a model in which the predicates are interpreted as Boolean combinations of recursively enumerable relations. Related open problems are canvassed.

A lesser-known but quite interesting contribution of Hilary Putnam to mathematical logic concerns the complexity of models of schemata of the predicate calculus, that is, first-order quantificational schemata. To frame his results, let me start by recalling what might be dubbed the Hilbert–Bernays Theorem, namely, that any satisfiable schema of quantification theory has a model over $\mathbb{N}$ (or a finite subset of $\mathbb{N}$) in which the predicate letters are interpreted as Δ_2 relations. (A Δ_2 relation is one that can be defined as both a Π_2 relation $\forall x \exists y Q$ and a Σ_2 relation $\exists x \forall y R$, where Q and R are recursive relations.) The proof is sketched in Hilbert and Bernays (1939), pp. 243–252; it proceeds by arithmetizing Gödel's completeness proof (1930). The argument is elaborated rigorously in Kleene (1952), pp. 389–394. Kleene's proof is very complex, but a much simpler one is now available (Ebbs and Goldfarb 2018).

In the 1950s, there was attention to the question of whether the Hilbert–Bernays Theorem could be improved: could one always find number-theoretic interpretations of lesser complexity than Δ_2? Kreisel (1953) and Mostowski (1953) independently showed that there were satisfiable schemata of quantification theory that could not be satisfied over $\mathbb{N}$ with recursive relations, that is, had no recursive models. Those proofs used set-theoretic means, but in Mostowski (1955) gave a more elementary argument that used only recursion-theoretic concepts.

Putnam's first result (1957) strengthened this: a restriction to relations that are either Σ_1 or Π_1 would also not suffice. (Σ_1 relations are existential quantifications of recursive relations, usually called "recursively enumerable" or "computably enumerable"; Π_1 relations are the complements of Σ_1 relations, and sometimes called "co-r.e". or "co-c.e".) The argument is simple and elegant. Putnam starts with a schema that has no recursive model, and constructs from it a schema that cannot

W. Goldfarb (✉)
Department of Philosophy, Harvard University, Cambridge, MA 02138, USA
e-mail: goldfarb@fas.harvard.edu

G. Hellman and R. T. Cook (eds.), *Hilary Putnam on Logic and Mathematics*,
Outstanding Contributions to Logic 9, https://doi.org/10.1007/978-3-319-96274-0_4

have a model when its predicate is either Σ_1 or Π_1. Assuming the predicates of the given schema that has no recursive models are dyadic $P_1, \ldots, P_n$, first introduce new constants $a_1, \ldots a_n, b_1, \ldots b_n$ and a triadic predicate Q, then replace each occurrence of $P_i(x, y)$ with $Q(x, y, a_i)$, and conjoin the resulting formula with

$$\forall x \forall y \left(\bigwedge_i Q(x, y, a_i) \leftrightarrow \neg Q(x, y, b_i) \right)$$

Since a Σ_1 set with Σ_1 complement is recursive, it follows that if Q is interpreted as either a Σ_1 or Π_1 relation, then it must in fact be recursive, which would then yield recursive interpretations of $P_1, \ldots, P_n$. So if the given schema has no recursive models, then the new schema cannot have a model with Q interpreted as either Σ_1 or Π_1.

Putnam obtained his other, more important, result a few years later; it is published as (1965). This is a positive theorem: he improved the Hilbert–Bernays Theorem by showing that a model *could* always be obtained using relations that were boolean combinations of Σ_1 relations. The notation he used for this class of relations was Σ_1^*. This yields a sharp characterization in terms of the Kleene arithmetical hierarchy: one-quantifier forms do not suffice, but boolean combinations of one-quantifier forms do.

As we shall see, Putnam's argument, unlike the original Hilbert–Bernays Theorem, applies only to quantification theory without identity, that is, what is called the "restricted functional calculus" in Hilbert and Ackermann (1928).

Putnam's argument is quite ingenious. Here I present a streamlined version in more modern notation. We are given a satisfiable schema S. In the interest of simplicity of exposition, Putnam invoked a well-known reduction due originally to Herbrand (1931), which allows him to assume that S contains only one predicate letter P (the method is a generalization of the reduction above of several dyadic predicate letters to one triadic letter). Suppose P is n-adic.

The first lemma needed for Putnam's proof is comes from the basic model theory of quantification theory. Suppose $\mathfrak{M}$ is a model for a schema with universe $\mathbb{N}$, and suppose $\varphi : \mathbb{N} \to \mathbb{N}$ is onto. Define $\mathfrak{N}$ also with universe $\mathbb{N}$ so that the interpretation of each predicate letter is the preimage under φ of its interpretation in $\mathfrak{M}$. Then $\mathfrak{N}$ is also a model for the schema. This elementary fact is easily shown by induction on the logical complexity of the schema. In the case we are considering, in which the schema contains one predicate letter P, the preimage interpretation is this: $P^{\mathfrak{N}}$ holds of an n-tuple $(i_1, \ldots i_n)$ iff $P^{\mathfrak{M}}$ holds of $(\varphi(i_1), \ldots \varphi(i_n))$. That φ is required only to be onto and not also one-one is what limits the scope of the lemma to quantification theory without identity; if "=" were allowed and φ were not one-one, the lemma would fail.

The ingenuity in the proof lies in a second lemma. Putnam started by showing that an n-adic number-theoretic relation R is Δ_2 iff there is an $(n+1)$-adic recursive function f such that $R(i_1, \ldots, i_n)$ holds iff the limit of $f(i_1, \ldots, i_n, y)$ as y goes to infinity is 1, and does not hold if that limit is 0. He called such a function a "trial and error function", although that name has not stuck. Nowadays, this result is considered a standard one in recursion theory. In fact it was already known before Putnam's paper, as it follows quickly from Shoenfield (1959). But in fact this result is not necessary for his theorem, because the simplified proof of the Hilbert–Bernays Theorem mentioned above establishes *first* that the predicate letters of any satisfiable quantificational schema can be interpreted as limits of trial-and-error functions, and only subsequently infers from this that they are Δ_2. Thus one can simply start with those functions and avoid the detour through Δ_2 (see Ebbs and Goldfarb 2018).

Now if f is an $(n + 1)$-adic trial-and-error function, then for all $i_1, \ldots i_n$ there is some k such that $f(i_1, \ldots, i_n, y)$ changes its value at most k times as y increases. Putnam's innovation was to reverse the quantifiers, that is, to consider recursive functions f with the property that there exists a k such that $f(i_1, \ldots, i_n, y)$ changes its value at most k times, no matter what $i_1, \ldots i_n$ are. Putnam called these "k-trial" functions, and the n-adic relations they define in their limit, "k-trial predicates". Several years later, k-trial functions also found application in the theory of tilings of the plane; see Hanf (1974).

Putnam then noted that, for any k, k-trial relations are boolean combinations of r.e. relations. The boolean complexity of the relation is straightforwardly dependent on k. For example, suppose $g(i, j, y)$ is a 2-trial function. The relation $R(i, j)$ that it defines in the limit can be specified thus: there are two numbers y at which $g(i, j, y)$ changes value, and at the greater of them the value is 1; or there aren't two numbers y such that $g(i, j, y)$ changes value at y but there is one number y at which $g(i, j, y)$ changes value, and then its value is 1, or there is no number y at which $g(i, j, y)$ changes value, and $g(i, j, 0)$ is 1. Each of these clauses can be expressed as either an r.e. relation or a co-re. relation.

Now let $\mathfrak{M}$ be a model for S for which $P^{\mathfrak{M}}$ is defined as the limit of a trial-and-error function f. Given the above, it suffices to find an onto mapping φ such that the φ-preimage of $P^{\mathfrak{M}}$ can be defined as the limit of a k-trial function for some k.

For any $i_1, \ldots, i_n$, an $(i_1, \ldots, i_n)$-*modulus* is any number b such that $f(i_1, \ldots, i_n, b) = f(i_1, \ldots, i_n, y)$ for all $y \geq b$. Since $f(i_1, \ldots, i_n, y)$ has a limit as y goes to infinity, there always exists an $(i_1, \ldots, i_n)$-modulus. For any i, an i-*modulus* is a number b that is an $i_1, \ldots, i_n$-modulus for all $i_1, \ldots, i_n \leq i$. Thus for every i there exists an i-modulus.

Let $\langle i, j\rangle$ be a standard primitive recursive one-one onto pairing function. Let φ be the one-place function such that, for all i and b, $\varphi(\langle i, b\rangle) = i$ if b is an i-modulus, and $= 0$ if not. It follows that φ is onto.

Now let $h(i, b, y) = i$ if for all z, $b \leq z \leq y$ and all $i_1, \ldots, i_n \leq i$,

$$f(i_1, \ldots, i_n, b) = f(i_1, \ldots, i_n, z)$$

otherwise let $h(i, b, y) = 0$. Note that if $h(i, b, y) = 0$ then $h(i, b, y') = 0$ for all $y' \geq y$. Thus for $i \neq 0$, $h(i, b, y)$ can change its value for increasing y at most only once: from i to 0. Moreover, for $i \neq 0$, $h(i, b, y) = i$ for every y iff b is an i-modulus. (For $i = 0$, $h(i, b, y)$ is always 0.) Hence

$$\varphi(\langle i, b\rangle) = \lim_{y\to\infty} h(i, b, y)$$

We now define an $(n+1)$-place function g, which we show to be a n-trial function and to define, in its limit, the φ-preimage of $P^{\mathfrak{M}}$. Let b_0 be any 0-modulus, that is, any number such that, for all $y \geq b_0$, $f(0, \ldots, 0, b_0) = f(0, \ldots, 0, y)$. Let $i_1, \ldots, i_n$ and $b_1, \ldots, b_n$ be any integers. Then let

$$g(\langle i_1, b_1\rangle, \ldots, \langle i_n, b_n\rangle, y) = f((h(i_1, b_1, y), \ldots, h(i_n, b_n, y), b_p)$$

where $p = 0$ if $h(i_m, b_m, y) = 0$ for each m, $1 \leq m \leq n$; and p is such that i_p is the largest among $i_1, \ldots, i_n$ for which $h(i_p, b_p, y) \neq 0$ otherwise.

Note first that g is an n-trial function. For as noted above, $h(i, b, y)$ can change its value at most once, from nonzero to zero. Hence the arguments of f on the righthand side of the definition of g can change their values at most n times.

Moreover, g defines, in the limit, just the φ-preimage of $P^{\mathfrak{M}}$. That is, as y goes to infinity the limit of $g(\langle i_1, b_1\rangle, \ldots \langle i_1, b_1\rangle, y)$ is the same as that of $f(\varphi(\langle i_1, b_1\rangle), \ldots, \varphi\langle i_1, b_1\rangle), y)$. For let y be any number large enough that, for all $y' \geq y$, $h(i_m, b_m, y') = \varphi(\langle i_m, b_m\rangle)$ for each $m \leq n$. Then, for all $y' \geq y$, $g(\langle i_1, b_1\rangle, \ldots \langle i_n, b_n\rangle, y') = f(\varphi(\langle i_1, b_1\rangle), \ldots, \varphi(\langle i_n, b_n\rangle), b_p)$, where $p = 0$ if all the $\varphi(\langle i_m, b_m\rangle)$ are 0, and otherwise p is such that i_p is the largest among $i_1, \ldots, i_n$ such that $\varphi\langle i_p, b_p\rangle) \neq 0$. In either case b_p is an i-modulus for some i such that $\varphi(\langle i_m, b_m\rangle) \leq i$ for each m, so that $f(\varphi(\langle i_1, b_1\rangle), \ldots, \varphi(\langle i_n, b_n\rangle), b_p) = f(\varphi(\langle i_1, b_1\rangle), \ldots, \varphi(\langle i_n, b_n\rangle), y')$ for all $y' \geq b_p$; hence the two limits are identical. This completes the proof.

Since g is an n-trial function, and the boolean complexity of the defined predicate depends only on n, we see that the boolean complexity of the Σ_1^* relations depends only on the number of argument places of the predicate letter and not on anything further about the logical structure of the schema S.

As mentioned above, this proof applies only to quantification theory without identity. Strictly speaking, the theorem is false for quantification theory with identity, since there are schemata which have only finite models, and hence no model over all of $\mathbb{N}$. If we restrict attention to those schemata that do have infinite models, then it is an open question as to whether the result holds.

Putnam's theorem is not widely known, I think, for two reasons. First, due to the growth of model theory, by the late 1960s logicians who studied quantification theory began taking quantification theory with identity as a more basic object of study than the "restricted functional calculus". (To see the shift, one need only compare the introduction of quantification theory in Kleene (1952) or Church (1956) with that in Enderton (1972).) Second, advances in recursion theory caused more emphasis to be put on Turing degree rather than position in the arithmetical hierarchy as a measure of complexity. Here the crucial result was the Low Basis Theorem of Jockusch and Soare (1972). This theorem states that any recursive infinite binary tree has an infinite path whose jump is Turing equivalent to K. By applying the same arithmetization of Gödel's completeness proof as Hilbert and Bernays had done, this yields – for quantification theory *with* identity – the same Turing degree for the interpretations of predicate letters in models over $\mathbb{N}$. This characterization lies athwart Putnam's characterization, since there are low sets that are Δ_2 but not Σ_1^*.

Nonetheless, I believe, there are open problems stemming from his result that are worth investigating. First, as mentioned above, there is the question of whether Putnam's result can be extended to quantification theory with identity. His technique certainly cannot be, and I am doubtful that the result holds there. It should, however, be settled one way or the other. Second, because the function φ used in the proof is ordinarily not recursive, one cannot extract from the Low Basis Theorem information about the Turing degree of the interpretation. Hence, the question remains: if we restrict ourselves to Σ_1^* interpretations, are there schemata which require relations that are Turing equivalent to K? Finally, in the model constructed for the Hilbert–Bernays Theorem, even though the predicates are not interpreted recursively, there are recursive Skolem functions (this is built in to Gödel's completeness proof). This is not preserved by φ-preimages, but then the question remains as to whether one can obtain recursive Skolem functions in Σ_1^* models, and if not, how great their computational complexity might have to be.[1]

References

Church, A. (1956). *Introduction to mathematical logic*. Princeton: Princeton University Press.

Ebbs, G., & Goldfarb, W. (2018). First-order logical validity and the Hilbert-Bernays theorem. *Philosophical Issues 28: Philosophy of Logic and Inference* (forthcoming)

Enderton, H. (1972). *A mathematical introduction to logic*. New York: Academic Press.

Gödel, K. (1930). Die Vollständigkeit der Axiome des logischen Funktionenkalküls. *Monatshefte für Mathematik und Physik*, *37*, 349–360. English translation in Gödel (1986), 102–123.

Gödel, K. (1986). In S. Feferman et al. (Eds.), *Collected works, volume I*. New York: Oxford University Press.

Hanf, W. (1974). Nonrecursive tilings of the plane. I. *Journal of Symbolic Logic*, *39*, 283–285.

Herbrand, J. (1931). Sur le probléme fondamental de la logique mathématique. *Sprawozdania z posiedze? Towarzystwa Naukowege Warszawskiego, Wydzial III*, *23*, 12–56. English translation in Herbrand (1971), 215–271.

Herbrand, J. (1971). In W. D. Goldfarb (Ed.), *Logical writings*. Dordrecht and Cambridge: Reidel and Harvard University Press.

Hilbert, D., & Ackermann, W. (1928). *Grundzüge der theoretischen Logik*. Berlin: Springer.

Hilbert, D., & Bernays, P. (1939). *Grundlagen der Mathematik* (Vol. 2). Berlin: Springer.

Jockusch, C. G., Jr., & Soare, R. I. (1972). Π_1^0 classes and degrees of theories. *Transactions of the American Mathematical Society*, *173*, 33–56.

Kleene, S. C. (1952). *Introduction to metamathematics*. Amsterdam: Van Nostrand Reinhold.

Kreisel, G. (1953). Note on arithmetic models for consistent formulae of the preciate calculus II. In *Proceedings of the XIth International Congress of Philosophy* (pp. 39–49). Amsterdam: North-Holland.

Mostowski, A. (1953). On a system of axioms which has no recursively enumerable model. *Fundamenta Mathematicae*, *40*, 56–61.

[1] I am grateful to Gary Ebbs and Carl G. Jockusch, Jr. for helpful comments.

Mostowski, A. (1955). A formula with no recursively enumerable model. *Fundamenta Mathematicae*, *43*, 125–140.
Putnam, H. (1957). Arithmetic models for consistent formulae of quantification theory. *Journal of Symbolic Logic*, *22*, 110–111.
Putnam, H. (1965). Trial and error predicates and the solution to a problem of Mostowski. *Journal of Symbolic Logic*, *30*, 49–57.
Shoenfield, J. R. (1959). On degrees of unsolvability. *Annals of Mathematics*, *69*, 644–653.

Warren Goldfarb is the Walter Beverly Pearson Professor of Modern Mathematics and Mathematical Logic at Harvard University. He was the editor of Jacques Herbrand's *Logical Writings* (1971), co-editor of Kurt Gödel's *Collected Works* vols. III–V (1995–2003), co-author with Burton Dreben of *The Decision Problem* (1979), and author of the textbook *Deductive Logic* (2003). He has published articles on mathematical logic and on the development of analytic philosophy, particularly Frege, Russell, Wittgenstein, Carnap, and Quine.

Chapter 5
Extendability and Paradox

Geoffrey Hellman and Roy T. Cook

Abstract In this essay we examine the *revenge* problem as it arises with respect to accounts of both the set-theoretic and the semantic paradoxes. First we review revenge as it arises in the set-theoretic setting – the Burali-Forti paradox – and outline its *modal-structural* resolution, highlighting the roles played by the logic of plurals, modal principles, and especially the extendability of models of set theory on this account. We then we turn to the semantic paradoxes, especially the Liar, and develop an analogy between the problems of expressive incompleteness and revenge affecting proposals to resolve the semantic paradoxes, on the one hand, and, on the other, the always incomplete and extendable nature of domains of sets on the modal structural approach to set theory. We then argue for a corresponding parallelism in resolutions of the set-theoretic and the semantic paradoxes. Focusing on recent accounts stemming from the work of Martin and Woodruff (Philosophia, 5(3):213–217, 1975) and Kripke (Journal of Philosophy, 72:690–716, 1975), we formulate a modal account of the extendability of languages on the Embracing Revenge account of the semantic paradoxes (see, e.g. Cook, Embracing Revenge: On the Indefinite Extendability of Language, 2007; Schlenker, Review of Symbolic Logic, 3(3):374–414, 2010) analogous to the formulation of extendability principles for set theoretic universes on the modal structural approach. Finally, however, we examine an interesting dis-analogy via a *meta-revenge* version of the Liar paradox that seems to have no analogue in the set-theoretic context, and we show how the solution to this puzzle also highlights even deeper connections between the modal-structural account of set theory and the Embracing Revenge account of truth and semantics.

5.1 Introduction

There is a broad consensus, amongst the majority of practicing set theorists and many philosophers of mathematics, that the famous set-theoretic paradoxes of Burali-Forti, Russell, and Cantor, which plagued the foundations of mathematics in the early

G. Hellman · R. T. Cook (✉)
University of Minnesota, Minneapolis, MN, USA
e-mail: roycookparadox@gmail.com

G. Hellman and R. T. Cook (eds.), *Hilary Putnam on Logic and Mathematics*,
Outstanding Contributions to Logic 9, https://doi.org/10.1007/978-3-319-96274-0_5

decades of the last century, were successfully resolved by systems of axiomatic set theory – especially those of Ernst Zermelo (Z and ZC) and of Zermelo and Abraham Fraenkel (ZF and ZFC). There is likely also a consensus, at least among philosophical logicians, that the semantic paradoxes, especially of truth and satisfaction, are without a generally accepted resolution. This is in spite of a spate of impressive contributions, beginning with the work of Martin and Woodruff (1975) and Kripke (1975) on fixed point languages containing their own truth predicates, leading to recent developments including Cook (2007, 2009), Field (2008), Schlenker (2010), and Tourville and Cook (2016).

The overarching theme of this paper is that the first consensus regarding the set-theoretic paradoxes – that the axiom systems of set theory are satisfactory for the purposes of pursuing mathematics – is too *optimistic*, since mere acceptance of one or the other of the axiomatic set theories in question is not really dispositive at the logico-philosophical level. The modal-structural (MS) approach to these paradoxes – implicit in the ground-breaking (Putnam 1967), developed in detail in Hellman (1989), and applied explicitly to the Burali-Forti paradox in Hellman (2011) – represents progress toward a dispositive, unified treatment, however – one both mathematically and philosophically satisfactory. Furthermore, our theme continues, the second point of consensus regarding the semantic paradoxes is too *pessimistic*, since resolutions closely analogous to those provided by the modal-structural approach in the set-theoretic case are now well on the way toward resolving the semantic paradoxes in a manner that may well be "best possible". In the final section of this paper, we will explore the extent of this analogy. Up to a point, it is fairly close; however, as we shall see, the "revenge" phenomenon is more resilient in the case of semantic paradoxes as compared with the set-theoretic; and the resolution of the semantic paradoxes thus requires correspondingly greater complexity.

5.2 The Burali-Forti Paradox

The Burali-Forti paradox is often referred to as the paradox of 'the largest ordinal', and can be obtained somewhat informally as follows: Let Ω be the class of all ordinals (say $\in$-well-orderings, beginning with the null set, following von Neumann's well-known construction). Then, since Ω represents the order-type of the well-ordering $<$ on the ordinals (i.e. $\in$ restricted to the ordinals), Ω itself qualifies as an ordinal. But then it has a successor, $\Omega + 1$, which is an ordinal and so must occur as a member of Ω, by definition of the latter as the class of all ordinals. But then we have that $\Omega + 1 < \Omega < \Omega + 1$; a contradiction. (Or, more directly, $\Omega \in \Omega$; whence $\Omega < \Omega$; contradiction.)[1]

How this is handled depends of course on how one formalizes set theory. On the standard set theory of contemporary mathematics – first-order ZFC – the "paradox"

[1] For some interesting history of the Burali-Forti paradox , see Ferreirós (2007), Menzel (1984), and Moore and Garciadiego (1981).

is blocked from the start as it cannot even be presented there, as the language of ZFC lacks any way of designating proper classes generally, hence no means of introducing Ω. Thus, ZFC resolves the Burali-Forti paradox in the same way that it resolves the Russell (the set of all and only non-self-membered sets) and Cantor (the set of all sets) paradoxes, by insisting that there simply cannot exist a *set* which contains all ordinals and hence dominates them (and itself) as an ordinal.[2]

While this might suffice for mathematical purposes, it should be clear that it is logically and philosophically unsatisfactory. By construction, $\in$ restricted to $\in$-well-orderings cannot be taken as a *set* of ordered pairs – that is, a well-ordered relation in the domain of the quantifiers (which only range over sets, and perhaps urelements). Nevertheless, equally obviously, there is such a well-ordered *relation* within any universe of objects satisfying the axioms of ZFC. At a minimum, expressed in the language of plural quantification, there exist all the ordered pairs of ordinals standing in the $\in$ relation (which is a well-ordering when restricted to $\in$) to one another. Hence this well-ordering (or plurality of pairs) *ought* to have an ordinal number representing it; after all, *that* is what ordinal numbers are *for*! But that leads straight to Ω, and then to the reasoning of the Burali-Forti paradox. In short, the 'resolution' offered by first-order ZFC is a paradigm of the ad hoc.

What happens if, instead, set theory is formalized as a two-sorted theory *á la* the set theory of von Neumann, Bernays, and Gödel (NBG), with explicit machinery for referring to proper classes and for distinguishing them from sets? Then Ω can be introduced via legitimate definition, but the definitional expression 'referring' to Ω denotes the (proper) class of all ordinals which are sets. So, even though NBG (unlike ZFC) recognizes the proper class of all ordered pairs constituting the ordering relation $<$ on set-ordinals, with Ω obviously representing its order-type, "$\Omega < \Omega$" is not even well-formed in NBG, since proper-class terms cannot occur in the first position of the $\in$-relation (and $<$ is just the restriction of $\in$ to ordinals). Once again, the reasoning of the Burali-Forti paradox cannot get off the ground, but only because the existence of certain ordinals (as objects) is blocked in an ad hoc manner.

The treatment of the Burali-Forti paradox in second-order systems such as ZFC2 is exactly analogous to that of two-sorted NBG. Again, however, while this is perhaps good enough for working mathematics, it is not good enough if one desires a genuine and philosophically satisfactory resolution of the Burali-Forti paradox, as (amongst other things) it violates the elementary principle that any ordinal have a unique successor. Indeed, it takes almost no mathematical inventiveness to 'generate', not only $\Omega + 1$; $\Omega + 2$; etc., but also $\Omega + \Omega$, $\Omega \times \Omega$; Ω^{Ω};the limit of $\Omega^{\Omega^{\cdot^{\cdot^{\cdot^{\Omega}}}}}$ and on and on, following the pattern already set with ω. The fact that such ordinals and the

[2]It is worth noting the oddness of the axiom of foundation in this respect. The presence of foundation, which rules out both the existence of self-membered sets and of infinitely 'descending' membership chains, is often invoked in explanations for why ZFC is not susceptible to the Russell, Cantor, and Burali-Forti paradoxes. But, in fact, foundation plays no role in blocking the paradoxes: it is not the fact that ZFC *proves* that the problematic sets do not exist, but rather the fact that ZFC minus foundation *does not prove* that these sets exist, that provides the explanation (if any) for the adequacy of ZFC as an alternative to naïve set theory. After all, adding an additional axiom – in this case, foundation – to a theory makes it more, rather than less, likely that the theory is inconsistent.

well-orderings they represent aren't countenanced in NBG or ZFC2 just shows that that both are arbitrarily restrictive, ad hoc theories. Philosophically, we are no better off with this 'resolution' than we were with first-order ZFC.

Let us now present the modal-structural reformulation and then resolution of the Burali-Forti paradox. First, we lay down some desiderata on a theory of ordinals. The first two are of a mathematical character; the third, which will command most of our attention, is more logico-philosophical:

1. It should be demonstrable that ordinals satisfy transfinite induction.
2. (A generalization of) Hartogs' theorem should hold, viz. that for any given ordinals, there exists a least strict cardinal upper bound of those ordinals (in the sense of having greater cardinality than that of any of the given ordinals).
3. Any well-ordered relation should be represented by a unique ordinal, in the sense that the pairs of the given well-ordered relation should be in one-one order-preserving correspondence with the pairs of ordinals strictly less than the representing ordinal (ordered by $<$).

Second, we introduce some familiar (and some less familiar) notation for plural quantification: Let us write:

- $x \prec yy$ as abbreviating "x is among the yy".
- $\alpha \prec yy$ as abbreviating "α is an ordinal among the yy".
- $x \prec \alpha\alpha$ as abbreviating "x is among the ordinals $\alpha\alpha$".
- $\alpha \prec \beta\beta$ as abbreviating "α is an ordinal among the ordinals $\beta\beta$".

In other words, we shall use Greek letters as abbreviations for quantification restricted to ordinals or pluralities of ordinals. Also, given a plural variable yy, $yy_<$ is a new plural variable ranging over the ordered pairs $\langle \alpha, \beta \rangle$ of ordinals α, β among the yy such that $\alpha < \beta$. Then desideratum (3) says that, for any given well-ordering R, construed as a plurality zz of pairs $\langle x; y \rangle$, there is a unique ordinal representing R – that is, there is a unique ordinal λ and bijective mapping f from the field of R to exactly the ordinals $\alpha\alpha < \lambda$ satisfying:

$$\langle x, y \rangle \prec zz \leftrightarrow \langle f(x), f(y) \rangle \prec \alpha\alpha_<$$

Note that this method of describing relations is general enough to apply both to relations as sets and relations as classes. But the advantage of plural variables (in the presence of an ordered pair function, which in the present context is supplied by ZFC2) is that they serve to replace variables ranging over objects of any given type.[3] Thus we have the means of expressing a theory of relations literally without treating relations as special objects distinct from the relata and pairs or n-tuples thereof.

Now let us demonstrate an elementary fact about ordinals on their natural well-ordering $<$ in the plural setting. Assume that the ordinals begin as do the von Neumann ordinals, with the null set $\varnothing$ representing the null well-ordering. Further, we will say that a plurality xx is *downward closed,* if and only if $\alpha < \beta$ and

[3]Talk of 'pluralities' is, as usual, strictly a *façon de parler*, not a reference to any special objects.

$\beta \prec zz$ entails $\alpha \prec zz$.[4] Then we have the following basic fact about ordinals as a good representation of well-ordered relations (e.g. the von Neumann ordinals):

Theorem 5.2.1 *If the $\beta\beta$ are downwards closed and represented by ordinal α; then α is the least strict upper bound of the $\beta\beta$, hence it is not the case that $\alpha \prec \beta\beta$.*

Proof By transfinite induction: The cases of the null well-ordering and of successor ordinals are trivial. For limits, λ, by hypothesis λ is the order-type of all its predecessors (under $<$); so if $\lambda \prec \beta\beta$, the order- type of the $\beta\beta$ has the predecessors of λ as a proper initial segment. Call the ordinals in this initial segment $\alpha\alpha$. But then the $\beta\beta$ cannot be order-isomorphic to the $\alpha\alpha$, since any bijective order-preserving map on the $\alpha\alpha$ to an extension thereof must be the identity map on the $\alpha\alpha$. This contradicts the assumption that λ represents the well-ordering of the $\beta\beta$.[5] □

The following is an immediate corollary:

Corollary 5.2.2 *Let T be an iterative set theory that defines a predicate 'ordinal' (e.g. à la von Neumann), where the ordinals index the ranks of a fixed cumulative hierarchy, along with a metatheory that assigns 'ordinal' a fixed, maximal extension over that hierarchy (as a proper class or plurality); then T does not satisfy Desideratum 3.*

Let T be as in the statement of the Corollary; then all the ordinals recognized by T, call them $\omega\omega$, are well-ordered by $<$. But if an ordinal represents this well-ordering, it would have to occur as one of the $\omega\omega$, which, by Theorem 5.2.1, is impossible. This is just the Burali-Forti 'paradox' in a new guise.[6]

What has emerged is the exact point at which we have a *choice*: Either we stick with the above instance of 'absolute generality' and give up on Desideratum 3, or we enforce the latter but deny the antecedent of Corollary 5.2.2 – that is, we deny that it makes sense to refer to 'absolutely all ordinals', or 'absolutely all well-ordered relations'. Standard set-theoretic presentations of the Burali-Forti paradox obscure this choice because they build in type distinctions which are then exploited in the ways reviewed above in order to block contradiction. But the formulation via plurals in effect collapses these type distinctions, allowing full generality in quantification over well-orderings. This not only avoids the ad hoc-ness detected in the traditional resolutions of the Burali-Forti paradox mobilized within standard set theory and reviewed above, it also opens up the choice just indicated. As a result, the question

[4]Note that a downwards closed plurality xx can 'contain' non-ordinals as well.

[5]That λ must be the *least* strict upper bound on the $\beta\beta$ follows similarly.

[6]Note that the proof of Corollary 5.2.2 (as well as the proof of Theorem 5.2.1, on which it depends) does not invoke Desideratum 2. The alert reader will have noticed that Desideratum 2 also leads to problems closely related to Burali-Forti. As usually stated, Hartogs' Theorem governs arbitrary sets of ordinals, but the generalization in the language of plurals applies to 'arbitrary collections', as it were. So again, if it makes sense to refer to 'absolutely all ordinals', then this would present an exception to Hartogs' Theorem, as there could then be no initial ordinal, cardinally – hence ordinally – greater than *all* ordinals! ('Initial ordinal' here just means 'ordinal cardinally greater than any earlier ordinal'.)

that now comes to the fore is how, without absolute generality, can we implement Desideratum 3?

The key idea goes back to Zermelo (1930): Quantification over ordinals (and sets generally) makes sense only relative to a hypothesized model of given set-theoretic axioms – in the case of Zermelo's focus, the axioms of ZFC2.[7] Hence, we cannot speak of 'all ordinals' in an absolute sense, only of all ordinals of/in a model $\mathcal{M}_1$: As is clear from the results above, the well-ordering of all such ordinals will not be represented by any ordinal of $\mathcal{M}_1$, but it will be represented by an ordinal of *any* proper extension $\mathcal{M}_2$ of $\mathcal{M}_1$, and Zermelo postulated that any model $\mathcal{M}_1$ of his axioms has such a proper extension $\mathcal{M}_2$ (indeed, a proper end-extension).[8]

In general, within any model $\mathcal{M}_1$, the well-ordered relations will outstrip the ordinals available to represent them in $\mathcal{M}_1$, but ordinals of any appropriate (i.e. 'large enough') proper extension $\mathcal{M}_2$ then become available for representing the well-ordered relations of $\mathcal{M}_2$. Of course, these extensions will then give rise to further well-ordered relations not representable by any ordinal of the least model in which they occur (as second-order relations or as pluralities of pairs), but they in turn will get represented in further extensions, and so on. This appears to be an attractive way to accommodate the spirit, if not the letter, of Desideratum 3 without falling prey to the Burali-Forti paradox.

But is it? Zermelo did not formalize the axioms of his (Zermelo 1930) paper, but if he had, he would have confronted a serious problem: had he stated the general comprehension principle for classes (and relations) usually taken as axiomatic in modern formalisms for second-order logic:

$$(\exists R)(\forall x_1)(\forall x_2)\ldots(\forall x_n)[R(x_1, x_2, \ldots x_n) \leftrightarrow \Phi(x_1, x_2, \ldots x_n)]$$

(where Φ does not contain R free) he would have seen that indeed the Burali-Forti paradox re-emerges on taking Φ to be the formula expressing that x is an ordinal in the domain of *some* model $\mathcal{M}$, for this collection of 'all ordinals contained in the domain of some model' would be well-ordered by a relation (loosely put, the union of the $<_{\mathcal{M}}$s where $<_{\mathcal{M}}$ is $\in$ restricted to ordinals in the domain of $\mathcal{M}$) having no ordinal as representative, violating Desideratum 3 after all. This is the Burali-Forti version of the 'revenge problem'.

---

[7] Here we mean 'model' in a broad sense, as any universe of discourse with an interpretation function assigning ordered pairs of elements of the universe to binary relation-symbol $\in$, along with standard interpretations of the plural quantifiers. We explicitly do not restrict 'model' to require that the domain be a set.

The use of second-order Replacement is essential in Zermelo's proofs of the quasi-categoricity of ZFC2 (see the next note).

[8] These conditions are best understood as only 'up to isomorphism'. A major achievement of Zermelo (1930) was to present a general proof of the quasi-categoricity of the ZFC2 axioms, viz. that given any two models (without urelements), one is an end-extension of a model isomorphic to the other.

It should also be noted that Hilary Putnam independently formulated an extendability principle very much in the spirit of Zermelo's in Putnam (1967), although it applied to models of Z rather than those of ZFC. Also, Putnam was the first to formulate it explicitly as a *modal* principal.

One way to preserve the core of Zermelo's idea while meeting Desideratum 3 is to introduce a *logico-mathematical* notion of modality, in terms of which we can formulate the following *extendability principle*[9]:

$$\mathsf{EP}_{\mathsf{MS}} : \Box(\forall \mathcal{M}_1)\Diamond(\exists \mathcal{M}_2)[\mathcal{M}_1 \subseteq_{\mathsf{end}} \mathcal{M}_2]$$

where the variables $\mathcal{M}_\alpha$ range over standard models of **ZFC2**, and:

$$\mathcal{M}_1 \subseteq_{\mathsf{end}} \mathcal{M}_2$$

indicates that $\mathcal{M}_2$ is an end-extension of $\mathcal{M}_1$. With the modal operators available, we have a new degree of freedom, and we can now adopt natural restrictions on the appropriate comprehension principles of second-order or plural logic, namely:

1. Only modal-free formulas are taken to define classes.
2. Classes or pluralities are guaranteed only to contain or encompass items existing within a given "world" (and similarly for relations). Thus, classes of objects drawn from the universes of different worlds are not recognized.[10]

Thus, for example, we do *not* write:

$$(\exists R)\Box(\forall x_1)\Box(\forall x_2)\ldots\Box(\forall x_n)[R(x_1, x_2, \ldots x_n) \leftrightarrow \Phi(x_1, x_2, \ldots x_n)]$$

or:

$$(\exists zz)\Box(\forall x)[x \prec zz \leftrightarrow \Phi(x)]$$

or even this preceded by $\Diamond$, let alone $\Box$; and we stipulate that the formula Φ occurring on the right-hand side of the comprehension schema:

$$(\exists R)(\forall x_1)(\forall x_2)\ldots(\forall x_n)[R(x_1, x_2, \ldots x_n) \leftrightarrow \Phi(x_1, x_2, \ldots x_n)]$$

or:

$$(\exists zz)(\forall x)[x \prec zz \leftrightarrow \Phi(x)]$$

contains no modal operators.[11] As a result, we cannot form, for example, 'the proper class of all possible ordinals' (or 'the plurality of all possible ordinals', or 'the property holding of all possible ordinals') – that is, the 'class of all ordinals of any possible model', nor can we even speak plurally of 'all possible ordinals'. Although

[9]For more on the modal structural approach generally, see Hellman (1989). The application of modal structuralism to the Burali-Forti paradox discussed here builds on and extends the account found in Hellman (2011).

[10]Officially, there is no need to quantify over worlds. Everything needed for pure and applied mathematics can be expressed directly via the modal operators.

[11]To get the effect of assertions of relations of structures across worlds, e.g. that two such are isomorphic, we can get by with additional assumptions of compossibility of models satisfying the relevant conditions. For further details, see chapter 1 of Hellman (1989).

of course we can speak of the class (or plurality) of all ordinals of a given hypothetical model, or all well-ordered relations on the domain of such a model, these totalities are subject to $\mathsf{EP_{MS}}$ and lead to no paradoxes or contradictions.

Are these restrictions on second-order comprehension really natural or are they ad hoc? We maintain the former. Indeed, they merely express a widely held *actualist* view of existence and modality. Put in terms of classes or collections, it makes no sense to speak of actual collections, or classes, of merely possible but non-actual things. At most we can speak of collections that *would* exist if there *were* such things. Similarly with plural formations: there are, in fact, only those items (or n-tuples of items) which *actually* exist; if there were or had been others, then there would be or would have been those items along with any (subpluralities) of them you like.[12]

How, though, does this block the cogency of speaking, not individually but plurally or collectively, of "any ordinals of any possible model"? The key here is to recognize that the the inference from "There might have been Φs and there might have been Ψs" – that is:

$$\Diamond(\exists xx)(\Phi(xx)) \wedge \Diamond(\exists yy)(\Psi(yy))$$

to "There might have been Φs and Ψs" – that is:

$$\Diamond(\exists xx)(\exists yy)(\Phi(xx) \wedge \Psi(yy))$$

is invalid. Of course, special cases of this inference do hold in the framework described here, but they depend on the compossibility of the Φs and the Ψs – an additional assumption. For example, in the modal-structural framework for set theory, where Φ is, say, "ordinal of a standard ZFC2 model of height κ", and Ψ is "ordinal of a standard ZFC2 model of height $\kappa' > \kappa$", $\mathsf{EP_{MS}}$ does in fact guarantee the requisite compossibility, and likewise for any finite generalization.[13] And transfinite generalizations can be (and are) licensed by stronger extendability principles – for example, via ensuring the existence of common proper extensions of a given α-sequences of models, where α is an ordinal of a given model. But nothing permits

[12]Indeed, these expressions of actualism involving plural constructions sound even more tautologous than those involving classes. In effect, we get instances of plurals comprehension such as:

There are only the things satisfying condition Φ that exist.

and, in counterfactual circumstances C:

There would only be those things satisfying Φ that would then exist.

[13]Note, however, that this is not guaranteed for arbitrary choices of Φ and Ψ: For example, if Φ is "pairs coding a bijective order-preserving map from all ordinals to all accessible ordinals", and Ψ is "pairs coding a bijective order-preserving map from all ordinals to all ordinals some of which are inaccessible", clearly there is no way for:

$$(\exists xx)(\exists yy)(\Phi(xx) \wedge \Psi(yy))$$

to be satisfied in a single (ZFC2) model.

generalizing to the compossibility of *all possible* models, and indeed such a generalization would contradict $\mathsf{EP}_{\mathsf{MS}}$. Now those in favor of an ontology with *possibilia* may remain unsatisfied with this resolution. The development of modal-structural interpretations, however, shows that their extravagances are just that, at least as far as recovering pure and applied mathematics is concerned.

It is important to notice, however – and this will become important in the comparison of MS and superficially similar fixed-point approaches to semantic paradox discussed below – that we *can* quantify *objectually* over all possible ordinals. For example, if we want to say that all ordinals whatsoever satisfy $\Phi(x)$, we can express this as:

$$\Box(\forall \mathcal{M})(\forall x)(x \in \mathsf{On}_{\mathcal{M}} \rightarrow \Phi(x))$$

where $x \in \mathsf{On}_{\mathcal{M}}$ expresses the claim that x is an ordinal (in model $\mathcal{M}$), and if we want to say that some ordinal satisfies $\Phi(x)$, we can express this as[14]:

$$\Diamond(\exists \mathcal{M})(\exists x)(x \in \mathsf{On}_{\mathcal{M}} \wedge \Phi(x))$$

Thus, we can speak of all possible ordinals, but not of the plurality of all possible ordinals, and hence not of the (or any) well-ordering on all possible ordinals. Since Desideratum 3, on the understanding developed here, only requires the existence of ordinals corresponding to well-orderings understood as (or as equivalent to) *existing* plurality or pairs, no Burali-Forti paradox emerges.

The Russell and Cantor paradoxes are handled, within the MS framework, along similar lines. First, it is worth noting that both paradoxes are blocked within more traditional approaches, such as those based merely on the adoption of NBG or ZFC2, via replacing naive comprehension:

For any predicate $\Phi(y)$ without x free:

$$(\forall x)(\exists y)(y \in x \leftrightarrow \Phi(y))$$

with Zermelo's *Aussonderung Axiom*:

$$(\forall z)(\exists x)(\forall y)(y \in x \leftrightarrow (y \in z \wedge \Phi(y)))$$

In short, there just is no set of all sets that are not members of themselves (Russell), or set of all sets whatsoever (Cantor). In ZF or ZFC, that is the end of the matter, as no other type of collections is recognized, while in NBG or ZFC2, there is another type of collection – the so called *proper classes* (which are explicitly the range of

[14]Actually, in some instances where we might wish to formalize informal claims of the form "some ordinal is Φ", stronger formalizations such as:

$$\Diamond(\exists \mathcal{M}_1)\Box(\forall \mathcal{M}_2)(\mathcal{M}_1 \subseteq_{\mathsf{end}} \mathcal{M}_2 \rightarrow (\exists x)(x \in \mathsf{On}_{\mathcal{M}}(x) \wedge \Phi(x)))$$

which express that there is an ordinal that satisfies $\Phi(x)$, and there will continue to be such an ordinal in any extension of the current model, will be more apt.

the second sort in NBG, and implicitly or indirectly the range of the second-order quantifiers in ZFC2) – which are not members of collections at all. This move blocks both the Russell and Cantor paradoxes, but, for much the same reasons as were presented above with regard to the Burali-Forti paradox, this account seems ad hoc and unsatisfactory from a logic-philosophical standpoint, since it violates the following analogue of Desideratum 3:

(3′) For any collection whatever, there should be a corresponding set whose elements are all and only the members of that collection.

The Russell and Cantor paradoxes show that this principle cannot be respected on the standard ways of formulating set theory. But we can once again apply Zermelo's insight: just as we cannot ask whether a set of all ordinals whatsoever exists *tout court*, but must instead ask whether such a collection exists relative to a model $\mathcal{M}$, we must similarly ask whether the set of all non-self-membered sets, or the set of all sets, exists relative to a particular model. And, just as the collection of all ordinals $\Omega_{\mathcal{M}}$ of a model $\mathcal{M}$ cannot exist in $\mathcal{M}$, but will exist in any end-extension of $\mathcal{M}$, the set of all non-self-membered sets of $\mathcal{M}$, and the set of all sets of $\mathcal{M}$, cannot exist in $\mathcal{M}$, but will exist in all end-extensions of $\mathcal{M}$.

Of course, as was the case with the Burali-Forti Paradox, this Zermelo-style resolution of the Russell and Cantor paradoxes breaks down when we consider the collection of all sets of any model. By definition, such a collection must be maximal, and hence lack any proper extension. And so we have the "revenge" of the Russell and Cantor paradoxes – that is, the paradox re-framed in terms of a maximal model built on the putatively unique maximal universe of 'all sets', construed in an absolute sense.

But once again, the MS framework comes to the rescue: relative to any given model $\mathcal{M}$, both a Russell collection and a Cantor collection (understood, say, as pluralities) arise, where neither is identifiable with a set in $\mathcal{M}$. But, given any such model, there is a *possible* extension of $\mathcal{M}$ and, in any such possible extension, that collection becomes representable as a set. There is no 'ultimate revenge' comparable to that facing Zermelo's non-modal framework, since there simply is not, and could not be, a collection of 'all possible sets', or of 'all possible non-self-members sets', just as there simply is not and could not be a collection or plurality of 'all possible well-orderings'. These locutions simply make no sense on a natural, actualist reading of the relevant modal operators in relation to the concept of *collection* (or *plurality*).

In sum, the modal-structural account of ordinals and well-orderings provides a natural way of meeting Desideratum 3 while blocking the Burali-Forti paradox (and corresponding moves applied to 'universal' sets and 'Russell' sets provide a similarly natural way of meeting Desideratum 3′). In Zermelo's terms, we must recognize limits on the tendency toward "all-embracing completeness", which seeks to recognize "ultimate infinities" – for example, "absolutely all ordinals", "absolutely all well-orderings", and the like. Instead, MS favors the opposing tendency of "creative progress", expressed in principles such as $\mathsf{EP}_{\mathsf{MS}}$. And then all the ordinals we ever could require become available for all the legitimate mathematical work they could be asked to perform.

Let us make a final comment on this way with the set-theoretic paradoxes: Should we think of it as 'embracing revenge'? In a sense, yes, since, at any stage in an ordinal progression of models of set theory, a transcending of any instance of one of the paradoxes is immediately undermined by an exactly analogous paradox at a higher level. Any such paradox can be transcended, but any such resolution is in turn shown not to be 'final' due to a higher-level paradox. But there is a sense in which the modal structural resolution does not embrace revenge: there simply is not and cannot be any 'ultimate' revenge, precisely because there cannot be any ultimate totality or plurality of 'absolutely all sets'.

5.3 Semantic Paradoxes

Martin and Woodruff (1975) and Kripke (1975) independently achieved a breakthrough by proving the existence of formal languages that, in a suitable sense, contain their own truth predicate via now-familiar fixed-point constructions. This constituted a major improvement on the then-prevailing formal theory of truth, due to Tarski (1944), according to which the semantics of a language $\mathcal{L}$ capable of expressing basic arithmetic cannot be wholly expressed within $\mathcal{L}$ itself, thereby engendering an infinite hierarchy of languages $\{\mathcal{L}_n : n \in \omega\}$, where each language $\mathcal{L}_{n+1}$ comes equipped with a 'new' truth predicate $\mathsf{T}_{n+1}(x)$ (or, equivalently, a satisfaction predicate $\mathsf{Sat}_{n+1}(x, y)$) that applies only to expressions in the preceding language in the hierarchy $\mathcal{L}_n$ (or, on some treatments, to all languages $\mathcal{L}_m$ such that $m \leq n$).[15]

Martin and Woodruff (1975) and Kripke (1975) both show how to improve on this by constructing formal languages (without any restrictions on self-reference or ungrounded reference chains) each of which contains a truth predicate that applies generally to sentences of the very same language, thereby avoiding semantic ascent to new levels of language.

Kripke (1975) illustrates the advantages of this move with a discussion of ordinary language contexts in which each of two parties (e.g. Richard Nixon and John Dean during the notorious Watergate hearings of the day) accuses the other of systematically lying, etc., and where it is indeterminate to which 'level of language' (i.e., to which of Tarski's languages $\mathcal{L}_n$) each assertion should be assigned. These advances over the Tarski hierarchy make use of non-classical logics which allow statuses other than *true* and *false* to be assigned to sentences – for example, the strong Kleene three-valued semantics, where a third status ρ_1 is assigned to sentences, and in particular to paradoxical sentences such as the Liar sentence[16]:

---

[15]As typically described, the hierarchy is defined only for languages indexed by finite ordinals, but it can be extended into the transfinite, a refinement inessential for our purposes here. The languages must be indexed by ordinals (or some other well-ordered collection of objects) where $\mathcal{L}_\alpha$ is an extension of $\mathcal{L}_\beta$ if and only if $\alpha > \beta$ – otherwise a version of the Yablo paradox arises, see Visser (2002).

[16]The use of the somewhat ambiguous term 'semantic status' is intentional, since one can read the additional value in Kripke's construction as a 'gap' (sentences assigned this value receive no genuine

This sentence is false.

In spite of this evident progress, however, these solutions were subject to a widespread phenomenon of 'revenge' - a phenomenon that seems to imply certain limitations in expressive power of the object language in question. The revenge phenomenon, in its simplest form, arises when we consider the *strengthened Liar*:

This sentence is either false or receives ρ_1 as its semantic value.

that seems to 'undo' the three-valued 'resolution' of the original Liar sentence (and related conundra such as the Curry and Yablo paradoxes). The strengthened Liar cannot be true, and cannot be false, and cannot be assigned ρ_1.

To see the connection between the revenge phenomenon and expressive resources, we need only consider the distinction between *choice negation*:

Φ	$\neg_{\mathsf{Ch}}\Phi$
$\top$	$\bot$
$\bot$	$\top$
ρ_1	ρ_1

and *exclusion negation*:

Φ	$\neg_{\mathsf{Ex}}\Phi$
$\top$	$\bot$
$\bot$	$\top$
ρ_1	$\top$

The construction found in Martin and Woodruff (1975) and Kripke (1975) requires that we use choice negation – hence, the intuitively true sentence:

The Liar sentence is not$_{\mathsf{Ex}}$ true.

(where the subscript indicates that we intend the negation to be understood as an informal natural language analogue of exclusion negation) cannot be expressed or represented in the formal object language $\mathcal{L}$. Further, if we extend the language via the addition of exclusion negation $\neg_{\mathsf{Ex}}$, then we can then express the strengthened Liar, and as a result the three-valued semantics is inadequate.

Thus, it appears that a higher-level Tarski-style meta-language, $\mathcal{L}'$, is required to express this obvious truth. As Kripke acknowledged at the end of his paper, "The ghost of the Tarski hierarchy is still with us." (Kripke 1975: 714)[17] In short, just as

truth value), a 'glut' (sentences assigned this value receive more than one genuine truth value), or in any number of other ways. Similar comments apply to the 'Embracing Revenge' semantics sketched below, which adds infinitely many such additional statuses. Different such readings will result in different classes of designated values, and hence different consequence relations. Although we prefer the 'gappy' reading, all of the arguments presented below affect only the semantics, and are thus compatible with consequence relations corresponding to 'glutty' readings (and other sorts of readings) of that semantics.

[17] The parallelism between this general reasoning and that of Tarski's proof of his celebrated theorem on the indefinability of arithmetic truth in the language of arithmetic is, of course, not accidental.

traditional (non-modal) solutions to the Burali-Forti paradox 'succeed' by refusing to acknowledge the existence of collections whose existence seems undeniable, the Kripke/Martin and Woodruff approach can only succeed via denying that we can express notions, such as exclusion negation, that we can clearly express.

Recognition of these difficulties led to a fresh round of interesting, original work on the semantic paradoxes, some leading examples of which have only recently appeared. Two bodies of work within this recent trend will command our attention here:

1. Hartry Field's *paracomplete* theory (in which the law of excluded middle fails in full generality).
2. Roy T Cook's *Embracing Revenge* (ER) approach, independently developed and pursued by Philippe Schlenker.[18]

Let us consider these in turn.

Field's work takes up where (Kripke 1975) left off, seeking to improve on the situation inherited from that seminal work, specifically seeking fixed-point languages that:

- Allow for the machinery of self-reference.
- Contain their own truth predicates.
- Respect Tarski's T-schema:

$$\mathsf{T}(\ulcorner \Phi \urcorner) \leftrightarrow \Phi$$

- Assign a truth-value gap to paradoxical and other *ungrounded* sentences.
- Contain a logically well-behaved conditional improving on the truth-functional conditional of classical systems and Kleene's three-valued ones.[19]

Field, via a rather complicated variation on the construction of Kripke (1975), proves the existence of fixed point languages meeting the desiderata listed above – see Field (2008) for the full details, which need not detain us unduly here. Irrespective of the criticisms leveled below, it is clear that Field's construction amounts to significant progress, and is a genuine improvement on, the simple fixed-point approach of Martin and Woodruff (1975) and Kripke (1975) upon which it builds.

There are, however, a number of problems and limitations affecting Field's fixed-point languages.[20] The main one of interest here concerns the intertwined issues

[18] Cook's most recent work on this topic has involved collaboration with Nicholas Tourville – see, e.g. Tourville and Cook (2016).

[19] Note that, in a Kripke/Martin-Woodruff style three-valued semantics, there is no *monotonic* binary truth-functional connective $*(A, B)$ such that $*(A, A)$ is always true (other than the trivial connective that outputs true for any arguments). Thus there is no truth-functional connective (other than the trivial one) $\rightarrow^*$ such that, for any expression Φ, $\Phi \rightarrow^* \Phi$ is a theorem. Hence, for the Tarski T-scheme to even be *expressible* in a fixed-point language, we require a better, non truth functional conditional. For good, general assessments of these sorts of issues within Kripke's fixed-point approach, see, Gupta (1982) and Hellman (1985).

[20] For a useful overview of these sorts of objections, see Priest (2010). We restrict our attention to objections relevant to the task at hand.

of expressive incompleteness and revenge. In brief, Field's fixed-point languages, like Kripke's (and like the view developed by Cook, Schlenker, and Tourville), are subject to the following general type of revenge argument: Let us suppose that $\mathcal{L}$ is a (Kripkean or Fieldian) fixed-point language with truth predicate $\mathsf{T}(x)$, and suppose $\mathcal{L}$ has the resources to express the following predicate:

> x (is a name of, or the Gödel code of, etc.) a sentence that receives some semantic value provided by the semantics for $\mathcal{L}$ other than that corresponding to $\mathsf{T}(x)$

Let us assume, for simplicity, that $\mathcal{L}$ expresses this predicate in terms of a (primitive or defined) exclusion negation and the truth predicate – that is, as $\neg_{\mathsf{Ex}}\mathsf{T}(x)$. Then, via standard diagonalization techniques, we can construct a sentence λ_{Sup} – the *super-Liar* – that is equivalent to:

$$\neg_{\mathsf{Ex}}\mathsf{T}(\ulcorner\lambda_{\mathsf{Sup}}\urcorner)$$

Surely the $\mathcal{L}$ truth theorist, who of course grasps the semantics for $\mathcal{L}$, finds the predicate in question – and hence the sentence λ_{Sup} – intelligible. But this sentence cannot be assigned any semantic value provided by the semantics of $\mathcal{L}$:

Theorem 5.3.1 *Let $\mathcal{L}$ be a fixed point language with the usual machinery of self-reference, and let $\mathcal{T}_{\mathcal{L}}$ be a truth theory for $\mathcal{L}$ respecting the Tarski T-scheme via allowing intersubstitutivity of S and $\mathsf{T}(\ulcorner S\urcorner)$ in all extensional contexts, and suppose that $\mathcal{L}$ is able to express exclusion negation $\neg_{\mathsf{Ex}}$. Then $\mathcal{T}_{\mathcal{L}}$ is inconsistent.*

Proof Let λ_{Sup} be the sentence $\neg_{\mathsf{Ex}}\mathsf{T}(\ulcorner\lambda_{\mathsf{Sup}}\urcorner)$. Suppose that λ_{Sup} is true; then by the Tarski T-scheme, we have that $\mathsf{T}(\ulcorner\lambda_{\mathsf{Sup}}\urcorner)$ is true; but by the construction of λ_{Sup}; we also have that $\neg_{\mathsf{Ex}}\mathsf{T}(\ulcorner\lambda_{\mathsf{Sup}}\urcorner)$. Contradiction. This establishes that λ_{Sup} must receive some semantic value other than the true, hence that $\neg_{\mathsf{Ex}}\mathsf{T}(\ulcorner\lambda_{\mathsf{Sup}}\urcorner)$ is true. By the construction of λ_{Sup}, this entails that λ_{Sup} is true. Contradiction. Hence $\mathcal{T}_{\mathcal{L}}$ is inconsistent. □

Therefore, as an immediate corollary, if the truth theory for $\mathcal{L}$ is indeed consistent, then $\mathcal{L}$ cannot express $\neg_{\mathsf{Ex}}$.

This strong version of the *revenge argument* shows the futility of seeking a *single* such language $\mathcal{L}$ and a theory of truth $\mathcal{T}_{\mathcal{L}}$ for $\mathcal{L}$ where:

- $\mathcal{L}$ is *expressively adequate* (in particular, $\mathcal{L}$ can express exclusion negation).
- $\mathcal{T}_{\mathcal{L}}$ is *consistent* (and hence *revenge proof*).

That is asking the impossible. The situation is similar to the one posed for Kripke's three-valued semantics – in fact, Kripke's comments about the "ghost of the Tarskian hierarchy" amount to nothing more than a recognition of a particular instance of this theorem, and the corresponding limitation within Field's system is another instance: neither semantics allows us to express exclusion negation for the given language in question.

Now Field seeks to improve upon this by adding a *determinately true* operator D to the language, so that $\neg_{\mathsf{Ch}}\mathsf{D}(\mathsf{T}(\ulcorner\Phi\urcorner))$ (equivalently, $\neg_{\mathsf{Ch}}\mathsf{D}(\Phi)$) is weaker than

$\neg_{\mathsf{Ch}}\mathsf{T}(\ulcorner\Phi\urcorner)$. In a sense, this does result in an improvement, for, as Field points out, the natural attempt to formulate a *determinately* strengthened Liar sentence λ_{D} equivalent to $\neg_{\mathsf{Ch}}\mathsf{D}(\mathsf{T}(\lambda_{\mathsf{D}}))$ is blocked since excluded middle fails for the complex negation operator $\neg_{\mathsf{Ch}}\mathsf{D}$. But even if *excluded middle* were accepted, paradox is blocked: although λ_{D} obviously entails $\neg_{\mathsf{Ch}}\mathsf{D}(\mathsf{T}(\lambda_{\mathsf{D}}))$, λ_{D} does not entail $\mathsf{D}(\mathsf{T}(\lambda_{\mathsf{D}}))$, hence no contradiction follows. In short, the analogue of the T-scheme with D, or, more carefully, $\mathsf{D}(\mathsf{T}(x))$, replacing $\mathsf{T}(x)$ does not hold. We can rest with λ_{D} being both true and not determinately true.

So far so good; but, although $\neg_{\mathsf{Ch}}\mathsf{D}(\mathsf{T}(\ulcorner\Phi\urcorner))$ is weaker than $\neg_{\mathsf{Ch}}\mathsf{T}(\ulcorner\Phi\urcorner)$, it still is not as weak as $\neg_{\mathsf{Ex}}\mathsf{T}(\ulcorner\Phi\urcorner)$, which says that Φ is something other than true (and therefore other than determinately true as well, since $\mathsf{D}(\mathsf{T}(\ulcorner\Phi\urcorner))$ entails $\mathsf{T}(\ulcorner\Phi\urcorner)$).[21] For, as we've just seen, if the language can express $\neg_{\mathsf{Ex}}$, we can construct a sentence λ_{Sup} equivalent to $\neg_{\mathsf{Ex}}\mathsf{T}(\ulcorner\lambda_{\mathsf{Sup}}\urcorner)$, as above, and then we have, not only that λ_{Sup} entails $\neg_{\mathsf{Ex}}\mathsf{T}(\ulcorner\lambda_{\mathsf{Sup}}\urcorner)$; but also that λ_{Sup} entails $\mathsf{T}(\ulcorner\lambda_{\mathsf{Sup}}\urcorner)$ (by the Tarski truth scheme), whence $\neg_{\mathsf{Ex}}\lambda_{\mathsf{Sup}}$, and with a few more steps, as above, we get a contradiction in the truth theory of the fixed point language itself. Thus, despite the success of determinacy operators, the revenge impulse still gets its way. (As Field and Priest both remark: It's a mean old world.)[22]

In light of the apparent inevitability of revenge, the second approach, developed in Cook (2007, 2009), Schlenker (2010), and Tourville and Cook (2016), builds this phenomenon into the theory, thereby seeking to make a virtue out of necessity.[23] This Embracing Revenge view, or ER, involves recognizing an endless hierarchy of fixed-point languages $\{\mathcal{L}_\alpha : \alpha \in \mathsf{On}\}$, each interpreted in terms of a semantics containing (at least) one new semantic value not utilized in the semantics for predecessor languages (the so-called *pathological* values ρ_α), and each containing a yet weaker approximation of exclusion negation – a *weak relativized* negation $\neg^{\mathsf{W}}_{\alpha}$ – inexpressible in predecessor languages. In contrast with the superficially similar Tarskian hierarchy, however, there is a single, univocal and transparent truth predicate $\mathsf{T}(x)$ that applies to all sentence in all languages in the hierarchy.

In common with Kripke (1975) and Field (2008), the truth theories of each of the fixed-point languages $\mathcal{L}_\alpha$ are consistent (at least relative to a rich mathematical

---

[21] Field goes further, and introduces a hierarchy of successively weaker negations $\neg_{\mathsf{Ch}}\mathsf{D}$, $\neg_{Ch}\mathsf{DD}$, $\neg_{Ch}\mathsf{DDD}$, etc. Although each of these is weaker than the next, none is equivalent to $\neg_{\mathsf{Ex}}$, on pain of contradiction.

[22] Actually, Field resists the dilemma posed by what we are calling exclusion negation for the given fixed-point language, arguing that, since the preferred logic for handling the paradoxes is non-classical and excluded middle does not hold (on his paracomplete approach), what we have been calling *exclusion negation* (for a given fixed-point language) is not really a legitimate notion. This is a natural result for a paracomplete theory of the semantic paradoxes, since the acceptance of the inter-substitutivity of Φ and $\mathsf{T}(\ulcorner\Phi\urcorner)$ forces a renunciation of excluded middle for any notion of negation expressible in the language. And exclusion negation, by definition, obeys excluded middle. As we will see momentarily, however, there is an attractive way of transcending this situation.

[23] The technical details differ among Cook (2007), Cook (2009), Schlenker (2010), and Tourville and Cook (2016). Here we will follow the most recent, and most powerful, formulation of the view – that found in Tourville and Cook (2016) – but all of the points made here also apply to the expressively weaker formal systems found in the earlier papers.

background, e.g. ZFC): each avoids assigning inappropriate semantic values to paradoxical sentences in a particular language $\mathcal{L}_\beta$ via assigning the most recently added pathological value (ρ_β) to such sentences, slaying the semantic dragons at that level. This works since each such language $\mathcal{L}_\beta$ is subject to the expressive incompleteness necessitated by the theorem proven above: the most recently added negation $\neg^{\mathsf{W}}_{\beta}$ fails to behave like $\neg_{\mathsf{Ex}}$ on ρ_β (but behaves like $\neg_{\mathsf{Ex}}$ on all other values). The next language $\mathcal{L}_{\beta+1}$, however, contains a better negation $\neg^{\mathsf{W}}_{\beta+1}$ that behaves exactly like $\neg_{\mathsf{Ex}}$ on all semantic values applied to sentences in $\mathcal{L}_\beta$ (including ρ_β), although this new vocabulary generates new semantic dragons, which must in turn be slain via the addition of a new value $\rho_{\beta+1}$. And so on. This back-and-forth goes on ad infinitum. On this view, there simply cannot be a single, all-embracing, expressively complete fixed-point language that can express all semantic notions any more than there can be a single, all-embracing, inextendable universe for set theory.

Reflection on this last point leads us to recognize that, in a sense, ER is, indeed, revenge-free: There just is no limit to possible semantic values, and literally no sense to be made of referring to *absolutely all* of them, in any possible language. In short, there is no *ultimate exclusion negation*, and hence no corresponding ultimate version of the Liar, since for any language interpreted in terms of some possible collection of semantic values, there is an extension of that language whose interpretation requires additional semantic values. In the metaphor of the Raphael painting, a detail of which appears on the dust jacket of Field's book, every possible dragon threatening our semantic account is slayable, indeed, each such dragon generated in a particular language is slain at that language, and anything we wish to say about such dragons (such as saying that they have a semantic value from the semantics for the language in which they are introduced other than the true) can be said in the next language.

However, although there may be no *ultimate revenge*, there is a worry regarding a kind of *meta-revenge* that arises by reflecting on the entire hierarchy of languages comprising the Embracing Revenge framework. The route to the *meta-Liar* and the problem that it poses begins with explicitly recognizing the similarity between the hierarchy of languages that is central to the ER account and the hierarchy of models of set theory central to the MS account discussed in the previous section. Just as the MS view involves a rejection of a single, all-encompassing universe of sets in favor of a (modalized) hierarchy of increasingly ontologically rich, but never richest possible, universes of sets, the ER view involves rejecting a single, all-encompassing language within which everything can be said in favor of a hierarchy of increasingly expressively rich, but never richest possible, languages. This suggests that the best way of understanding the ER view is in terms of an *extendability principle* for languages along the lines of the extendability principle $\mathsf{EP}_{\mathsf{MS}}$ introduced for models of set theory in the previous section – that is, in terms of the following:

$$\mathsf{EP}_{\mathsf{ER}} : \Box(\forall \mathcal{L}_1)\Diamond(\exists \mathcal{L}_2)[\mathcal{L}_1 \subseteq_{\mathsf{sem}} \mathcal{L}_2]$$

where the variables range over the languages in the hierarchy described above, and:

$$\mathcal{L}_1 \subseteq_{\mathsf{sem}} \mathcal{L}_2$$

expresses the claim that $\mathcal{L}_2$ is an extension of $\mathcal{L}_1$ in the sense outlined above – that is, $\mathcal{L}_2$ can express all notions required to describe the semantics of $\mathcal{L}_1$ (including the inclusion of the restricted exclusion negation $\neg^{\mathsf{W}}_{\alpha+1}$ that behaves exactly like the illegitimate $\neg_{\mathsf{Ex}}$ on sentences of $\mathcal{L}_1$).

Something like $\mathsf{EP}_{\mathsf{ER}}$ seems to capture exactly what is at issue in the ER account. Further, the obvious similarity to the original set-theoretic $\mathsf{EP}_{\mathsf{MS}}$ also seems to capture an interesting and important connection between ER as a response to the semantic paradoxes and semantic versions of the revenge phenomenon on the one hand, and MS as a response to the set theoretic paradoxes and set theoretic versions of revenge (i.e. Burali-Forti) on the other.

The (purported) trouble with $\mathsf{EP}_{\mathsf{ER}}$ arises as follows[24]: If $\mathsf{EP}_{\mathsf{ER}}$ is a legitimate way to describe the hierarchy of languages described informally in, for example, Tourville and Cook (2016), then it (and all of the resources required to express it, including both the relevant modal notions and quantification over languages) must occur in some possible language $\mathcal{L}_\alpha$ in the hierarchy – after all, one of the main claims of ER is that anything that can be said can be said in some such $\mathcal{L}_\alpha$. But then that $\mathcal{L}_\alpha$ has the resources to express the following predicate[25]:

$$\Diamond(\exists \mathcal{L}_\alpha)(\exists \neg^{\mathsf{W}}_\beta \in \mathcal{L}_\alpha)(\neg^{\mathsf{W}}_\beta(\mathsf{T}(x)))$$

That is, any language $\mathcal{L}_\alpha$ that can express $\mathsf{EP}_{\mathsf{ER}}$ can also express the predicate that applies to (the name or Gödel code of) a sentence if and only if the application of *some* possible approximation $\neg^{\mathsf{W}}_\beta$ of exclusion negation to (the application of the truth predicate to the name or Gödel code of) that sentence is true. But then, via standard diagonalization techniques, we can construct a sentence λ_{Meta} – the *meta-Liar* – that is equivalent to:

$$\Diamond(\exists \mathcal{L}_\alpha)(\exists \neg^{\mathsf{W}}_\beta \in \mathcal{L}_\alpha)(\neg^{\mathsf{W}}_\beta(\mathsf{T}(\ulcorner \lambda_{\mathsf{Meta}} \urcorner)))$$

It would seem that λ_{Meta} cannot have any truth value introduced in the hierarchy described by ER however: If λ_{Meta} were true, then there would have to be some language $\mathcal{L}_\beta$ and $\neg^{\mathsf{W}}_\beta \in \mathcal{L}_\beta$ such that $\neg^{\mathsf{W}}_\beta(\mathsf{T}(\ulcorner \lambda_{\mathsf{Meta}} \urcorner))$, and hence $\neg^{\mathsf{W}}_\beta(\lambda_{\mathsf{Meta}})$, is true. But, for every β, $\neg^{\mathsf{W}}_\beta(\Phi)$ is true only if Φ fails to be true. Contradiction. So

---

[24] We are indebted to Kit Fine for this observation.

[25] The alert reader might notice that the notation of this purported meta-liar is potentially problematic: the predicate should attribute truth to the result of concatenating the βth-level negation (construed as a *syntactic* object) with the formal truth-predicate (followed by a free variable). As written, however, the negation sign is being *used* rather than merely *mentioned*, and, as will soon emerge, this turns out to be the key to ER's immunity to any purported meta-Liar. Now, in the case where $\beta \leq \alpha$ (and hence $\neg^{\mathsf{W}}_\beta \in \mathcal{L}_\alpha$), the predicate:

$$\neg^{\mathsf{W}}_\beta(\mathsf{T}(x))$$

is equivalent to the explicit formulation just described via an application of the relevant Tarski T-sentence. In such a case ($\beta \leq \alpha$), the βth level exclusion negation is available in $\mathcal{L}_\alpha$ for *use*, so the disquotation is harmless. But of course, disquotation is entirely illegitimate (in fact, impossible!) when $\beta > \alpha$, as explained below.

λ_{Meta} fails to be true. So λ_{Meta} is either false or receives a pathological value ρ_γ as its semantic value. But then there is a language $\mathcal{L}_\beta$ and $\neg^{\mathsf{W}}_\beta$ such that $\neg^{\mathsf{W}}_\beta(\mathsf{T}(\ulcorner\lambda_{\mathsf{Meta}}\urcorner))$, and hence $\neg^{\mathsf{W}}_\beta(\lambda_{\mathsf{Meta}})$, is true – we need merely pick any β such that $\beta \geq \gamma$ (or $\beta = 1$ if λ_{Meta} is false). But then λ_{Meta} is true. Contradiction.

One way to understand what is going on is to note that the predicate described above *appears* to be an predicate-analogue of exclusion negation – that is, loosely put, within any language $\mathcal{L}_\alpha$ that can express $\mathsf{EP}_{\mathsf{ER}}$ in the first place, and given any sentence Φ, the following equivalence *seems* to hold:

$$\neg_{\mathsf{Ex}}(\Phi) \leftrightarrow \Diamond(\exists\mathcal{L}_\alpha)(\exists\neg^{\mathsf{W}}_\beta \in \mathcal{L}_\alpha)(\neg^{\mathsf{W}}_\beta(\ulcorner\Phi\urcorner))$$

If this is right, then no $\mathcal{L}_\alpha$ in the ER hierarchy – and thus no language at all – can express the right hand side of this equivalence, and hence no language can express $\mathsf{EP}_{\mathsf{ER}}$.

At first glance this apparent restriction on the expressive resources available to the ER theorist might seem unsurprising: after all, the ER approach is predicated on the idea that no possible language can be semantically and expressively complete, and hence for every language $\mathcal{L}_\alpha$ there is a further, more expressive language $\mathcal{L}_{\alpha+1}$. On the other hand, if there can be no such expressively complete language, and no language that allows for quantification over all languages (and hence over all negations) in the way required to formulate $\mathsf{EP}_{\mathsf{ER}}$, then the ER approach is vitiated, since the account seems to rule out the expressibility of notions required to adequately describe the view in the first place.

This way of characterizing the problem also points to the solution, however. We need merely ask: if, according to the ER account, there is no language within which we can quantify over all possible languages $\mathcal{L}_\beta$, or all possible weak relativized negations $\neg^{\mathsf{W}}_\beta$, or all semantic values ρ_β, then what language have we been using when describing the possible languages $\mathcal{L}_\beta$, the possible weak relativized negations $\neg^{\mathsf{W}}_\beta$, and the semantic values ρ_β? In short, within which $\mathcal{L}_\beta$ is this very paper written?

This question is a variant of a general puzzle that plagues any view that rejects absolutely general quantification: If we can't quantify over all objects, then how can we express the claim that we can't quantify over all objects? The answer to this more general puzzle, applied to the ER account, was already given in the first essay on the topic:

> How can we claim that we can never talk about all truth-values at once, so the criticism goes, when we obviously quantified over all of them in the formal account given in the previous section? The easy answer to this question is that it misrepresents what exactly the formal model is doing. In particular, this objection confuses *describing* a language and *using* that very language.
>
> The formal semantics [...] is a description (i.e. a model, in the intuitive sense of "model") of a sequence of possible language extensions. No semantic predicates are used in describing this mathematical structure – the account is (or, can be reformulated) within first-order set theory. As a result, the formal model (can) occur in our (actual) base language corresponding to $\mathcal{L}_0$ (and, in fact, this is the proper place for such theorizing). Thus, the account of the formal semantics does not occur within a language that *uses* all of the semantic notions which it *describes* as occurring in the hierarchy. (Cook 2007: 46–47)

In short, all of the prose regarding possible languages, negations, semantic values, and so on in both that paper and the present one should be interpreted as occurring within whatever set theory is contained in our initial language $\mathcal{L}_0$ (or in the purely set-theoretic portion of any extension of it $\mathcal{L}_\beta$), and the definitions, proofs, etc., given above are not really directly about the (entire) hierarchy of languages itself, but are instead about either (i) our current language $\mathcal{L}_\alpha$ (that, after all, is what the transparent truth predicate $\mathsf{T}(x)$ is for) or (ii) the formal, set-theoretical models of initial segments of the hierarchy of languages that are (or can be) constructed in our current language $\mathcal{L}_\alpha$. In short, whenever we say things like "Every semantic status such-and-such" or "Every language such-and-such" we are not actually talking directly about the entire hierarchy of languages (not even in this very sentence), but are instead either (i) talking directly about the current language of set theory within which such claims are formulated (or sub-languages thereof), or, in the more interesting case, (ii) talking directly about our set-theoretic model(s) of an initial segment of the hierarchy of languages as formulated within our current set theory (and hence, only indirectly speaking about the corresponding initial segment of the hierarchy of languages themselves).[26]

This brings up an important point. The languages, and expressive resources, that can be modeled (hence mentioned) in a particular language $\mathcal{L}_\alpha$ will, in general, far outstrip the linguistic resources used in $\mathcal{L}_\alpha$ itself. For example, we can construct models of all languages $\mathcal{L}_\beta$ for β less than the first strong inaccessible cardinal in $\mathcal{L}_0$ if $\mathcal{L}_0$ is (classical) $\mathsf{ZFC2}$. As a result, within any $\mathcal{L}_\alpha$ in the hierarchy (or within any formal model of such an $\mathcal{L}_\alpha$) there will be many expressions that we can mention in $\mathcal{L}_\alpha$ (since, loosely put, we can model them) but not use in $\mathcal{L}_\alpha$ (since they 'belong' to later languages in the hierarchy).

With this in mind let us return to the problematic predicate that, via diagonalization, generated the meta-Liar:

$$\Diamond(\exists \mathcal{L}_\alpha)(\exists \neg^{\mathsf{W}}_\beta \in \mathcal{L}_\alpha)(\neg^{\mathsf{W}}_\beta(\mathsf{T}(x))).$$

Recall that this predicate must, in some sense, be formulable in some $\mathcal{L}_\alpha$ if $\mathsf{EP}_{\mathsf{ER}}$ is expressible. We have now seen two distinct ways that we might read the quantifiers ranging over negations and over languages in this predicate: as ranging over expressions (and collections of expressions) that can be used in $\mathcal{L}_\alpha$, and as ranging over expressions (and collections of expressions) that can be mentioned (or modeled) in $\mathcal{L}_\alpha$.

On the first reading the quantifier ranging over languages becomes superfluous, hence the quantifier ranging over negations just ranges over those negations in $\mathcal{L}_\alpha$, and the supposed meta-Liar is nothing more than a strengthened Liar sentence that must receive ρ_α as its semantic value. Note further that $\mathsf{EP}_{\mathsf{ER}}$ is false on this interpretation of the quantifiers in question, since from the perspective of $\mathcal{L}_\alpha$ there is a maximal last possible language – $\mathcal{L}_\alpha$ itself.

---

[26]Hartry Field makes a similar move in Field (2003), from which we have brazenly stolen the idea (see also Field 2008).

Much more interesting, however, is the second reading, where the quantifiers range over the languages (and negations) that we can model, and hence mention, in $\mathcal{L}_\alpha$. On this reading $\mathsf{EP}_{\mathsf{ER}}$ is true (assuming, at least, that the set theory contained in $\mathcal{L}_\alpha$ proves that there is no last ordinal), since $\mathsf{EP}_{\mathsf{ER}}$ only requires that we *mention* the languages in question, and not that we can *use* them in $\mathcal{L}_\alpha$ (the claim in question is analogous to the claim that there are many non-classical logics that can be modeled in classical ZFC). But the predicate used in the construction of the meta-Liar is, on this reading, not well-formed, since it *uses* the negation(s) in question, and the fact that we can *model* or *mention* a negation within $\mathcal{L}_\alpha$ no more entails that we can use that negation in $\mathcal{L}_\alpha$ than the fact that we can *model*, and hence *mention*, non-classical negations in classical ZFC entails that we can use non-classical negations when reasoning in classical mathematical theories.[27]

Thus, the extendability principle $\mathsf{EP}_{\mathsf{ER}}$ is both expressible and true in each language $\mathcal{L}_\alpha$ – that is, in each such language we can express the (true) claim that there is no last (formal model of a) language $\mathcal{L}_\beta$ describable in terms of the resources of $\mathcal{L}_\alpha$ – but no such language $\mathcal{L}_\alpha$ quantifies over all such languages $\mathcal{L}_\beta$. Instead, the

[27]Of course, we could obtain something like a meta-Liar for $\mathcal{L}_\alpha$ by *adding* an operator $\mathsf{meta}_\alpha(x)$ such that:

- $\mathsf{meta}_\alpha(x)$ is true of x if and only if there is some weak negation $\neg^{\mathsf{W}}_\beta$ that can be modeled in $\mathcal{L}_\alpha$ such that $\neg^{\mathsf{W}}_\beta\mathsf{T}(\langle x\rangle)$ is true of x.
- $\mathsf{meta}_\alpha(x)$ applied to x is false otherwise.

(Note that β need not be less than α). The sentence that results from diagonalizing on such a notion would express a version of the meta-Liar – albeit one relativized to $\mathcal{L}_\alpha$ – and would not be interpretable in terms of the semantics for $\mathcal{L}_\alpha$. The $\mathsf{meta}_\alpha(x)$-Liar *would*, however, be interpretable on the semantics for some $\mathcal{L}_\beta$ for sufficiently large β – large enough such that any negation that can be modeled/mentioned in $\mathcal{L}_\alpha$ can be used on $\mathcal{L}_\beta$. But this still does not get us a genuine meta-Liar, since we can repeat the construction, adding a new $\mathcal{L}_\beta$-relative notion $\mathsf{meta}_\beta(x)$ such that:

- $\mathsf{meta}_\beta(x)$ is true of x if and only if there is some weak negation $\neg^{\mathsf{W}}_\gamma$ that can be modeled in $\mathcal{L}_\beta$ such that $\neg^{\mathsf{W}}_\gamma\mathsf{T}(\langle x\rangle)$ is true of x.
- $\mathsf{meta}_\beta(x)$ applied to x is false otherwise.

But now we can construct a new $\mathcal{L}_\beta$-relativized meta-Liar, which (assuming that the set theory in $\mathcal{L}_\beta$ guarantees the existence of $\aleph_\gamma$ for any $\gamma < \beta$) cannot be interpreted in the semantics for $\mathcal{L}_\beta$, but can be interpreted on the semantics for any language high enough in the hierarchy such that it allows one to use any negation that can be mentioned/modeled in $\mathcal{L}_\beta$. And so on.

This construction is notable in that it allows us to make 'large' jumps in the hierarchy. On the standard revenge construction, adding the resources required to describe the semantics of a particular $\mathcal{L}_\alpha$ 'pushes' us up to $\mathcal{L}_{\alpha+1}$. Adding $\mathsf{meta}_\alpha(x)$ to $\mathcal{L}_\alpha$, however, will 'push' us to an $\mathcal{L}_\beta$ where β is significantly larger than $\alpha + 1$. In particular, β must be greater than any ordinal γ whose existence is guaranteed by the set theory contained in $\mathcal{L}_\alpha$. This is yet another reflection of the fact that the languages $\mathcal{L}_\beta$ whose semantics can be set-theoretically modeled in a particular $\mathcal{L}_\alpha$ far outstrip $\mathcal{L}_\alpha$ itself.

It is important to note, however, that the difference between the $\mathsf{meta}_\alpha(x)$ construction and more familiar revenge-style constructions is, in a sense, a difference of degree, not of kind – while $\mathsf{meta}_\alpha(x)$-constructions ascend the hierarchy of languages more quickly, the ascent is of the same sort, and in the end this is just one more sequence of increasingly strong revenge Liars.

quantifiers used in $\mathcal{L}_\alpha$ to range over languages in our formulation of $\mathsf{EP}_{\mathsf{ER}}$ should be interpreted as ranging over the languages we can *model* within $\mathcal{L}_\alpha$.

A few pages later in the same essay Cook provides some further clarification of this point, including some prescient points foreshadowing the connections between MS and ER explored here[28]:

> Since we can never formulate a set theory which implies the existence of all possible ordinals, we can never formulate a formal semantics for our account which implies the existence of all possible extensions of our language (and corresponding truth-values) in some absolute sense of the word 'all'. While no single set theory implies the existence of all ordinals, however, there seems to be no reason to doubt that, for any ordinal, there is a set theory that implies its existence. As a result, for any possible extension of our language, we can formulate a semantics for it (by utilizing a suitably strong set theory in the base theory).
>
> Earlier we drew an analogy between the indefinite extensibility of the concept *ordinal* and the indefinite extensibility of the concept *language* (and the corresponding indefinite extensibility of the concept *truth-value*). The previous few paragraphs suggest, however, that there is more to this than just an analogy – in fact, the indefinite extensibility of our language just *is* the indefinite extensibility of the ordinals. This insight promises fruitful connections between the semantic and set-theoretic paradoxes. (Cook 2007: 48)

Here the connection we have been drawing out between MS and ER becomes explicit (even if unintended in Cook 2007): as we move from one set theory to another, with a different, and more expansive, universe of sets and ordinals, as MS and its extendibility principle $\mathsf{EP}_{\mathsf{MS}}$ guarantees we can, then we can reinterpret all of this semantic theorizing as occurring within a new, more extensive initial segment of the set theoretic hierarchy – one that can represent or model a more extensive collection of languages $\mathcal{L}_\beta$. Thus, for any initial segment of languages $\mathcal{L}_\beta$ we might be 'talking about' (i.e modeling within a particular model of set theory $\mathcal{M}_1$), we can ascend to a richer, more extensive collection of languages (i.e. those that are modeled in a proper extension of $\mathcal{M}_1$), and then another even more extensive model, ad infinitum.

This observation – that the sequence of languages $\mathcal{L}_\beta$ that we can model is relative to the set theory we are currently employing, and that richer set theories provide models of more extensive initial segments of the hierarchy of languages involved in the ER view – provides us with an alternative formulation of the extendability principle for languages. Instead of formulating this principle *sui generis* as $\mathsf{EP}_{\mathsf{ER}}$, we can instead understand the extendability of languages $\mathcal{L}_\beta$ as piggy-backing, so to speak, on the extendability of the set theoretic universe. We need only supplement the modal structuralist extendability principle for set-theoretic universes $\mathsf{EP}_{\mathsf{MS}}$ with the following *existence principle for languages*:

$$\begin{aligned}\mathsf{EPL} :& \Box(\forall\mathcal{M})(\forall\alpha \in \mathsf{On}_{\mathcal{M}})(\forall\beta \in \mathsf{On}_{\mathcal{M}})\\ &(\alpha \leq \beta \rightarrow (\exists\mathcal{L}_\alpha \in \mathcal{M})(\exists\mathcal{L}_\beta \in \mathcal{M})(\mathcal{L}_\alpha \subseteq_{\mathsf{sem}} \mathcal{L}_\beta))\end{aligned}$$

In short, in order to capture the never-ending nature of the hierarchy of languages involved in the ER account, we merely need to combine the modal structuralist view

[28]We assure you, the author of Cook (2007) did not have any explicit links to the MS view in mind when writing these passages, although that seems almost hard to believe in retrospect!

with the claim that, necessarily, given any set theoretic universe and any two ordinals α and β ($\alpha < \beta$) contained in that universe, there are languages $\mathcal{L}_\alpha$ and $\mathcal{L}_\beta$ such that the latter is an extension of the former in the sense relevant to ER. We leave it to the reader to verify that $\mathsf{EP}_{\mathsf{MS}}$ plus EPL entails $\mathsf{EP}_{\mathsf{ER}}$.

As a result, we can understand the always extendable nature of the Embracing Revenge view along lines analogous to the extendability of set theoretic universes on the MS account. We cannot, on pain of the meta-Liar, however, formulate the extendibility principle for ER as applying directly to the natural languages that in some sense make up the actual extendable hierarchy of languages. Instead, an adequate account of the never-ending nature of the hierarchy of languages on the ER view must be formulated in terms of the formal models of these languages that we construct within set theory.[29] As a result, the Embracing Revenge account can be understood as parasitic on a prior understanding of the extendable nature of the set-theoretic universe as codified within the modal structuralist account. In light of these deep and surprising connections, let us conclude by suggesting that the appeal of each of these accounts lends support to the other.[30]

References

Beall, J. (Ed.). (2007). *Revenge of the liar: New essays on the liar paradox*. Oxford: Oxford University Press.

Cook, R. (2007). Embracing revenge: On the indefinite extendability of language. In Beall (2007), pp. 31–52.

Cook, R. (2009). What is a truth value, and how many are there? *Studia Logica*, *92*, 183–201.

Ferreirós, J. (2007). The early development of set theory. In *The stanford encyclopedia of philosophy*. http://plato.stanford.edu/entries/settheory-early/.

Field, H. (2003). A revenge-immune solution to the semantic paradoxes. *Journal of Philosophical Logic*, *32*(2), 139–177.

Field, H. (2008). *Saving truth from paradox*. Oxford: Oxford University Press.

Gabbay, D., & Guenthner, F. (Eds.). (2002). *The handbook of philosophical logic* (Vol. 11). Dordrecht: Kluwer Academic.

Gupta, A. (1982). Truth and paradox. *Journal of Philosophical Logic*, *11*, 1–60.

Hellman, G. (1985). Review of Martin and Woodruff, 1984, Kripke, 1984, Gupta, 1984b, and Herzberger, 1984. *Journal of Symbolic Logic*, *50*(4), 1068–1071.

Hellman, G. (1989). *Mathematics without numbers: Towards a modal-structural interpretation*. Oxford: Oxford University Press.

Hellman, G. (2011). On the significance of the Burali-Forti paradox. *Analysis*, *71*(4), 631–637.

Kripke, S. (1975). Outline of a theory of truth. *Journal of Philosophy*, *72*, 690–716.

Martin, R., & Woodruff, P. (1975). On representing true-in-L in L. *Philosophia*, *5*(3), 213–217.

[29]One interesting question is whether the relationship is parasitic or symbiotic. In other words, is there some natural way to understand the modal structuralist account as parasitic on the embracing revenge account via obtaining $\mathsf{EP}_{\mathsf{MS}}$ as a consequence of $\mathsf{EP}_{\mathsf{ER}}$ plus some principle governing which set-theoretic structures exist relative to each language $\mathcal{L}_\alpha$? We set aside examination of this question for another time.

[30]We would like to thank Harty Field, Kit Fine, and Nicholas Tourville for helpful conversations on the issues raised in this paper.

Menzel, C. (1984). Cantor and the Burali-Forti paradox. *The Monist, 67*, 92–107.
Moore, G., & Garciadiego, A. (1981). Burali-Forti's paradox: A reappraisal of its origins. *Historia Mathematica, 8*, 319–350.
Priest, G. (2010). Hopes fade for saving truth. *Philosophy, 85*(331), 109–140.
Putnam, H. (1967). Mathematics without foundations. In Putnam (1975), pp. 43–59.
Putnam, H. (1975). *Mathematics, matter, and method: Philosophical papers* (Vol. I). Cambridge: Cambridge University Press.
Schlenker, P. (2010). Super liars. *Review of Symbolic Logic, 3*(3), 374–414.
Tarski, A. (1944). The semantic conception of truth. *Philosophy and Phenomenological Research, 4*(3), 341–376.
Tourville, N., & Cook, R. (2016). Embracing the technicalities: Expressive completeness and revenge. *Review of Symbolic Logic, 9*(2), 325–358.
van Heijenoort, J. (Ed.). (2002). *A sourcebook in mathematical logic: 1879–1931*. Cambridge: Harvard University Press.
Visser, A. (2002). Semantics and the liar paradox. In Gabbay and Guenthner (2002), pp. 149–240.
Zermelo, E. (1930). Über Grenzzahlen und Mengenbereiche: Neue Untersuchungen über die Grundlagen der Mengenlehre. *Fundamenta Mathematicae, 16*, 29–47.

Geoffrey Hellman received his AB and Ph.D. (1973) from Harvard. Having published widely in analytic philosophy and philosophy of science, he has, since the 1980s, concentrated on philosophy of quantum mechanics and philosophy and foundations of mathematics, where he has, following the lead of his adviser, Hilary Putnam, developed modal-structural interpretations of mathematical theories, including number theory, analysis, and set theory. He has also worked on predicative foundations of arithmetic (with Solomon Feferman) and pluralism in mathematics (with J. L. Bell). In 2007, he was elected as a fellow of the American Academy of Arts and Sciences.

Roy T. Cook is CLA Scholar of the College and John M. Dolan Professor of Philosophy. He is the author of *Key Concepts in Philosophy: Paradoxes* (2013) and *The Yablo Paradox: An Essay on Circularity*, as well as numerous articles and essays in the philosophy of mathematics, the philosophy of logic, and the aesthetics of popular art.

Chapter 6
The Metaphysics of the Model-Theoretic Arguments

Kate Hodesdon

Abstract This paper presents a exposition of Putnam's model theoretic arguments in the context of his broader philosophical position. I argue that Putnam used the arguments not just to undermine metaphysical realism, but to reveal that the philosophical debate between metaphysical realism and internal realism is dialectically problematic in that the metaphysical realist defence cannot "count against" (Putnam in Philosophical Topics: The philosophy of Hilary Putnam 20(1):355, 1992c) the converse position. Putnam's response is that this is a debate that we should simply undercut.

Putnam's model theoretic arguments have posed challenges of interpretation since their publication. In this article I shall make two claims about the arguments that add to this debate. One is to clarify the arguments' target: while it is clear that the arguments are designed to refute the position of metaphysical realism, it is less clear just which hypothesis is at stake. I present a thesis that Putnam takes to be constitutive of metaphysical realism and targets with the model-theoretic arguments. This is the posit of *epistemic humility*, which states that it is possible that there is an aspect of the world that is epistemically inaccessible as a matter of principle. The second aim of this paper is to suggest a new direction in which to seek justification for the notorious 'just more theory' response that Putnam gives to critics of the model-theoretic arguments.

Metaphysical or "external" realism and Putnam's own internal realism are broad positions, comprising two "philosophical temperaments" (1981, p. 49) or "tendencies" (1980, p. 474). They are also foundational: Putnam hints that their consequences affect almost every area of philosophy (1981, p. 49), particularly scepticism ("the question of 'Brains in a Vat' would not be of interest if it were not for the sharp way in which it brings out the difference between these philosophical perspectives"; *Ibid.*) The distinction between the two positions is inspired by Kant, with internal realism representing Kant's own position (1992d, p. 114, 1987, pp. 36–37), although the relationship between Kant and Putnam's views is complicated. Metaphysical real-

K. Hodesdon (✉)
Department of Philosophy, University of Bristol, Bristol, UK
e-mail: kate@hodesdon.com

G. Hellman and R. T. Cook (eds.), *Hilary Putnam on Logic and Mathematics*, Outstanding Contributions to Logic 9, https://doi.org/10.1007/978-3-319-96274-0_6

ism, on the other hand, is a traditional position in philosophy. Putnam defines it as committed to the following three broad claims about the world and the way that we refer to it:

> [T]he world consists of some fixed totality of mind-independent objects. There is exactly one true and complete description of 'the way the world is'. Truth involves some sort of correspondence relation between words or thought-signs and external things and sets of things. I shall call this perspective the externalist perspective, because its favorite point of view is a God's Eye point of view. (Putnam 1981, p. 49)[1]

The first claim to which metaphysical realism is committed is that *ontology is "fixed once and for all"* (1989, p. 352), independently of what our theories of the world might be, or indeed what our conceptualisation of the world might be (1982). The members of this ontology are philosophically privileged: to talk of the world in terms of them is to speak of things as they are *in themselves* (Putnam 1981, p. 50, 1995a, p. 303).

According to the second claim, there is just one true and complete theory of the world. Of course, humans speak a number of different, but inter-translatable, languages, so the uniqueness of this one theory must be due to the concepts that it employs relative to a language, rather than the particular words it uses. Specifically, the uniquely true theory of the world is the one that describes the fixed totality of objects singled out in the first claim. As Putnam explains, if there is just one true theory of the world, then there is automatically one privileged ontology: the one in terms of which the theory is given.[2] This is the non-perspectival ontology of things in themselves.

> One true theory requires a *ready-made* world—the world itself has to have a "built-in" structure since otherwise theories with different structures might correctly "copy" the world (from different perspectives) and truth would lose its Absolute (non-perspectival) character. (Putnam 1982, p. 147).

The third posit of metaphysical realism tells us what it means for a theory to be true: it must *correspond correctly with the world.* The nature of the correspondence relation is not specified, but it is intended to be *unique*, and to yield a bivalent semantics (1989, p. 352, 1991, p. 110). As an example of a reference relation, Putnam typically uses the thesis that reference is causally mediated. Putnam has argued specifically

---

[1]While Putnam's characterization of his own position, 'internal realism', shifted during the period that he endorsed it, Putnam used 'metaphysical realism' to capture more or less the same thesis throughout. See, for instance, (1989, p. 352, 1999, p. 18, n. 41).

[2]The claim is held by at least one contemporary metaphysician. In his recent book, *Writing the Book of the World*, Ted Sider argues for precisely this metaphysical realist thesis. He claims that not only is there one true account of the world as it is in itself, but that there is precisely one correct language for properly describing it.

> In order to perfectly describe the world, it is not enough to speak truly. One must also use the right concepts—including the right logical concepts. One must use concepts that 'carve at the joints', that give the world's *structure*. There is an objectively correct way to 'write the book of the world'.

against causal theories of reference (1982, 1989, pp. 358–360, 1992b, pp. 61–79), but in the model-theoretic arguments, a causal chain theory functions as a toy example of a physical constraint on reference: what goes for a correspondence based on causal chains goes equally for a correspondence based on any other relation. A correspondence theory of truth rules out other norms of rational inquiry, such as verifiability, which was a central part of Putnam's own position, internal realism, until he abandoned it in 1990.

Together, these three posits of metaphysical realism entail that there is precisely one theory (up to notational variance) which must correspond in the correct way, to the correct objects in order to be true.

> For the externalist philosopher [...] the truth of a theory does not consist in its fitting the world as the world presents itself to some observer or observers (truth is not 'relational' in this sense), but in its corresponding to the world as it is *in itself*. (Putnam 1981, p. 50, emphasis added)

The three posits also entail a deep and seemingly inescapable problem for the metaphysical realist. The problem is that *our* theories of the world are unavoidably perspectival. The describe objects that are informed by what Putnam calls our "epistemological position" in the world, which is determined both by our sensory experiences and by the conceptual schemata we employ. But the "fixed totality" of things in themselves, on the other hand, may be epistemically inaccessible to us. Consequently, it may be the case that even our best theory of the world—and Putnam is careful to spell out exactly what we mean by 'best theory' (1980, p. 473)—fails to describe the privileged totality of objects, and represents only appearances. In this case, our best theory of the world is false.

6.1 Epistemic Humility

We have seen that metaphysical realism amounts to a traditional epistemological duality thesis that posits an underlying noumenal world of things in themselves, which may possibly be very different from the appearances that our theories describe.[3] In this way, the epistemic duality thesis is coupled with a form of *epistemic humility*, which says that it is possible that as a matter of principle, we know nothing about a certain aspect of reality. Indeed, Putnam tells us that "the sharp distinction between what really is the case and what one judges to be the case is *precisely what constitutes metaphysical realism*" (Putnam 1981, p. 50, emphasis added). Elsewhere he writes:

> The realist—or, at least, the hard-core metaphysical realist— [...] wishes it to be the case that what, e.g., electrons are should be distinct (and possibly different from) what we believe

[3]The duality is also presented by Putnam's differentiation in "Brains in a Vat" between "vats *themselves*" and"*vats-in-the-image*" (1981, Ch. 1).

> them to be or even what we would believe them to be given the best experiments and the epistemically best theory. (Putnam 1980, p. 472)

It is not sufficient for epistemic humility just that we are, as a matter of fact, ignorant of some aspect of the world. After all, irrespective of any stance on things in themselves, we are almost certainly ignorant about the distant past, and remote regions of space. Epistemic humility says that this ignorance is *as a matter of principle*, which is to say that even in the most epistemically ideal conditions we could remain ignorant of the aspect of reality in question.

The metaphysical realist's epistemic duality between things as they are for us and things in themselves, together with humility about the latter, is a traditional thesis. It is tacitly assumed by many philosophers, but its most explicit defence comes from Kant. There are (at least) two ways of understanding the distinction in Kant: as demarcating between *two kinds of object* in terms of how we know them, or between *two ways of knowing* any given object. Pippin has called the first of these interpretations the "two worlds" or "two realms" view (1982, p. 196 ff.), and it is this that Putnam seems to have in mind for the metaphysical realist's predicament. Another place that epistemic humility finds expression—this time postdating Putnam's arguments against it—is in the work of Lewis (2009). Lewis argued, on grounds entirely different from Kant's, for an epistemic humility thesis regarding the identities of fundamental *properties*. Lewis named the position Ramseyan Humility, after its Ramseyan justification. Specifically, Ramsey argued that scientific theories describe only the nomological structure of the world and the roles of the fundamental properties within it. However, there are multiple ways in which the properties may realize the roles introduced by a theory. And quidditism, which Lewis also defends (2009, §4), says that a permutation of properties between their theoretical roles yields a distinct, but qualitatively identical, possibility. A permutation of two fundamental properties is the result of replacing all occurrences in space and time of one of them with those of the other, and vice versa.[4] From Ramsey's thesis and quidditism, Ramseyan humility follows: since we cannot distinguish between the distinct possibilities that result from such permutations, we are ignorant of the identities of the properties that may be permuted.[5]

However, there is one significant respect in which metaphysical realism differs from most other forms of epistemic humility, most notably the reading of Kant that Langton defends as 'Kantian humility' (1998). This is the centrality of scepticism. Putnam has often remarked that metaphysical realism can be *characterized* by its tolerance of scepticism (1980, p. 473, 1981, Ch. 1, Ch. 3, 1989, pp. 352–354). But, at least according to Langton, Kantian humility does not lead to scepticism: "it allows plenty of knowledge of the real world, but denies knowledge of things as they are in themselves." (2004, p. 134) Likewise, Lewis's easy acceptance of the predicament

---

[4]Of course, the properties must be both of the same logical "category" in order to be permuted—both monadic, or both relational magnitudes, for instance—but Lewis thinks that there are such multiply-instantiated categories of properties.

[5]Since Lewis does not believe haecceitism, which is the thesis analogous to quidditism concerning individuals instead of properties, his argument does not establish ignorance of things.

Ramseyan humility places us in indicates that, for him, it falls short of scepticism: "Who ever promised me that I was capable in principle of knowing everything?" (2009, p. 211).

6.2 The Model-Theoretic Picture

The epistemic duality of metaphysical realism makes it possible for Putnam to use model theory to give an analogy for the metaphysical realists thesis, and then, by carrying over results from model theory to the metaphysical realist account of the world, to undermine it. This is the job of the model-theoretic arguments.

In a model, we have a domain of objects, $\mathcal{D}$, an object language, $\mathcal{L}$, and an assignment function, π, that assigns terms of $\mathcal{L}$ to objects and sets of objects of $\mathcal{D}$. Models in this sense are the basis of the branch of mathematical logic known as model theory, but models in the more general sense, including physical models, like replicas, are constructed in a similar way. What is crucial for the analogy with metaphysical realism is that an interpreted model does not only introduce a correspondence, π, between names and objects; it also determines a correspondence between *two domains of objects*. This is by virtue of the fact that its object language, $\mathcal{L}$, is already at least partially interpreted: it includes terms that are meaningful and which, in the metalanguage, refer to objects and sets of objects that are generally independent of the model. If the object language did not have such an interpretation—that is, if it were just meaningless symbols—then the model could not serve its purpose of *representing* the particular state of affairs described in $\mathcal{L}$ that it was designed to portray. As Putnam has said, a model is only a model in so far as it represents *something* by somebody (1981, p. 5). For an example of the two domains in practice, consider a set-theoretic model of arithmetic. The model's assignment function will induce a correspondence between the *sets* in the domain of the model (say, the von Neumann ordinals) and the *natural numbers* described in the language of arithmetic.

Thus, an interpreted model picks out a correspondence between two discrete sets of objects: the domain elements (which are doing the representing) and those named by the object language (which are being represented). In some cases, these two domains will be structurally similar. For instance, scale models preserve the same relative distances between domain elements and the objects they represent. On the other hand, a Rutherfordian model of the atom made of plastic beads will almost certainly misrepresent the relative distances between subatomic particles, but will preserve their relative *positions* as inside or outside the nucleus. The presence of structural similarities like these are what make a model *apt* as a representation. However, although there may be such structural similarities,[6] in general the domain

[6] I do not want to claim that *all* representation trades on shared structure between the represented objects and the representing objects (domain elements). In particular, the Löwenheim–Skolem theorems provide plausible counterexamples.

elements of a model will have quite different properties from those of the objects being represented. To use Putnam's example from "Models and Reality", in a model of set theory a set that is non-constructible from the perspective external to the model may well represent a constructible set in the model.

This two-level ontology associated with a model is analogous to the metaphysical realist's "two realms" of things for us and things in themselves. The analogy justifies Putnam's application of theorems from model theory that show that there is no unique way to model a theory, to the "models" whose domain elements are the things in themselves, and whose objects are things for us. Just as there is no unique way to model a theory by pairing up domain elements with the objects they represent, neither is there a unique mapping from things for us onto things in themselves. But, just such a mapping is required by the single unique reference relation posited by the metaphysical realist's correspondence theory of truth. Therefore, by analogy, we can infer that there is no such unique reference relation to things in themselves. In this way the model-theoretic arguments establish the conditional thesis that, "assuming a world of mind-independent, discourse-independent entities [. . .] there are [. . .] many different 'correspondences' which represent possible or candidate reference relations (infinitely many, in fact, if there are infinitely many things in the universe)." (Putnam 1981, p. 47) Consequently, and as I shall claim, the model-theoretic arguments lead Putnam to reject the notion of an epistemic duality as well as the notion of things in themselves.

6.2.1 Reductio ad Absurdum

We have seen that model theory provides a suitable language for talking, by analogy, about metaphysical realism. Let us now turn to the model-theoretic arguments in closer detail to see how model theory undermines the position.

The arguments are intended to be a *reductio ad absurdum* of metaphysical realism (Putnam 1993, pp. 280–281, 1995a, p. 303). But what is the absurd conclusion that they establish? Perhaps the most obvious candidate is just what the permutation argument (1981, Ch. 2) establishes: that *reference is radically underdetermined.* Although this may be the most straightforward form for a *reductio* to take, the reading doesn't sit well with the characterization of the metaphysical realist just given. To really appreciate the problem with metaphysical realism, we shall have to turn to the other model-theoretic argument: the 'Skolemite' argument.

There are two reasons why the *reductio* of metaphysical realism requires more than simply showing that it incurs radical indeterminacy of reference. The first problem is that in order to derive referential indeterminacy from metaphysical realist posits *by contradiction,* the metaphysical realist must be implicitly or explicitly committed to the converse: not necessarily that reference *is* determinate, but at least that it is not radically underdetermined. However, as we have seen, the metaphysical realist already lives with the threat of scepticism, and so believes that it is possible that her theories are radically false of the world. Therefore, while the metaphysical realist might *hope* that reference is fixed via causal chains, or something similar, the fixity of reference cannot be an assumption of her position, since it is inconsistent with her scepticism about knowledge. For all the metaphysical realist knows, she might be a brain in a vat. If she were a brain in a vat, she would still believe that causal chains link her talk of trees with trees *themselves*, even when, in reality causal chains link her talk only to trees "in the image". If the model-theoretic arguments show that the metaphysical realist has no guarantee that her theories correctly refer, and consequently, that she has no guarantee that they are true of reality, then this just supplies her with *one more* sceptical hypothesis. It is simply more grist for her mill.

There is a second problem with understanding Putnam's *reductio* of metaphysical realism to have even the *logical form* of a proof by contradiction. The problem is how to reconcile a proof by contradiction with Putnam's frequent claims that the model-theoretic arguments reveal metaphysical realism to be *incoherent*, or *unintelligible* (Putnam 1978, p. 126, 1980, p. 474, 1992a, p. 85, n. p. 173, 1992c, p. 355). Generally, a *reductio* only licenses us to conclude that at least one of the assumed premises is false. We could interpret the claim of metaphysical realism's incoherence as meaning only that its posits cannot all be true together. But there are other forms of absurdity that are not outright logical contradictions. And, as Putnam later claimed, his charge was never that metaphysical realism is logically inconsistent. Instead, he says it is incoherent.

> What is consistent or not is a matter of pure logic; what is coherent, or intelligible, or makes sense to us, and what is incoherent, or unintelligible, or empty, is something to be determined not by logic but by philosophical argument. (Putnam 1989, p. 354)

For the remainder of this paper, I will focus attention on Putnam's objection that metaphysical realism is incoherent in the sense that it is "empty" (1995a, p. 303):

> [M]etaphysical realism cannot even be intelligibly stated [. . .] attempts at clear formulation never succeed in capturing the content of 'metaphysical realism' because there is no real content there to be captured. (1992c, p. 353)

The emptiness claim is justified by the model-theoretic arguments—specifically, by the Skolemite argument of "Models and Reality", together with the 'just more theory' rejoinder. The argument rests on what is taken to be an extension of Skolem's historical 'paradox', that all first-order formal theories, including those like the theory of real analysis that we take to deal with non-denumerably many objects, have denumerable models. While this theorem is no longer considered paradoxical, it does indicate something about the inability of such theories to pin down (even the cardinality of) their models. Putnam's Skolemite argument shows that something similar is true of the theory consisting of our total science:

> [E]ven a *formalization of total science* (if one could construct such a thing), or even a *formalization of all our beliefs* (whether they count as "science" or not), could not rule out [. . .] *unintended* interpretations (Putnam 1980, p. 466; see also Putnam 1989, p. 353)

The Skolemite argument requires the notion of an epistemically ideal theory, T_I. This is a theory (together with an interpretation) that meets operational and theoretical constraints, defined as follows.

Operational constraints constrain the theory from contradicting facts that can be confirmed observationally. The constraints are imposed by the stipulation that, first, the theory T_I be given in a language containing an observational vocabulary sufficient to name every one of the countably many things or events we could possibly observe. (1980, p. 472) Second, T_I is given a partial interpretation $\mathcal{OP}$ such that, for all terms $\bar{t}$ in the language of T_I that denote observable things and events, and for all predicate and relational symbols R in the language of T_I that denote observable properties or relations, if the object denoted by $\bar{t}$ *can be observed to have* the property denoted by R then the valuation $\mathcal{OP}$ makes $R\left(\bar{t}\right)$ true. This guarantees that T_I "does not lead to any false predictions" (1980, p. 473) about observable events. In sum, operational constraints ensure the following conditions:

> [I]f 'there is a cow in front of me at such-and-such a time' belongs to T_I, then 'there is a cow in front of me at such-and-such a time' will certainly *seem* to be true—it will be 'exactly as if' there were a cow in front of me at that time. [. . .]
>
> On the other hand, if 'there is a cow in front of me at such-and-such a time' is *operationally* 'false' (falsified) then 'there is a cow in front of me at such-and-such a time' is [false in the model]. (Putnam 1978, p. 126)

Theoretical constraints are extra-empirical constraints. They ensure that T_I possesses all epistemic virtues that would make it rational for scientists to accept T_I in the limit of human inquiry. Of course, these virtues include *consistency* (*Ibid.* p. 473)—along with "simplicity, elegance, subjective plausibility" (1989, p. 35). It is important that these virtues capture what is "*epistemically* ideal for *humans*" (1980, p. 472); although they are relativized to an ideal limit of inquiry, they pin down a theory that *we* would accept as best, given our epistemological position.

Operational and theoretical constraints are the best yardstick that we can reliably use to measure a theory's success. A theory that meets them will make no predictions that are falsified by what we can observe, since, by virtue of the operational constraints, all atomic sentences describing observable things and events will be theorems of the theory. But the constraints do not guarantee that a theory is true, in the sense of the correspondence theory of truth. As Putnam explains, the Skolemite argument was intended to put pressure on the notion that a theory could meet operational and theoretical constraints yet fail to be true:

> Example: an ideal theory might say that there are intelligent extraterrestrials somewhere in space-time, although in fact there aren't any. There might be overwhelming evidence that there are intelligent extraterrestrials (somewhere, some time), evidence for laws according to which the probability that such never did, don't, and never will exist is less than one in a trillion, let us say (which would certainly justify believing that intelligent extraterrestrials exist in spacetime), when, in fact, ours is a universe in which the one in a trillion chance that they don't exist is realized. This is an example of the way in which "correspondence truth" can differ from even idealized verifiability. *The purpose of the model-theoretic argument was to cast doubt on the very intelligibility of this very plausible set of beliefs.* (Putnam 2012, p. 75, emphasis added)

We are now in a position to state Putnam's Skolemite argument. Let T_I be an epistemically ideal theory.

6.3 The Skolemite Argument

P1. It is possible that T_I is *false* of reality (1980, p. 473). This is just the metaphysical realist's thesis of epistemic humility.
P2. Let T_I be false. This assumption is justified by **P1**.
P3. T_I has models by the completeness theorem. (*Ibid.*)
P4. Let $\mathcal{M}$ be a model of T_I whose domain contains the countably-many macroscopically observable things and events, and whose interpretation agrees with the partial interpretation $\mathcal{OP}$.
P5. A theory is *true* if it is true in the intended model. (*Ibid.* p. 474)
P6. Since T_I meets *all* the constraints that we can impose on a theory, the model $\mathcal{M}$ is an "intended" model.
C1. T_I is true. (*Ibid.* p. 474)
C2. Since **C1** contradicts **P2**, **P2** must be false. But **P2** follows from **P1**, so **P1** must be false. This contradicts epistemic humility..

P1 is true by definition of metaphysical realism, and **P2** follows from it. **P3** is also uncontroversial, as a theorem of first-order logic. **P4** can also be justified without much trouble, since it asserts the existence of a model that we can directly construct. However, the justification for the remaining premises has been the subject of much debate—for a sample, see Douven (1999), Bays (2001, 2008), Hale and Wright (1997).

P5 implies that truth can be equated with truth in a (specific) model. This equivalence is supported by the analogy discussed earlier: that the metaphysical realist's world picture of a uniquely privileged ontology, one true theory about it, and one correspondence relation making the theory true can be thought of as a single model.

P6 appears to equivocate on the notion of "intended", and so gives the metaphysical realist some leeway to reject the argument's conclusion. While the metaphysical realist will insist that the previously-described model is the one she "intends", **P6** asserts that $\mathcal{M}$, which is an arbitrary model of the epistemically ideal theory T_I, is in fact intended. The reason why, for Putnam, $\mathcal{M}$ *is* an intended model is that the theory it makes true satisfies operational and theoretical constraints. He asks, rhetorically, "what else could single out a model as 'intended' than this?" (*Ibid.* p. 473). However, for the metaphysical realist, the model $\mathcal{M}$ has to be unintended because it satisfies T_I. For, T_I is false. The point of contention therefore comes down to rival theories of truth: to the question whether a theory which is as epistemically ideal as we can possibly measure may be false, by virtue of its failure to describe some epistemically inaccessible part of reality.

We might well wonder why Putnam believes that operational and theoretical constraints on truth yield the only possible ways to determine reference. As Button (2013, §4.3) has convincingly shown, it is clear that Putnam considers these to be the limit of *naturalistic* constraints on reference. Button draws attention to Putnam's repeated characterisation of any constraints that go beyond them as "magical"—in other words, nonnatural. And while Putnam may have a reputation for changing his mind on central topics, he has *always* been a scientific realist (Putnam 2012, p. 52ff.), so rules nonnatural methods of reference-fixing out of the question: "the suggestion that metaphysical realism might be *nonempirically* true is a possibility I did not—and still do not—take seriously." (1995a, p. 304).

Putnam's own positive account of reference (1975) on the other hand firmly links successful reference with our own, human practice of inquiry. To use his example, whether or not a substance can be referred to as 'gold' depends on whether it has properties privileged by *our* theory of chemical composition. Indeed, as Putnam wryly remarked: "Kripke expressed dissatisfaction with "The Meaning of Meaning" precisely on the ground that the notion of the "essence" of a natural kind I employ there is *not* independent of scientific practice" (1992c, p. 349). This, then, is why $\mathcal{M}$ is an intended model: because it makes true a theory that satisifies the limit of constraints that we can put on truth without contradicting naturalism.

6.4 Just More Theory

If Putnam is right that the metaphysical realist cannot supply a naturalistic account of reference that supports her view of truth, then his argument raises a dilemma for the metaphysical realist. Either she must abandon metaphysical realism altogether for antirealism about truth, or else bolster her realism with a nonnatural account of intentionality (1980, pp. 474–475). There is a commonly-raised objection to the first disjunct. The objection proceeds by simply stating that reference is fixed by some relation in particular, typically by causation. According to this objection, *contra* the Skolemite argument, $\mathcal{M}$ is not intended at all unless its assignment function picks out this causal relation.

The 'just more theory' reply is the response that Putnam gives to interlocutors who make this objection. The reply claims that since the causal theory of reference is part of our overall best theory of the world, the occurrences of 'cause' and other terms in the theory are subject to Skolemite reinterpretation. Thus, the model-theoretic arguments cannot be dismissed using a causal theory of reference. And the same goes for any other reference-fixing relation. As stated, it might seem that the 'just more theory' move is an objection to any reference fixing constraint whatsoever. However, the point is only supposed to be aimed at reference-fixing constraints supplied by the metaphysical realist. In particular, the ability to pick out one reference relation is incompatible with the metaphysical realist's epistemic humility. This is suggested by Putnam's rebuke that in proposing some referencing fixing constraint "the philosopher is *ignoring his own epistemological position*" (1983, p. xi, emphasis added). Recall that the metaphysical realist's epistemological position is perspectival. She cannot rule out the hypothesis that the way that the world is really—in terms of things in themselves—is radically different from her best theory of the world.

Before continuing to examine the 'just more theory' move, let us look briefly at the second disjunct of the metaphysical realist's dilemma: that she opt for a non-natural theory of reference. Lewis (1984, pp. 232–233) asked why Putnam offered the metaphysical realist this way out, given that doing so missed an opportunity to generalize his argument fully. We have already seen that Putnam's naturalism means that he would not take non-natural constraints on reference seriously; to him, a non-natural constraint on reference does not consitute a legitimate way out of the dilemma at all. However, why permit it as an option for the metaphysical realist at all? Douven (1999, p. 490) answered this question with the claim that only naturalistic theories are vulnerable to the 'just more theory' rejoinder. We saw that the just more theory move requires us to suppose that the theory of any reference-fixing constraint offered is false. But according to Douven, the metaphysical realist is only committed to fallibilism for naturalistic theories. So, if the metaphysical realist held a theory of a non-natural referential constraint, she could simply reject the just more theory move when directed at this theory.

Button gives an alternative explanation for the restriction of the model-theoretic arguments to naturalistic theories of reference. He argues that, given metaphysical realism, the only constraints actually capable of fixing reference—and thus being more than mere theory—are non-empirical (2013, p. 31). So, the 'just more theory' move can *only* apply to empirical accounts of reference, in other words, naturalistic ones. The reason for this is that the metaphysical realist endorses what Button calls a Cartesian Principle, which says that "even an ideal theory might be radically false" (p. 10).

> [T]he external realist must accept that her attempts to constrain reference are without empirical content. Whatever her view of empirical content, her Cartesian Principle sets up a sceptical veil between herself and the world, and between her words and the world" (Button 2013, p. 53, see also p. 58)

Button adds to the model-theoretic arguments the premise that, according to metaphysical realism, beliefs have narrow content. In other words, the belief that I am seeing a cat requires a certain kind of cat-like sense data, but not necessarily any cat itself. In fact, the idea can be generalized beyond merely sensory data: the metaphysical realist has a full system of "constructions" in terms of which her theories of the world are given, that are over and above the objects themselves. In this way, Button's metaphysical realist posits something much like the epistemic duality I have described here. When the metaphysical realist makes a claim that goes beyond her constructed ontology, then she is talking about an "unconstructed world, made up of objects that are largely mind-, language- and theory-independent" (p. 37).

Where I disagree with Button on Putnam is his claim that "the empirical content of any claim is exhaustively accounted for *within* the construction system itself" (*Ibid.*) Button uses this thesis to justify the just more theory move: if all claims with empirical content can be given solely in terms of the constructions, then simply by virtue of being an empirical claim, any theory about reference will be 'just more theory' in so far as it talks only about the constructions, and not about anything beyond the veil. Certainly, we can imagine a dual ontology comprised of, on the one hand, objects in terms of which *all empirical theories* can be given, and on the other hand, whatever objects are left. Just think, as Button suggests, of Carnap's distinction between 'internal' and 'external' questions. But I am not convinced that the metaphysical realist's dual ontology *is* like this: her ontology of constructions, or appearances, contains (roughly) things like cat-like sensory impressions or objects based on appearances of cats, but her ontology of things in themselves is supposed to be made up of cats, vats, and so on—all very much empirical things. This is, of course, unless her sceptical hypothesis is true, and we are being radically deceived—and in this case, things in themselves are indeed *who knows what* and constructions are al we have to go on. In short, I don't see how to justify the idea that empirical claims are exhausted by appearances or constructions, *without* assuming that the metaphysical realist's sceptical hypothesis is correct. While Button's exegesis of the internal realist and metaphysical realist positions is convincing, his account of the 'just more theory' move appears to assume that the metaphysical realist is *in fact* veiled off from the reality beneath appearances—which is just what the model-theoretic arguments were supposed to establish.

But if the metaphysical realist's belief that *it is possible that* she is epistemically isolated from reality doesn't justify the 'just more theory' move, what does? And what should we make of Putnam's remark, quoted earlier, that the move is justified by the metaphysical realist's epistemological position? I want to suggest that we can justify the 'just more theory' manoeuvre on the basis simply of the metaphysical realist's model-theoretic picture of the world and our theories' relationship to it, without any assumption regarding whether or not the metaphysical realist is in fact epistemically isolated from the world. But first I will flesh out the 'just more theory' move a little more. My interpretation here is guided by an account that Putnam has given in just some of his later discussions of the model-theoretic arguments, most fully in a 1992 special edition of *Philosophical Topics*, in which he responded to his critics (1992c, see also Putnam 1995a). There Putnam urged that the problem with metaphysical realism was a problem of *demarcation*: the metaphysical realist cannot articulate a theory that rules out the antirealist position that equates truth with satisfaction of operational and theoretical constraints. Putnam explained that,

> The main point [of "Models and Reality"] was that metaphysical realism cannot even be intelligibly stated. I expressed this by saying that metaphysical realism is 'incoherent'. I did not mean by that it is inconsistent in a deductive logical sense, but rather that when we try to make the very vague claims of the metaphysical realist precise, we find that they become compatible with strong forms of 'antirealism'. (Putnam 1992c, p. 353, see also 1995a, p. 303)

In this remark, Putnam cannot mean that metaphysical realism and antirealism are compatible in the sense of being jointly consistent, since clearly these theories say different things—about what makes a sentence true, for example. Their compatibility, as Putnam explains it, is due to the fact that the metaphysical realist's theory about how reference gets fixed, and consequently how sentences are made true, can *itself* be made true in the antirealist sense of truth. This is because the core posits of metaphysical realism, given at the start of this paper, satisfy operational and theoretical constraints.

This is true even of claims to which the metaphysical realist is committed that are directly contradicted by antirealism, such as the claim that reference is fixed by causal connections. As long as the claim that 'causation fixes reference' does not contradict anything observable, and thus does not violate operational constraints, nor violate any theoretical constraints, the antirealist may accept it. It is precisely because operational and theoretical constraints couple truth to appearances that the antirealist can accept that causation fixes reference. Putnam continues,

> Granting that one can *say* [. . .] that reference is fixed by some physical relation, say, 'causal connection' [. . .], the question is whether these statements (assuming they make sense) express what the metaphysical realist 'wants to say'. After all, the claim that 'reference is fixed by causal connection' will, if true, be satisfied by all intended models. It will satisfy operational and theoretical constraints. Its truth does not, by itself, count against [the antirealist] conception of truth" (Putnam 1992c, p. 355)

Putnam's remark that the metaphysical realist's claim can be made true in a model, without *counting against* the theory external to the model, recalls the demonstration at the beginning of "Models and Reality" that V = L can be made true in a model of set theory even when "in reality" V = L is false. The remark also puts the 'just more theory' manoeuvre in a somewhat different context. It emphasizes a similarity between the treatment of the metaphysical realist's claims by her interlocutor and the way that a formal theory is treated model-theoretically, which is to say, with reference that is only fixed up to isomorphism. This suggests that the problem lies with the model theoretic nature of the metaphysical realist relationship between a successful theory and reality. We have seen that the analogy between metaphysical realism and the way that models represent theories makes the Skolemite argument possible (it justifies **P5** in particular). It is this model-theoretic picture that permits the metaphysical realist's interlocutor to treat her words as 'mere theory' with no uniquely privileged correspondence to any particular objects or relations. If it is merely the existence of this strong similarity that justifies the 'just more theory' move, then there is no need to assume that the metaphysical realist's theories do not, in fact, correspond with things in themselves: we can remain agnostic on this matter.

Putnam's remark that "the question is whether [causal theories of reference] express what the metaphysical realist 'wants to say'" suggests that Putnam's goal is not to show that metaphysical realism is false,[7] but rather that it cannot *get an edge* on antirealism—cannot "count against it". Quite simply, when the metaphysical realist makes a claim, she can't guarantee that her interlocutor, who holds a different theory of truth, will interpret her claim as she intends. Describing this dialectic, we could say that if two opposing sides in a debate don't agree on the notion of truth then there is no hope in either side getting one up on the other. With this understanding of the justification of the 'just more theory' reply, the antirealist's treatment of metaphysical realism—simply reinterpreting its claims—is indicative that the dispute between metaphysical realist and antirealist, who each hold different theories of truth, is seriously fraught. It is not the kind of dispute that can give way to a mutual resolution.

[7] Putnam says as much in (1995a): "clear attempts at a formulation of [metaphysical realism] never succeed—because there is no real content to be captured. My aim [. . .] therefore, was not to argue for the truth of a counter-thesis (one which could be identified with the negation of metaphysical realism but rather simply to provide a reductio ad absurdum of metaphysical realism by teasing out the consequences of its own presuppositions." (p. 303).

This observation has also been made by Maximilian de Gaynesford, who, in a talk at a conference at Harvard in 2010 for Putnam's 85th birthday, drew attention to Putnam's use of the term "antinomy" to describe the model-theoretic arguments. De Gaynesford presented what he called the "antinomy picture" of the model-theoretic arguments, according to which, whichever party begins the argument invariably wins it; neither is outright victorious, and neither succeeds in convincing the other.

As I understand it, Putnam's attempted resolution of the model-theoretic arguments was simply to undercut this dispute. This is seen in the origins of Putnam's own position, internal realism. When Putnam first introduced 'internal realism', in (1978), it was the name of a position intended as a form of scientific realism that could be endorsed by metaphysical realists and antirealists about truth alike (1992c, p. 352, see also 2012, pp. 55–56). It was an empirical theory (1978, p. 130) about how scientific theories referred. The position said nothing about what truth *was*, and thus could be seen as a kind of Carnapian 'internal question', circumventing the debate between metaphysical realism and antirealism. Later on, when Putnam published his own verificationist position on truth, distinct from the account based on operational and theoretical constraints (Putnam 1992c, p. 353), he found that his readers took that position to be the one named 'internal realism', and decided that "it seemed easiest to me to go along with this, as I did in *Reason, Truth and History*" (*Ibid.*).

While Putnam may have initially intended to undercut the metaphysical realist-antirealist debate by staying neutral on the topic of truth, he then moved into this debate with his defence of verificationism. Given this, it is worth identifying how Putnam's own position avoided the pitfalls of the model-theoretic arguments. Putnam's strategy for avoiding the problem himself is, in part, to deny that an epistemically ideal theory might be false, in favour of the alternative verificationist thesis that truth coincides with epistemic ideality, at least in the limit of inquiry. But there is another significant difference between Putnam's position and the metaphysical realist's: internal realism does away with the model-theoretic picture of truth by denying any kind of dichotomy between things for us and things in themselves. And, as we have seen, a dichotomy between things in themselves and things for us is essential to the model-theoretic arguments. So, whereas a distinction between appearances and things in themselves was "precisely what constitutes metaphysical realism" (Putnam 1981, p. 50), Putnam tells us that, *"[T]he adoption of internal realism is the renunciation* of the notion of the 'thing in itself'. (1987, p. 36, emphasis added). He continued:

> Internal realism says that the notion of a 'thing in itself' makes no sense; and *not* because 'we cannot know the things in themselves'. [...] Internal realism says that we don't know what we are talking about when we talk about 'things in themselves'. (Putnam 1987, p. 36)

6.5 The Renunciation of the Notion of the 'Thing in Itself'

Putnam defended the claim that there can be no such thing as a thing in itself throughout his internal realist period (1995a, p. 302, 1978, p. 6, 1987, p. 41, 1982, p. 163), and continued to do so even after abandoning internal realism. It is a goal of his 2005 book, *Ethics Without Ontology*, to refute the project of "describing the world as it is 'in itself'" (2005, p. 24). In this final section, I'l highlight some areas of Putnam's broader philosophical thought to which this thesis is foundational.

One place where Putnam's rejection of the dichotomy between things for us and things in themselves is particularly apparent is in his critique of ontological relativity. Ontological relativity is the thesis that "there is no absolute sense in speaking of the ontology of a theory" (1969, p. 48). It was established by Quine's *Gavagai* argument, which aimed to show that there is no possible answer to the question whether a given object is a rabbit, versus an undetached rabbit-part, and so on, since no difference in behaviour can be detected between the case in which 'rabbit' refers to one or the other.

Putnam's disagreement with Quine consists in fact that, whereas Quine would be "willing to put up with the slack" (*Ibid.* p. 45) that the *Gavagai* argument reveals between language and world, Putnam is already convinced that there is no slack. If there were, then we would have an epistemic duality of the kind that the model-theoretic arguments refute. As Putnam sees it, Quine just accepts that there is a world of objects out there, and we cannot know to which one 'rabbit' refers. Quine thus retains a realm of epistemically inaccessible objects just like those posited in the metaphysical realist's sceptical hypotheses. Consequently, as for the metaphysical realist, only a magical theory of reference would allow us to refer to these objects: "it is magical, in Quine's view, to think that science can do more than fix the structure of the world up to isomorphism" (Putnam 1989, n. 20) Ultimately, its commitment to things in themselves makes ontological relativity as untenable as metaphysical realism:

> What am I to make of the notion of an X which is a table *or* a cat *or* a black hole (or the number three *or*...)? An object which has *no* properties at all in itself and any property you like 'in a model' is an inconceivable *Ding an sich*. The doctrine of ontological relativity avoids the problems of medieval philosophy (the problems of classical realism) but it takes on the problems of Kantian metaphysics in their place. (Putnam 1983, p. xiii)

Putnam therefore takes the argument to show that an "alternative" is needed to the entire metaphysical picture it presupposed. Quine's *modus ponens* is Putnam's *modus tollens* (Putnam 1993, p. 280).

The untenability of ontological relativity is a surprising consequence of Putnam's attack on metaphysical realism, given the extent to which Quine's work in this area influenced Putnam's. Indeed, the permutation argument was devised during what Putnam described as a period of "intense interaction" (1978, p. ix) with Quine's views, and introduced as extending Quine's work in *Ontological Relativity* "in a very strong way" (1981, p. 34, see also 1992a, p. 112). And, as Putnam says, it is true that both the permutation argument and the *Gavagai* argument were designed to refute the thesis that "words stand in some sort of one-one relation to (discourse-independent) things and sets of things." (1981, p. 41) Putnam's rejection of ontological relativity is also surprising, given his defence of a quasi-structuralist position about mathematics (1967), in which he argued that there are multiple ontologies for mathematical objects, each equally good, but suited for different purposes. This sounds remarkably like ontological relativity restricted to mathematics.[8]

Internal realism's "renunciation of the notion of the'thing in itself'" is also ostensibly hard to reconcile with Putnam's repeated claims that Kant *himself* was an internal realist (1981, p. 60, 1987, p. 43), given that Kant famously posited the existence of a noumenal domain. But this is not the interpretation of Kant that Putnam endorses. During his internal realist period Putnam defended an interpretation of Kant's transcendental realism according to which there is no bijection between the noumenal and phenomenal realms (1981, p. 61), which is to say, no correspondence between things in themselves and things for us. In fact, Putnam went on to claim that "almost all of the *Critique of Pure Reason* is compatible with a reading in which one is not at all committed to a Noumenal World, or even [...] to the intelligibility of thoughts about noumena (1987, p. 41). In light of this reading of Kant, the Kantian heritage of Putnam's thought is much clearer.

To conclude, while the model-theoretic arguments defy easy characterisation, they do provide a refutation of the metaphysical thesis of an epistemic duality, a concept aptly illustrated by the ontologies involved in modeling. Moreover, I believe that we can find justification for the 'just more theory' manoeuvre—the most troubling step in the arguments for many of their critics. The manouvre is justified when made against the metaphysical realist by someone already committed to the view that truth is satisfaction of operational and theoretical constraints. However, this justification also reveals the problematic nature of the debate between metaphysical realism and rival metaphysical theories.

References

Bays, T. (2001). On Putnam and his models. *The Journal of Philosophy*, *98*(7), 331–350.

Bays, T. (2008). Two arguments against realism. *The Philosophical Quarterly*, *58*, 193–213.

Braddon-Mitchell, D., & Nola, R. (Eds.). (2009). *Conceptual analysis and philosophical naturalism*. Cambridge: MIT Press.

Button, T. (2013). *The limits of realism*. Oxford: Oxford University Press.

Douven, I. (1999). Putnam's model-theoretic argument reconstructed. *Journal of Philosophy*, *96*(9), 479–490.

Gunderson, K. (Ed.). (1975). *Language, Mind, and Knowledge. Minnesota Studies in the Philosophy of Science, Vol. VII*. Minneapolis: University of Minnesota Press.

Hale, B., & Wright, C. (1997). Putnam's model-theoretic argument against metaphysical realism, pp. 427–457.

Hodesdon, K. (2014). Mathematical representation: Playing a role. *Philosophical Studies*, *168*(3), 769–782.

[8]An example that Quine also considered (1969, pp. 58–60).

Hodesdon, K. (forthcoming). Structure, symmetry and semantic glue. *Philosophy in an age of science: Themes from the philosophy of Hilary Putnam, Berger A*. Oxford: Oxford University Press.
Langton, R. (1998). *Kantian humility*. Oxford: Oxford University Press.
Langton, R. (2004). Elusive knowledge of things in themselves. *Australasian Journal of Philosophy, 82*(1), 129–136.
Lewis, D. (2009). Ramseyan humility. *See Braddon-Mitchell and Nola*, pp. 203–223.
Lewis, D. K. (1984). Putnam's paradox. *Australasian Journal of Philosophy, 62*(3), 221–236.
Pippin, R. (1982). Kant's theory of form. *An Essay on the Critique of Pure Reason, New Haven, London*.
Putnam, H. (1967). Mathematics without foundations. *The Journal of Philosophy, 64*(1), 5–22.
Putnam, H. (1975). The meaning of meaning. See Gunderson, pp. 131–193.
Putnam, H. (1978). *Meaning and the moral sciences*. Boston: Routledge and Kegan Paul Ltd.
Putnam, H. (1980). Models and reality. *The Journal of Symbolic Logic, 45*(3), 464–482.
Putnam, H. (1981). *Reason, truth, and history*. Cambridge: Cambridge University Press.
Putnam, H. (1982). Why there isn't a ready-made world. *Synthese, 51*(2), 141–167.
Putnam, H. (1983). *Philosophical papers: Volume 3, realism and reason*. Cambridge: Cambridge University Press.
Putnam, H. (1987). *The many faces of realism: The Paul Carus lectures*. La Salle: Open Court.
Putnam, H. (1989). Model theory and the 'factuality' of semantics. See Putnam (1995b), pp. 351–375.
Putnam, H. (1991). *Representation and reality*. Cambridge: MIT Press.
Putnam, H. (1992a). *Realism with a human face*. Cambridge: Harvard University Press.
Putnam, H. (1992b). *Renewing philosophy*. Cambridge: Harvard University Press.
Putnam, H. (1992c). Replies. *Philosophical Topics: The philosophy of Hilary Putnam, 20*(1), 347–408.
Putnam, H. (1992d). Why is a philosopher? See Putnam (1992a), pp. 105–109.
Putnam, H. (1993). Realism without absolutes. See Putnam (1995b), pp. 279–294.
Putnam, H. (1995a). The question of realism. See Putnam (1995b), pp. 295–312.
Putnam, H. (1995b). *Words and life*. Cambridge: Harvard University Press.
Putnam, H. (1999). *The Threefold cord: Mind, body and world, the John Dewey essays in philosophy*. Irvington: Columbia University Press.
Putnam, H. (2005). *Ethics without ontology*. Cambridge: Harvard University Press.
Putnam, H. (2012). In M. Di Caro, & D. Macarthur (Eds.), *Philosophy in an age of science*. Cambridge: Harvard University Press.
Quine, W. (1969). *Ontological relativity and other essays*. New York: Columbia University Press.

Kate Hodesdon is research explores how it is possible to pick out and refer to mathematical objects. She is interested in what mathematical logic itself, particularly model theory and set theory, tells us about what mathematics is all about.

Chapter 7
Normativity and Mechanism

Timothy McCarthy

Abstract This paper presents a new perspective on the Lucas-Penrose arguments, which attempt to connect Gödel's incompleteness theorems to the thesis of mechanism in the philosophy of mind. I begin by taking a close look at Hilary Putnam's own response to Lucas-Penrose, which is widely taken to be decisive. I shall largely concur, but there is more to be learned. I will suggest that certain non-monotonic models of mathematical reasoning significantly alter the philosophical context for these arguments. I go on to describe a structural constraint on the rational coherence of the alternative cognitive evolutions allowed by these models. In the presence of that constraint, it is shown that if the evolutions allowed by such a model are mediated by an effective rule of revision, then the model is incapable of capturing certain inductive inferences of the most elementary kind.

Roger Penrose's book *The Emperor's New Mind* appeared in 1989. A year or so thereafter, the book came up in a conversation between the late George Boolos, Michael Detlefsen and myself. Mic led off by asking George what he thought of Penrose's attempt to give a new slant to an old argument. The argument, which had been circulating in various forms since the late 1950s, achieved a sort of plateau in a paper of John Lucas in 1961.[1] The argument purports to show that the human mind is 'non-algorithmic' on the basis of the Gödel incompleteness theorems. The claim is roughly speaking that Gödel's results show that the human mind cannot be 'represented' by a Turing machine.

[1] J. Lucas, Minds, Machines and Gödel, *Philosophy* **36**: 112–127.

T. McCarthy (✉)
Departments of Philosophy, University of Illinois, Urbana, IL, US
e-mail: tgmccart@illinois.edu

G. Hellman and R. T. Cook (eds.), *Hilary Putnam on Logic and Mathematics*, Outstanding Contributions to Logic 9, https://doi.org/10.1007/978-3-319-96274-0_7

George wanted no part of it – either the claim or, for that matter, the entire discussion. The whole thing had been disposed of once and forever, he said, by Hilary Putnam in "Minds and Machines" (1960). He held that Putnam had knocked the whole spectrum of arguments of this type out of contention in a couple of paragraphs.[2]

I hedged. Yes, Putnam's rejoinder was, on its own terms, decisive. But, I said, I'd never been able to escape the impression that there is something behind these arguments which hasn't been understood. This paper takes as its point of departure Boolos's claim which expresses, I believe, the dominant reaction to arguments of the Lucas–Penrose type among logicians and philosophers of mathematics.[3]

7.1 Putnam's Rejoinder

Let's start by looking at Putnam's argument. Here it is in its entirety:

> Let T be a Turing machine that 'represents' me in the sense that T can prove just the mathematical statements I can prove. Then the argument ... is that by using Gödel's techniques I can discover a proposition that T cannot prove, and moreover *I* can prove this proposition. This refutes the assumption that T 'represents' me, hence I am not a Turing machine.

That is intended to be a précis of the target argument. Putnam responds:

> The fallacy is a misapplication of Gödel's theorem, pure and simple. Given an arbitrary machine T, all I can do is find a proposition U such that *I* can prove
>
> (1) If T is consistent, U is true,
>
> where U is undecidable by T if T is in fact consistent. However, T can perfectly well prove (1) too! And the statement U, which T *cannot* prove (assuming consistency), *I* cannot prove either (unless I can prove that T is consistent, which is unlikely if T is very complicated)!

Let's unpack this a bit. The sentence U is, of course, the Gödel fixed point, a sentence provably equivalent, in elementary number, theory, to an arithmetical sentence expressing its own unprovability. A bit more explicitly, if $Pr_T(x)$ is the standard arithmetic provability predicate for T, the sentence

---

[2]Putnam's paper is reprinted in *Mind, Language and Reality: Philosophical Papers, v. 2*, Cambridge University Press, 1975. Boolos makes essentially the same claim about Putnam vis-à-vis Lucas–Penrose type arguments in his introductory note to Gödel's Gibbs Lecture in *Collected Works, v. III*, p. 295. Note that Putnam's 'response' was published *before* Lucas's paper! Its actual target was the related discussion of the significance of the incompleteness theorems in the concluding sections of Nagel and Newman's book *Gödel's Proof*, reissued by New York University Press in 2001.

[3]I should mention, of course, that Penrose has not been silent on these matters in the intervening twenty-five years. For example, his later book *Shadows of the Mind* (Oxford, 1994), Penrose gives a variant of the argument concerning soundness rather than consistency. We shall touch on this variation below. Solomon Feferman has drawn attention to a number of technical problems in Penrose's argument in Penrose's Gödelian Argument, *Psyche* **2**, 7 (1996), some of which play a role below.

(2) $U \leftrightarrow \neg Pr_T([U])$

is a theorem of, say Robinson's arithmetic Q and hence, on the assumption that T 'represents' me, of T (assuming I accept Q; here $[U]$ is the 'Gödel numeral' for U, a standard numerical term for its code number). From this it is very easy to see how the consistency of T allows us to establish the unprovability of U. Suppose that U *is* provable in T. Then some particular integer p that we could effectively calculate codes a proof of U in T. Thus if $Prf_T(x,y)$ represents the proof relation for T in Q, the sentence $Prf(\mathbf{p}, [U])$ will be provable in Q, hence in T. But since $Pr([U])$ is just the sentence

$$(\exists x) Prf_T(x, [U]),$$

this statement will be provable in T as well. But since T proves (2), if U were provable in T, $\neg Pr([U])$ would also be provable in T which is impossible *if T is consistent*.

If we could show that T is consistent, then, we could show that U is unprovable in T, and thus that $\neg Pr_T([U])$ is true and thus, via (2), that U is true. And so there would be at least one sentence, namely U, that we can prove to be true which T cannot prove. But how, Putnam asks, are we supposed to prove that T is *consistent*? T, Putnam supposes, 'represents me in the sense that T can prove just the number-theoretic sentences I can prove'. There are two things to attend to here, 'represents' and 'I can prove'. Putnam is reasonably explicit about the fact that he has in mind only a weak notion of representation: T 'represents' my number-theoretic capacity just in the sense that the output of T extensionally coincides with the arithmetic sentences that I can prove to be true. It is not required that T represent me in any more fine-grained sense. I am going to suggest that this is not at all what we normally mean when we talk about 'computational representations of cognitive capacities', but set that aside for now. What exactly is it that is *represented*? Putnam says that it is my capacity to *prove* number theoretic sentences, not, obviously, in the sense of 'provability' in a formal system, but in the informally rigorous sense appropriate to ordinary mathematical practice. This is a normative concept of demonstrative epistemic justification: the question is whether the number theoretic sentences that can be justified in this way coincide with the output of a Turing machine, or, equivalently, whether the codes of such sentences are computably enumerable.

If you look at the machine T just as a syntactic engine, and the problem of providing a consistency proof for T in a purely combinatorial way, it is not at all clear that I can supply such a proof. If T is just a Turing machine that happens to generate just the number-theoretic sentences I can prove, there is no reason at all to suppose this. And Putnam's description of T does not suggest or require that we look at it in any other way. Thus far, Putnam.

7.2 Normatively Reflective Representations

I said that we might want to add something to our description of the assumed representational properties of T. The thesis of mechanism in the philosophy of mind is not simply the assertion that the potential outputs of the mind can be coded in some way that makes them computably enumerable. It is the thesis that the mind, in its cognitive functioning, is *algorithmic*. In the sense normally intended in psychology, a computational model "represents" a portion of that functioning when the mind is a *realization* of that model; and what *that* means, very roughly speaking, is that there is an interpretation mapping data in the model to mental states under which the computational structure of the model describes the causal structure of the mind, what the mind is actually doing in generating cognitive outputs.

This characterization, while vague, seems to me correct as far as it goes. But it is not quite what is needed for the above application, the one concerning my capacity to give *proofs*. That capacity consists in part in my ability to situate the products of my mathematical activity in a normative context of justification. And so what the computations generated by the algorithm must reflect is not an order of causes per se but an order of reasons. The algorithm generates representations of *proofs*, and a proof is a sequence of inferences mediated by normative rules, drawing on contents which are either antecedently proved or warranted without proof. In this way one can arrive quite naturally at the idea of a formal axiomatic theory as a natural effective expression of the normative framework surrounding the notion of proof.

Placed in this context, then, Putnam's question becomes why a formal axiomatic theory T that "represents" me in the normative sense just indicated should be provably consistent *by me*. Here is an attempted argument that it should. This argument is closely related to the argument from soundness in Penrose's later formulation of his position.[4] The argument leans heavily on the assumption that such a theory introduces axiomatic contents and rules of inference that are *available* to me in a context of demonstrative justification; in particular, that I am *warranted* in appealing to these data in justifying number-theoretic beliefs. And the fact that these axioms and rules have this epistemic status has some strong consequences. First, if I am warranted in appealing the axioms in a proof, then I must be warranted in accepting those axioms. That does not mean that I must be in a position to produce an informative justification of the axioms, but it does require that (a) I can *recognize the axioms as being true*. Similarly, if I am warranted in appealing to a rule of inference in a mathematical proof, then (b) I can recognize that rule is *truth-preserving*. Using the standard inductive definition of proof in T, then, I can prove that any sentence formally provable in T is true (the assumption that the axioms of T are true provides the basis step, and the assumption that the rules of T are truth-preserving provides the induction step). Since I can recognize that $0=1$ is not true, I can prove that $0=1$ is not derivable in T. However, I can also recognize, by finitary mathematical means, that Gödel's arithmetic consistency statement Con_T holds if and only if $0=1$ is not

[4] *Shadows of the Mind, op, cit*. What follows is a reconstruction, not a rendition, of Penrose's later argument.

provable in T. Thus I can prove that Con_T is true. But since T meets the conditions of Gödel's Second, Con_T is not provable in T. Thus there is at least one number-theoretic statement, namely Con_T, that I can prove to be true that is underivable in T.[5]

I don't know of anywhere in the many pages which have been written attempting to connect Gödel's theorems to the thesis of mechanism in the philosophy of mind where *exactly* this argument has surfaced, but it closely connected to a number of arguments familiar from the literature.[6] I will call the present form of it the *Soundness Argument*. I am going to consider a response to this argument which was anticipated by Gödel himself in his Gibbs Lecture to the American Mathematical Society in 1951.

7.3 Escaping the Soundness Argument

There is a fairly obvious flaw in the Soundness Argument as it has just been presented. It has to do with a scope ambiguity in the condition (a), that 'I can recognize the axioms of T to be true'. We can distinguish two ways in which this claim can be understood:

(1) I can recognize each axiom of T to be true;
(2) I can recognize that each axiom of T is true.

The Soundness Argument involves an inductive proof that every consequence of T is true, and that requires (2) (I need to know *that each* axiom of T is true to deploy the basis of that induction); but it is only (1) that is required to ensure that T generates mathematically justifying proofs (for this purpose I just need to know that the axioms of T are true one by one). And of course (1) does not imply (2)!

Can the gap between (1) and (2) block the Soundness Argument? Let's look at the circumstances under which (1) *would* ensure (2). One such circumstance is that the set of axioms is finite: for in that case, via the Tarski schema the axioms jointly imply the truth of the (finite) conjunction of the axioms, and that is enough to ensure (2). And so the Soundness Argument works if we can establish the finiteness assumption.

It is plausible on general psychological grounds that a finite agent can hold only finitely many sentences to be true at any one time. However, a problem for the argument arises when we consider the possibility that an idealized finite cognizer **Ag** may accumulate infinitely many axioms over time. In order to undermine the Soundness Argument this would have to happen in such a way that

(i) The overall collection AX of axioms is computably enumerable;

[5] Observe that since the sentence U above is a consequence of Con_T in elementary arithmetics, we can also prove that U is true.

[6] For more or less similar arguments, see, for example, Stewart Shapiro, Truth and Proof – Through Thick and Thin, *Journal of Philosophy* **93**, 493–521; Neil Tenant, Deflationism and the Gödel Phenomena, *Mind* **111**, 551–582; and J. Ketland, Reply to Tennant, *Mind* **114**, 75–88. The issue here concerned specifically the "thickness" of the notion of truth required for something like this argument to go through.

(ii) The sentences formally derivable from *AX* coincide with the sentences that **Ag** can prove to be true;
(iii) **Ag** *cannot* recognize by mathematical means that every member of *AX* is true. A bit more precisely, there is no computable enumeration of *AX* that **Ag** can recognize to generate only true sentences.

And, as it happens, in the Gibbs lecture Gödel considers just this possibility:

> However, as to subjective mathematics,[7] it is not precluded that there should exist a finite rule producing all its evident axioms. However, if such a rule exists we with our human understanding could certainly never recognize it to be such, that is, we could never know with mathematical certainty that all propositions it produces are correct, or, in other terms, we could perceive to be true only one proposition after the other, for any finite number of them. The assertion, however, that they are all true could at most be known with empirical certainty, on the basis of a certain number of instances or by other inductive inferences.[8]

Conditions (i) and (ii) are required by the assumption that **Ag**'s mathematical competence is algorithmically representable; but condition (iii) must be assumed if the Soundness Argument isn't going to resurface. But now the difficulty is this. As above, we want a computational specification of *AX* that reflects the rational genesis of the axioms, and this means a computational description of the cognitive procedures which control axiomatic accumulation. If these procedures are thought of as *inferences* from the collection of axioms accepted at a given stage to the axioms accepted at the next stage, it would again seem that the application the procedures cannot *justify* me in believing such an axiom if I cannot recognize them to be truth-preserving. And if I can recognize the procedures to the truth-preserving, then by another inductive argument, using the truth of the axioms accepted at the initial stage as a basis, I can prove that every member of *AX* is true, in contradiction to (iii).

It is worth mentioning that the only *example* Gödel gives in the Gibbs lecture of a strategy for obtaining additional *evident* axioms fits this pattern exactly. Call a first-order theory *apodictic* if it is evident that each of its axioms are true. Since it is evident that the standard rules of inference preserve truth, it is potentially evident that the usual syntactic consistency claim for an apodictic first-order theory holds. Gödel's second example of what he calls the "inexhaustibility" of mathematics concerns precisely extending a given theory by adding the consistency statement for it, and this extension is mediated by an inference than we can recognize to be truth-preserving by mathematical means.

---

[7]The collection of sentences that **Ag** can prove to be true.

[8]Some Basic Theorems on the Foundations of Mathematics and Their Implications, in Gödel, *Collected Works, III, Unpublished Essays and Lectures* ed. by S. Feferman et al., Oxford University Press, 1986, p. 309.

7.4 Ampliative Accumulation: Gödel

The only model that we have for (i) – (iii) is the conception of infinitely many independent, yet somehow effectively generated, apodictic insights. It is difficult to know what to make of this suggestion in the overall context of Gödel's discussion; as indicated, it isn't illustrated anywhere. The oddness is further underscored by the first example Gödel gives of the thesis of the 'inexhaustibility of mathematics', strong axioms of infinity, or what we now call large cardinal axioms. The trouble is that this example does *not* fit into the pattern just rehearsed, because Gödel doesn't believe that these axioms are evident at all. Rather, the way they get introduced fits two non-apodictic cognitive strategies for obtaining axiomatic extensions that Gödel discusses in the Gibbs lecture:

(a) New axioms can be introduced on inductive or 'quasi-empirical' grounds[9];
(b) New axioms can be epistemically grounded on the basis of *perception of concepts.*

The axioms introduced in these two ways are *not* generally evident: both (a) and (b) are *essentially ampliative* and *epistemically risky*. I am going to suggest that this stance toward axiomatic accumulation forces us to entirely reconceive the whole problem of what it means for our framework for mathematical justification to be 'algorithmically representable'. But let's first see briefly how Gödel introduces both of these possibilities.

There is an intriguing passage in the Gibbs lecture where Gödel suggests that realism with respect to mathematics should allow the application of inductive reasoning, and he speculates that the abhorrence of the use of quasi-empirical methods in mathematics might be "due to the very prejudice that mathematical objects somehow have no real existence" (p. 313). Gödel writes:

> If mathematics describes an objective world just like physics, there is no reason why inductive methods should not be applied in mathematics just as in physics. The fact is that in mathematics we still have the same attitude today that in former times one had toward all science, namely, we try to derive everything by cogent proofs from definitions (that is, in ontological terminology, from the essences of things). Perhaps this method, if it claims monopoly, is as wrong in mathematics as it was in physics. (313)

Gödel understands the notion of 'inductive method' rather broadly, to include not only simple enumerative inductions (for example, inference of a Π_1 number theoretic statement on the basis of verification of instances), but analogical and explanatory inferences as well. Obviously, such inferences are ampliative and defeasible.

The second cognitive process identified in the Gibbs lecture as driving axiomatic discovery is what Gödel called 'perception of concepts'. In Gödel's middle-period realism, mathematical propositions are held to be objectively true or false in virtue of 'relations between concepts' (and so – in a certain sense – 'analytic'). Gödel says

[9] An idea subsequently vigorously defended by Hilary Putnam in, for example, Mathematical Truth, in *Matheamtrics, Matter and Method, Philosophica Papers, v. 1* (1975).

that concepts "form an objective reality of their own, which we cannot create or change, but only perceive and describe" (p. 320), and he held a view of perception of concepts on which concepts are rather analogous to ordinary perceptual objects that can be apprehended from different points of view and with varying degrees of distinctness. Perception of concepts, like perception of objects, is epistemically risky, a point brought out in the following remarkable passage:

> This concept of analytic is so far from meaning "void of content" that it is perfectly possible that an analytic proposition might be undecidable (or decidable only with ||a certain|| probability). For our knowledge of the world of concepts might be as limited and incomplete as our knowledge of ||the|| world of things. It is certainly undeniable that this knowledge not only is incomplete, but even indistinct. This occurs in the paradoxes of set theory, which are frequently alleged as a disproof of Platonism, but I think quite unjustly. Our visual perceptions sometimes contradict our tactile perceptions, for example, in the case of a rod immersed in water, but nobody in his right mind will conclude from this fact that the outer world does not exist. (321)

And so 'perceptual' judgments about concepts, like perceptual judgments about objects, are not apodictically grounded, in part because the presentations grounding them can fail to cohere with one another.[10] A particularly radical case of this is afforded by presentations of the concept of set underlying paradoxical instances of the comprehension schema.

In the epistemology of the Gibbs lecture, then, each of the two sorts of warrants that new axioms are said to possess are defeasible, or epistemically risky. Neither lead to evident axioms, and so neither can serve as a model for the possibility that Gödel initially suggested. But that possibility – the idea of an infinitely many apodictic insights, independent of one another, but somehow recursively organized – is just left hanging in the air. In the end, I think the whole idea of apodictic axiomatic extensions is supplanted in Gödel's thought by an alternative model derived from (a) and (b). And that picture of axiomatic evolution puts the whole question of the algorithmic representability of the concept of provability in a radically new light. If, as Gödel supposes, axioms evolve non-monotonically, what does it mean to say that a statement is 'provable' for or by me? If *proof* requires demonstrative justification, the question becomes: demonstrable *from what?* In the ordinary (normatively charged, informal) sense, a sentence is provable for me in a given epistemic situation if it is

---

[10]Charles Parsons has pointed out that it is not clear from Gödel's writings in this period that talk about 'perception' of concepts is not to be taken metaphorically (see Platonism and Mathematical Intuition in Kurt Godel's Thought, *Bulletin of Symbolic Logic* **1**, 1 (1995)). In the 1960s, Gödel came to subscribe to Husserl's phenomenological analysis of intentional consciousness, which makes available a more general characterization of the sort of epistemic access he was attempting to describe. It seems likely that Gödel would have accepted an analysis of 'perception' of concepts in terms of the noesis/noema correlation, on which concepts as ell as ordinary perceptual objects, are presented in a potentially infinite multiplicity of noetic acts through a corresponding multiplicity of noematic contents. For Gödel the (second-order) noema of the concept of set has an inexhaustibility analogous to that of the full noema of a concrete object. Gödel's remarks suggest the possibility of novel applications of Husserl's account of the intuition of concepts, and also some possible extensions of it. See Tieszen, *After Gödel* (Oxford University Press, 2013), for more on the Gödel–Husserl connection.

provable from sentences I accept without proof, and I am mathematically justified in accepting without proof, in that situation. These, one might say, function as my axioms in that situation, But if axioms may come and go, it is not at all clear what it means to say that a statement is 'provable for me' absolutely.

7.5 Effectively Presented Non-monotonic Structures

The lesson of all this is that the problem of algorithmically representing the notion of epistemic justification underlying my number theoretic competence must somehow be reframed in dynamic terms that allow axioms to accumulate non-monotonically. Beginning with Hilary Putnam's paper on trial and error predicates (once again, Putnam led the way!), a number of investigators have considered this problem. In the situation Gödel has described, there is a finitely axiomatized arithmetic theory T_0, taken to represent an initial epistemic situation. The problem is to describe the epistemically admissible evolutions stemming from that situation, and to specify a global notion of provability in terms of the whole ensemble of such histories. Such a description will qualify as 'algorithmic' if the evolutions are effectively generated.[11]

I am now going to sketch a framework for describing this sort of situation. Let K be a set of finite theories in the language of the initial theory T_0, which we think of as representing possible states of information expressible in the language of T_0. The finite epistemic evolutions stemming from T_0 form a subset of $K^{<\omega}$ generated by a revision relation R on K. For each point $\mathbf{p} \in K$, $\{\mathbf{q} \in K \mid \mathbf{q} R \mathbf{p}\}$ is the collection of all admissible immediate successors of $\mathbf{p}$ in F.

Here are a few definitions: the structure $\mathbf{F} = (K, T_0, R)$ is called a *frame.* If $\mathbf{p}$ and $\mathbf{q}$ are points of K, $\mathbf{q}$ is said to be *accessible from* $\mathbf{p}$ in $\mathbf{F}$ if there is an R-chain linking $\mathbf{q}$ to $\mathbf{p}$ in **F.** **F** will be called *effective* if the revision relation R is computable, *solvable* if for any two points in K there is a point accessible from each, and *deterministic* if R is the graph of a partial function on K. Effective frames are dynamic analogues of axiomatic theories, in that there is an algorithm generating the alternative admissible cognitive evolutions starting from the initial theory T_0. When there is a unique such evolution, whether effectively determined or not, the frame is deterministic (so "effectively deterministic" is not, in this context, a pleonasm). Finally, solvability refers to the circumstance that even though the evolutions from the initial state are not uniquely determined, conflicting states within alternative evolutions share a common refinement. (In the literature the term "confluence" is sometimes used to describe this idea.)

It is natural to say that a sentence is *assertable* in a frame when it is a first-order consequence of sentences which are *admissible axioms* in the frame: assertability in an effective frame is then a non-monotonic analogue of provability in an axiomatic

---

[11] What follows is an adaptation of self-referential phenomena in non-monotonic systems in my Self-Reference and Incompleteness in Non-Monotonic Structures, *Journal of Philosophical Logic* **23**:4 (1994).

theory. And so the question is what it means to say that a sentence is potentially axiomatic in the frame. The difficulty, as Gödel observed, is that axioms may come and go. However, within a frame there is a local analog of epistemic stability: say that a sentence is *indefeasible* at a point **p** in a frame **F** when it occurs at every point accessible from **p** in **F**, and let us say that a sentence is *acceptable at* **p** if it is indefeasible at some point accessible from **p**. Globally, the potentially axiomatic sentences in **F** may be taken to be the sentences which are acceptable at every point of **F**. The assumption that **F** is solvable allows one to simplify this definition: a sentence is an admissible axiom for **F** just in case it is indefeasible at some point in **F**. The usual notions from the metamathematics of number theory are extended to frames in the obvious way: for example, if *S* is a set of sentences in L(*PA*), a frame **F** is said to be *complete* for *S* if for any $\varphi \in S$ either φ or $\neg\varphi$ is assertable in **F**, and to be *sound* for *S* if a sentence of *S* is assertable in the frame only if it is true in the standard model. **F** is *consistent* when no sentence and its negation are assertable in **F**. Here then are a few quick facts about frames:

1. There exists no consistent, deterministic and Π_1 complete effective frame.
2. There exists no consistent, solvable and Π_2 complete effective frame.
3. There does exist a consistent, solvable Π_1 complete effective frame.[12]

7.6 A Limitative Argument

Models and the facts about them of the sort sketched in Sect. 7.5 can be applied and misapplied in a variety of ways. One application regards an effective frame as a computational description of the mathematical capacities of an actual cognizer. This is not the application I intend. There are in the first place the familiar obstructions to regarding any actual system as literally instantiating an algorithm with an infinite domain of definition; and the equally familiar observation that even an idealized system may perfectly realize an aberrant algorithm, an algorithm that systematically generates *mistakes.*[13] In the application I intend a frame is a normative model of a concept of epistemic justification.

Jeroslow developed the special case of frame models covering deterministic, effective frames, which Jeroslow called 'experimental logics', where we recall that

[12]For 1 see R. Jeroslow, Experimental Logics and Δ_2 Theories, *Journal of Philosophical Logic* **4** (1975), 253–267;. For 2, 3 see McCarthy *op. cit.* A result related to 3 ois proved in P. Kugel, Induction Pure and Simple, *Information and Control* **35**, 4 (1977) 276–336. The terminology of McCarthy (1994) and Jeroslow (1975) differs slightly from the above. In particular in McCarthy (1994) a frame is called a 'non-monotonic structure' and 'indefeasible' appears as 'stable'. A sentence is said to be *assertable* in a non-monotonic structure if every state has an extension at which the sentence is stable. A non-monotonic structure is first-order closed if the collection of sentences assertable in it is first-order closed. In the present treatment, any sentence assertable in a non-monotonic structure is treated as potentially axiomatic.

[13]On both points, see Kripke, *Wittgenstein on Rules and Private Language*, Harvard University Press, 1982.

'effective' refers to the decidability of the successor relation of the frame, whereas 'deterministic' means that each state in the frame has exactly one successor. In a general frame, the successor relation is many-one, and on the normative stance adopted above, is naturally taken to represent a notion of epistemic *permissibility*: the relation of a state to its successors represent the epistemically permissible moves for an agent in that state. On the normative construal a state of information in an experimental logic has only one permissible successor and so the successor relation describes a concept of epistemic *obligation*.

Now consider an idealized agent with the full complement of mathematical capabilities Gödel described in the Gibbs lecture. Following tradition, we shall call the agent 'Karl'. When does a frame constitute a normative characterization of Karl's number-theoretic competence? The natural minimal suggestion is that a frame *characterizes* Karl's arithmetic competence iff the sentences of L(*PA*) that Karl can know be true on mathematical grounds are just the sentences of L(*PA*) derivable in the frame.

I shall now argue that if Karl's competence includes the cognitive procedures explicitly recognized by Gödel, then no solvable, effective frame can characterize his arithmetic competence. First, we may assume that if a frame **F** characterizes Karl's number theoretic competence, then **F** is consistent; and if Karl is minimally competent, Robinson's arithmetic *Q* is derivable in **F**. Secondly, given Gödel's favorable remarks, canvassed above, about the admissibility of inductive inferences in mathematics, I think it is at least likely that he would have regarded the true Π_2 sentences as being within the scope of the inductive methods available to idealized number-theoretic practice. In any case, it seems to me that he should have so regarded them: consider a true Π_2 sentence, say (a) $\forall x B(x)$ where $B(x)$ is the formula $\exists y A(x,y)$, $A(x,y)$ primitive recursive. For any natural number n, there exists a k for which the statement $A(\mathbf{n},\mathbf{k})$ is provable in Robinson's Arithmetic Q, and thus for any n the sentence $B(\mathbf{n})$ is provable in Q. If via Q Karl can prove *every* instance of $B(x)$, then Karl can provide apodictic proofs for *every* bounded restriction of the sentence (a). If Karl's inductive method is not hopelessly weak, by proving sufficiently many of these he will infer (a). Such an inference is not apodictic, but it is *apodictically grounded*: the premises for it are provable in Q, and thus from evident axioms.[14]

Putting this together, then, if Karl's inductive method is reasonably robust, then any frame that characterizes Karl's arithmetic competence must be both consistent and Π_2 complete. By the second quick fact, then, no such structure can be both

[14] I assume that the axioms of Q are evident if anything is. The consistency condition in conjunction with the provability of the instances of B ensures that any projection of the negation of such in instance is unstable. Karl may even from time to time counter-project (a). Consider a true properly Π_2 statement of the form (a). Then the function $f(n) = \mu k A(n,k)$ is not majorized by any primitive recursive function; else, the variable y in (a) can be primitive recursively bounded, so that (a) equivalent to a Π_1 form. For a particular n, Karl may infer $\neg\exists y A\mathbf{n},y)$ inductively by reference to a long run of k's such that $\neg A(n,k)$ and thus temporarily project the negation of (a). But such a projection is never stable: eventually an integer k will be found such that the sentence $A(\mathbf{n},\mathbf{k})$ is verified in Q. Again, if Karl's inductive method is not hopelessly timid, Karl will frame the metainduction that such a projection is never stable and that the function f is in fact everywhere defined but not majorized by any p.r. function.

effective and solvable. Thus if effective frames are our all-purpose candidates for algorithmic representations of Karl's total mathematical competence, both apodictic and non-apodictic, then the possibility of such a representation is incompatible with the solvability constraint. The status of that constraint then becomes the question at issue.

7.7 Normative Status of Solvability

A frame **F** is solvable iff for each pair of points **p**, **q** in **F**, there is a point accessible from both **p** and **q** in **F**. If **F** is effective, the revision or successor relation of **F** is decidable and will be called a *rule of revision*. If we think of an effective frame **F** as describing a collection of alternative cognitive histories generated from an initial situation by application of the rule of revision, solvability requires that application of that rule will lead to a common refinement of any chosen pair of stages in these histories. That refinement need not be a part of either history, but it represents an epistemic situation that is a rationally permissible alternative from the standpoint of both of the initial points. Solvability, then, is a minimal principle of epistemic confluence.

The normative significance of this principle is that the frame in question allows conflicts between the epistemic situations represented in the frame to be evaluated from an epistemic perspective accessible from both of the initial situations. Such a conflict could appear in the form of incompatible partial states of information at two points which are indefinitely retained under iterated application of the rule of revision. Thus, suppose that **p** and **q** are points in the frame **F** that incorporate incompatible number theoretic hypotheses. These states contain apodictic evidence bases $e(\mathbf{p})$ and $e(\mathbf{q})$ which for concreteness in the number-theoretic case will be identified with finite sets of sentences provable in *PA*. The normative requirement is that there is a superposition of **p** and **q** in **F** in which these evidence bases are combined and the conflicting claims of **p** and **q** are evaluated. This process need not have a unique result: there may be several conflicting states incorporating $e(\mathbf{p}) \cup e(\mathbf{q})$ that constitute admissible refinements of **p** and **q**. But to deny that there exist any at all is to impute a sort of epistemic incommensurability to these states of information. It is tantamount to saying that when confronted with the combined evidence base the inductive method represented by the frame is powerless to generate a story which is rationally permissible from the point of each of the initial states of information.

Suppose that the frame **F** describes the global inductive structure of Karl's possible epistemic states. Let R be the successor relation associated with **F**. If Karl instantiates **p** at a time m, the picture one has of the diachronic structure of Karl's epistemic states is that of a sequence $\langle \mathbf{p}_n \rangle$ generated from the initial state by applying a selection function Φ to the collection of admissible successors of any stage. Thus if $\mathbf{p}_0$ is the initial state, for any n we have

$$\mathbf{p}_{n+1} = \Phi(\{\mathbf{h} \mid \mathbf{h} R \mathbf{p}_n\}),$$

where $\mathbf{p}_m = \mathbf{p}$. Similarly **q** arises from the initial state by iterated application of a selection function Ψ, generating a sequence$<\mathbf{q}_n.>$. These two sequences constitute deterministic substructures of the frame **F**. Let us first look at a simple case this situation that provides an unproblematic illustration of the operation of the solvability requirement.

Consider again a true Π_2 statement

(a) $\forall x\, \exists y\, A(x, y)$,

where $A(x,y)$ is a primitive recursive condition. Suppose the state **p** projects (a) on the basis of proofs of finitely many sentences

$$A(\mathbf{0}, \mathbf{k}_0), \ldots, A(\mathbf{n}, \mathbf{k}_n)$$

in Q; **q** is an incompatible state that projects the negation of (a) on the basis of a prior projection, for a fixed m, of the sentence

(b) $\forall y\, \neg A(\mathbf{m}, y)$

on the basis of refutations of the sentence $A(\mathbf{m}, \mathbf{i})$ for each i falling below a bound k_m. By suitably enlarging the computational evidence base, **p** and **q** can be sequentially extended, via suitable functions Φ and Ψ as above, to states $\mathbf{p}+$and $\mathbf{q}+$such that $\mathbf{q} \subseteq \mathbf{p}+$and $\mathbf{p} \subseteq \mathbf{q}+$. For suppose that $n < m$. Then by accumulating the computations in Q refuting the sentences $A(\mathbf{m}, \mathbf{i})$ for k_m a chain of situations obtained from **p** by applying Φ may be led to project (b) and retract (a); if $\mathbf{p}+$is the last term in this chain, $\mathbf{p}+$will be a situation accessible from both **p** and **q**. On the other hand, for each $i \leq m$ eventually an integer k will be found such that $A(\mathbf{i}, \mathbf{k})$ holds. If Ψ operates by accumulating computations verifying these statements, a chain generated by Ψ beginning with **q** may be led to retract (b) and project (a); if $\mathbf{q}+$is the terminal member of this chain, it will again be a state accessible from both of the initial states **p** and **q**. Up to a point, this process can go back and forth, sometimes favoring **p**, sometimes **q**. I pointed out in *n*. 14 above after sufficiently many retractions of sentences of the form (b) this process should form enough inductive confidence to frame a stable projection that (a) is true. Thus **p** is favored in the long run.

In cases of this sort, confluence of incompatible states is a byproduct of the accumulation of apodictic data; the rules of selection embodied in the selection functions Φ and Ψ operate in essentially the same way, revising states to restore consistency in the light of new apodictic data. But other cases are not so straightforward, and may involve very different principles of selection. In such cases, these functions may give rise to permanently conflicting projections along the separate paths. For a case of this sort, we imagine that **p** and **q** are extensions of ZF that are embedded in very different but familiar traditions defined by contrasting methodological perspectives. The first is controlled by some form of the *Maximize* rule, the idea that any set that can be consistently allowed to exist should be allowed to exist. The second is the contrasting idea that sets that do not have to exist shouldn't (the *Minimize rule*). These alternative heuristic pictures exist in many forms.

Now suppose that the functions Φ and Ψ above respectively reflect *Maximize* and *Minimize*, where we take both initial theories $\mathbf{p}_0$ and $\mathbf{q}_0$ to be a finite formulation of ZFC such as the system NGB, and let us start the process off by setting

$$\mathbf{p} = \Phi(\mathbf{p}_0) = \text{NGB} + \text{Ramsey cardinals exist;}$$
$$\mathbf{q} = \Psi(\mathbf{q}_0) = \text{NGB} + V = L.$$

These two hypotheses extending NGB are jointly incompatible in ZFC. Iterating Φ might take us through a sequence of large cardinal existence assumptions of increasing strength; iterating Ψ, on the other hand, might take us through a sequence of negative existence assumptions of increasing strength against the background of ZFC + $V = L$. The epistemic advantages of the first sequence lie in the systematic unification made possible by the presence of larger cardinals of various classes, a familiar argument used to motivate *Maximize*.[15] Whereas the second strategy has been a minority view in contemporary set theory, Jensen has argued that $V = L$ may be justified in terms of its mathematical fruitfulness and ontological economy, "a limited form of Ockham's Razor."[16] And so **p** and **q** might serve as models of stages in the early development of these traditions. However, the point is that neither hypothesis is currently regarded as rationally inadmissible by advocates of the other.[17] Thus each of the positions represented by **p** and **q** should be accessible from the other in the sense of the frame **F** if **F** adequately describes the rational structure of current practice.

Accessibility in a frame is supposed to represent a relation of relative epistemic possibility. Neither hypothesis above, then, is acceptable anywhere in the frame **F** if both hypotheses are epistemically possible everywhere in **F**. From the standpoint of the two methodological perspectives represented by the functions Φ and Ψ above, assertability in such a frame is an extremely strong epistemological requirement. It is natural to ask, from either perspective, not which sentences are assertable in **F**, but which sentences are assertable in substructures of **F** representing all situations which are epistemically admissible *within those perspectives.* These substructures would be generated by restrictions of the successor relation R of the wide frame that reflect the requirements of these divergent methodological standpoints. The experimental logics generated by Φ and Ψ constitute deterministic substructures of this sort, but in the general case there may be more than one successor to a given state of information that is admissible under the requirements of these perspectives. Thus we do not generally expect these methodologically reflective substructures to be deterministic.

A violation of the solvability requirement in the present case could arise only if an extension **p+** of **p** can be found at which the axiom of constructibility is not regarded as a rational alternative at all, and similarly an extension **q+** of **q** can be found at

[15]The label '*Maximize*', along with a number of important arguments for the position it represents, are due to Maddy, *Naturalism in Mathematics*, Oxford University Press, 1997.

[16]R.B. Jensen, Inner models and large cardinals, *Bulletin of Symbolic Logic*, 1 (1995): 393–407, p. 398.

[17]Jensen, *op. cit.* p. 400.

which the existence of Ramsey cardinals is regarded as rationally impermissible. Thus, for example, suppose that **p+** represents a maximalist position incorporating a catalog of epistemic successes so thick that $V = L$ comes to be regarded as having been conclusively refuted, not only in terms of the epistemic criteria appropriate to *Maximize* but in terms of the rule of revision of the wide frame. However, the minimalist position should be able to assess those successes in terms of those same wide criteria. Even if a different conclusion is reached in the light of conflicting data, such an evaluation should render **p** epistemically *permissible* from an extension of **q** in which the progression leading from the initial position to **p** to **p+** can be surveyed. But then the assumption of Ramsey cardinals is not, after all, rationally impermissible in **q**. In the case that the minimalist position **q+** incorporates a comparable set of epistemic successes, in occupying the evaluating state we face a situation of symmetrical underdetermination; but in such a situation each of the relevant extensions of ZFC should again be epistemically permissible from the standpoint of the other. The solvability requirement, then, has a very natural methodological motivation, and appears to lack counterexamples in current practice.

7.8 Concluding Remarks

The Lucas–Penrose query with which we began, "Is human mathematical competence algorithmically representable?", has now been sharpened considerably. We are now asking whether some very basic inductive capacities can be represented in effective frames, which are effectively presented structures that describe the permissible epistemic moves available to a hypothetical idealized cognizer. The observation of Sect. 7.6 showed that no frame **F** in which a minimal amount of elementary number theory is assertable can simultaneously satisfy the following conditions:

(a) **F** is effective;
(b) **F** is solvable;
(c) Each true Π_2 sentence is assertable in **F**.

Implicit at the conclusion of Sect. 7.6 was the following variant of condition (c). We observed above that if $\forall x B(x)$ is a true Π_2 statement, where $B(\mathrm{x})$ is the formula $\exists y A(x,y)$ for some primitive recursive predicate A, then for each n we have $\vdash_Q B(\mathbf{n})$. Moreover, since **F** is solvable, a sentence is assertable in **F** if it is acceptable at some point in **F**. The following condition therefore ensures the assumption (c):

(c#) If for each $n \vdash_Q B(\mathbf{n})$, then $\forall x B(x)$ is acceptable at some point of **F**.

(c#) is a weak condition of inductive closure: it says that if Karl can provide an apodictic proof of the most elementary kind for *every* instance of a generalization, and consequently never stably projects a counterinstance, then he will eventually project that generalization to be true and that projection will be stable under further evaluation. I am inclined to say that if Karl can effect any epistemically stable ampliative inferences, represented as indefeasible projections at some point in the frame **F**, than this ought be one of them.

Looked at in one way, the present observation simply adds to a line of limitative results, beginning with Putnam, Gold and Kugel,[18] concerning obstructions to computational models of induction. However, there are good reasons to suppose that the above form of the result may be the best possible. Certainly, it cannot ne strengthened in either of the two directions that naturally present themselves. First, the assumptions (a)–(c) (or (a)–(c#)) do not ensure the Π_1 incompleteness of the frame **F**: there are examples of effective, solvable frames in which each true Π_1 statement is assertable.[19] Secondly, it is the assumption of solvability that allows us to exhibit a true Π_2 unassertable sentence. But by making use of another technical result we can identify a true Π_3 unassertable sentence *independently of the solvability of* **F**.[20] The instances of that sentence are true Σ_2 statements, and so may themselves be products of non-monotonic inferences; in any case they will not in general be provable from evident sentences. Thus an inductive justification of such a Π_3 sentence will not be apodictically grounded. But so what?

The answer is that in the absence of apodictic groundedness the informal inductive case for the projected universal sentence is no longer very compelling. To see the significance of the requirement of apodictic groundedness, consider a Π_3 sentence, say

$$\forall x\, \exists y\, \forall z\, A(x, y, z),$$

where $A(x, y, z)$ is primitive recursive. For each n, an m can be found such that the Π_1 statement $\forall z A(\mathbf{n}, \mathbf{m}, z)$ holds; and, although each such statement is justifiable by an apodictically *grounded* induction, it need not itself be apodictically provable.

We are then faced with a standard obstruction to enumerative induction: to inductively justify a universal generalization $\forall x P(x)$, we require a suitable sample. But if the instances of the predicate $P(x)$ over that sample are individually uncertain, their finite conjunction may be too risky to ground the inference; and in the typical case the risk grows with the sample size. If, on the other hand, the inference is apodictically grounded, every finite conjunction of the instances of $P(x)$ is potentially free of epistemic risk, and therefore fully available to ground the inductive inference. The obstruction presented to an inductive inference of a number-theoretic generalization by the epistemic risk attached to instances over the relevant sample which are not apodictically justified may be remediable or it may not be; but in any case such an inference is not as convincing as one in which the premises are apodictically provable.

The assumption (c#), then, is a direct reflection of a compelling normative picture of inductive inference. The condition (b), on the other hand, reflects an equally com-

---

[18]Putnam, *op. cit.*, E.M. Gold, Limiting Recursion, *J. Symbolic Logic* **30**: 27–48 (1965), P. Kugel, Induction Pure and Simple, *Information and Control* **33**: 276–336 (1977).

[19]See fact 3 of Sect. 5 and the references there.

[20]McCarthy (1994), Corollary 9.

pelling normative picture of rational evaluation, requiring that a common descendent can be reached, by two chains of epistemically admissible moves, from any two initial epistemic positions. Finally, the condition (a) says simply that there is an algorithm (i.e., a *rule*) that tells what from the standpoint of any position the admissible moves *are*. What we have exhibited, then, is a conflict between three global constraints on a normative description of the epistemic structure of Karl's states of thinking. To say that, of course, is not to pinpoint where the fault lies, and in this paper I will not finalize an answer to that question. But it is, to my mind, something tantamount to an antinomy of reason that the combination of these three seemingly compelling constraints should self-destruct in this way.

Postscript

Uncharacteristically, I completed the above essay a few weeks before the deadline, and the editors were able to forward a draft to Hilary Putnam for comment in late 2015. In December, he drafted a reply. Apparently this reply is among the last things he wrote; I append it below. Although apparently unfinished, it is of considerable interest. It is a running commentary on my paper with a number of valuable observations both historical and philosophical. It ends with a simple pointed question: viewed as a contribution to the literature on Penrose et al, my piece delivers no verdict. It shows only that the combination of three structural constraints is unsustainable in the context of a model of the sort I described above. Hilary wanted to know: to what extent, if any, does such an argument constitute a vindication of Penrose? He was clearly unhappy with the idea that it might, and he encouraged me to tack on an answer to this question. I want to briefly oblige that request.

The assumption (a) in my text is a minimal expression of cognitive mechanism. It reflects the idea that Karl operates with an effective procedure for discriminating the admissible immediate successors to any given epistemic situation. In my text I simply assumed that this assumption is grounded in the idea that Karl applies a *rule* in making such discriminations. To violate that assumption, we must either make sense of cognitive discriminations which are not rule-governed or make sense of the idea of a non-effective rule. Neither of these options seems palatable, for reasons which are by now familiar. There are still two other places to go. You can deny that the frame describing Karl's epistemic alternatives is solvable, or you can accept the conclusion that apparently compelling inductive inferences are not captured by the frame. If we view the representability of such inferences as a basic normative constraint, that means that we must give up solvability. But what does *that* mean?

A frame is solvable if for any two states of information in it there is a point accessible from each. I interpreted this above as expressing a minimal sort of epistemic confluence: no matter how seriously alternatives may diverge, a common position can be reached by a chain of epistemic moves which are acceptable from the standpoint of both alternatives. We think of these confluent paths as representing a rational conversation between the given epistemic positions that leads to a common accessible position. This does not imply that the disagreement between them is ultimately resolvable, since there could be two such rational conversations, leading to a pair of conflicting alternatives; but it does imply, in one clear sense we can give to the phrase, that the initial alternatives are *epistemically commensurable*. Unless it violates this sort of commensurability, an effectively presented frame will be unable to accommodate certain inductive inferences of the most elementary kind. And this means that if capturing these inferences is a basic normative requirement, and effectiveness is a basic methodological one, we must accept the idea that what is epistemically possible for Karl depends upon his contingent epistemic situation in a particularly radical way. In the case described above from set theory, for example, we might have to say that at a certain point in the development of *Maximize*, minimizing moves loose all rational force for the maximizer. On the other hand, if solvability is retained as a normative constraint, and the effectiveness assumption (a) is in place, one can exhibit apodictically grounded inductive inferences that Karl *cannot make*; these inferences will have the status, for Karl's inductive method, of illusions of reason.

How then do we answer Putnam's question? What solace is there in all this for Penrose? The answer is: not much. We are free to accept the modest mechanist thesis (a) as a basic methodological constraint, and I am inclined to do so. But we must give up the conjunction of (b) and (c). In that case, what appeared as a joint inconsistency between formal expressions of mechanism, commensurability and inductive completeness becomes instead a tension between commensurability and inductive completeness alone. Against the background of a minimal mechanism, then, two ideals of reason stand in opposition. The first is an ideal of epistemic reconciliation, the second an ideal of epistemic closure. The loss of either is regrettable; the loss of one is necessary.

Appendix: Reply by Hilary Putnam

Reply to Tim McCarthy

Tim McCarthy's paper connects two parts of my work: my criticism of arguments by Ernest Nagel, John Lucas, and Roger Penrose that were supposed to show that our mathematical capacities outrun those of a Turing Machine – of *any* Turing machine - and my work on limiting recursion. The connection had not been seen by me, and the reflections on it initiated by McCarthy's paper are likely to be open-ended – which is great! Nagel, Lucas and Penrose all claimed that the Gödel Theorem shows that the mathematical power of the human mind outrun those of an computer, and I have contended that it shows no such thing. As I understand him, McCarthy does not disagree. His aim is not to defend the Nagel–Lucas–Penrose position, but to point out that the assumption that our mathematical powers could be represented by an algorithm for listing sentences that we, or an idealized version of ourselves, could "prove", the assumption to which Nagel, Lucas and Penrose assume their opponent to be committed, is not one that a sophisticated logician or philosopher of mind should accept. (Positions on this issue are not simply reflections disagreements about naturalism: Ernest Nagel was a naturalist and anti-dualist, and Penrose, while maintaining that present day physics is inadequate to describe the brain's functioning, is probably a naturalist as well, while Lucas was a traditional Anglican. What they agreed on is the negative claim that "the mind is not a Turing Machine".)

McCarthy writes:

> Putnam is reasonably explicit about the fact that he has in mind only a weak notion of representation: T 'represents' my number-theoretic capacity just in the sense that the output of T extensionally coincides with the arithmetic sentences that I can prove to be true. It is not required that T represent me in any more fine-grained sense. [The reason I 'had in mind" this weak notion of representation, is that it was the notion to which Nagel, Lucas's, and Penrose's original arguments[21] depended-H.P.] I am going to suggest that this is not at all what we normally mean when we talk about 'computational representations of cognitive capacities', but set that aside for now. What exactly is it that is *represented*? Putnam says that it is my capacity to *prove* number theoretic sentences, not, obviously, in the sense of 'provability' in a formal system, but in the informally rigorous sense appropriate to ordinary mathematical practice. This is a normative concept of demonstrative epistemic justification: the question is whether the number theoretic sentences that can be justified in this way coincide with the output of a Turing machine, or, equivalently, whether the codes of such sentences are computably enumerable.
>
> If you look at the machine T just as a syntactic engine, and the problem of providing a consistency proof for T in a purely combinatorial way, it is not at all clear that I can supply such a proof. If T is just a Turing machine that happens to generate just the number-theoretic sentences I can prove, there is no reason at all to suppose this. And Putnam's description of T [I would say "the description provided by Nagel, Lucas, and early Penrose" – H.P.] does not suggest or require that we look at it in any other way. Thus far, Putnam.

The next section of McCarthy's paper, "Normatively Reflective Representations," introduces an exploration, comprising the rest of the paper, of what happens if we "look at it" in a very different way, and brings in a very rich menu of ideas, including some from Gödel's famous Gibbs Lecture.[22] A key idea is that the thesis of mechanism should not just be that the output of the (idealized) mathematicians' mind is recursively enumerable (can be generated by a computer), but that the structure of the ability to recognize *proofs* should have an algorithmic description.

McCarthy writes,

> [The capacity to give proofs] consists in part in my ability to situate the products of my mathematical activity in a normative context of justification. And so what the computations generated by the algorithm must reflect is not an order of causes per se but an order of reasons. The algorithm generates representations of *proofs*, and a proof is a sequence of inferences mediated by normative rules, drawing on contents which are either antecedently proved or warranted without proof. In this way one can arrive quite naturally at the idea of a formal axiomatic theory as a natural effective expression of the normative framework surrounding the notion of proof.
>
> Placed in this context, then, Putnam's question becomes why a formal axiomatic theory T that "represents" me in the normative sense just indicated should be provably consistent *by me*. Here is an attempted argument that it should. This argument is closely related to the argument from soundness in Penrose's later formulation of his position.

[21] Penrose ref.

[22] Some basic theorems on the foundations of mathematics and their implications, *op. cit.*

Tim McCarthy does not claim that this "argument from soundness"[23] was employed by Nagel or Lucas, so from this point on we are dealing only with Penrose, or, more precisely, an argument "closely related" to arguments McCarthy finds in Penrose. I will not repeat McCarthy's first formulation of the argument; the critical point is that, in that formulation, the argument assumes that the mathematician ("Ag" for "agent"), *can recognize the axioms of T to be true.* If the axioms implicitly assumed by Ag are apodictic, and their number is finite this may be unproblematic, but in the Gibbs lecture, McCarthy reminds us, Gödel takes seriously the possibility that the idealized mathematician add axioms over the course of time (idealized as unbounded); the number of such axioms is potentially infinite. Moreover, new axioms may strike mathematicians as compelling for "quasi-inductive" reasons. Can such an open-ended intellectual procedure be represented with the tools of recursion theory? McCarthy's answer is that I myself had provided a way of doing this in my work on trial-end-error predicates. He writes.

> The lesson of all this is that the problem of algorithmically representing the notion of epistemic justification underlying my number theoretic competence must somehow be reframed in dynamic terms that allow axioms to accumulate non-monotonically. Beginning with Hilary Putnam's paper on trial and error predicates (once again, Putnam led the way!), a number of investigators have considered this problem. In the situation Gödel has described, there is a finitely axiomatized arithmetic theory T_0, taken to represent an initial epistemic situation. The problem is to describe the epistemically admissible evolutions stemming from that situation, and to specify a global notion of provability in terms of the whole ensemble of such histories. Such a description will qualify as 'algorithmic' if the evolutions are effectively generated.

McCarthy shows how to formalize descriptions of this kind by means of what he calls "effective frames." A property of such frames that he regards has highly desirable is *solvability*, explained thus:

If we think of an effective frame **F** as describing a collection of alternative cognitive histories generated from an initial situation by application of the rule of revision, solvability requires that application of that rule will lead to a common refinement of any chosen pair of stages in these histories. That refinement need not be a part of either history, but it represents an epistemic situation that is a rationally permissible alternative from the standpoint of both of the initial points. Solvability, then, is a minimal principle of epistemic confluence.

At this point let us cut to the chase:

Now consider an idealized agent with the full complement of mathematical capabilities Gödel described in the Gibbs lecture. Following tradition, we shall call the agent 'Karl'. When will a frame can constitute a normative characterization of Karl's number-theoretic competence? The natural minimal suggestion is that a frame *characterizes* Karl's arithmetic competence iff the sentences of L(*PA*) that Karl can know be true on mathematical grounds are just the sentences of L(*PA*) derivable in the

[23]The argument from soundness, in its simplest form, is that "the axioms of T are true, the rules of inference preserve truth, so all the theorem of T are true. But '1 = 0' is not true, so it is not a theorem of T. So T is consistent."

frame. I shall now argue that if Karl's competence includes the cognitive procedures explicitly recognized by Gödel, then no solvable effective frame can characterize his arithmetic competence.

What McCarthy does not tell us is, to what extent is this negative result a vindication of Penrose? I hope he will extend his paper to a discussion of this question.

Timothy McCarthy is a Professor in the Departments of Philosophy and Linguistics at the University of Illinois at Urbana-Champaign and a faculty affiliate in the Center for Applied Collaboration on Human Environments in the College of Engineering. He has written extensively on the philosophy of logic, philosophy of mathematics, and the theory of meaning; he also has written on social choice theory and risk. He is the author of Radical Interpretation and Indeterminacy and co-editor of Wittgenstein in America. His current research centers on inductive and analogical inference in mathematics and generalizations of Gödel's incompleteness theorems. In addition to his philosophical work, in recent years he has participated in projects related to inequality, STEM education, and skilled labor migration.

Chapter 8
Changing the Subject: Quine, Putnam and Waismann on Meaning-Change, Logic, and Analyticity

Stewart Shapiro

Abstract Hilary Putnam's views on analyticity, synonymy, and meaning-change loom large in his writing on logic, mathematics, and science. In "The analytic and the synthetic" (Scientific explanation, space, and time, Minnesota studies in the philosophy of science. University of Minnesota Press, Minneapolis, pp. 358–397, 1962), Putnam argues that (i) Quine is wrong in claiming that there just is no analytic-synthetic distinction, but (ii) Quine is right in arguing that analyticity plays no significant role in the philosophy or science (except, perhaps, linguistics). In some interesting ways, Putnam's views on these matters connect with those developed in Friedrich Waismann's "Analytic-synthetic", published serially in Analysis (Analysis 10:25–40, [1949], Analysis 11:25–38, [1950], Analysis 11:49–61, [1951a], Analysis 11:115–124, [1951b], Analysis 13:1–4, [1952], Analysis 13:73–89, [1953]), around the same period as Quine's "Two dogmas of empiricism" (Philosophical Review 60:20–43, 1951). Waismann provides a rich and subtle conception of analyticity and meaning, and the role that analyticity and synonymy play in linguistic interpretation (see also Waismann in Proceedings of theAristotelian Society, Supplementary 19:119–150, [1945]).

Hilary Putnam's views on analyticity, synonymy, and meaning-change loom large in his writing on logic, mathematics, and science. "Is logic empirical" (1968) and, to a lesser extent, "Mathematics without foundations" (1967), raise the issue of when two expressions in the same or different, formal or informal, languages are strictly synonymous or, in other words, whether they have the *same meaning*. In those papers, the focus is on logical terminology and on terms that figure in mature scientific and mathematical theories. Putnam is particularly interested in the question of whether a given term, like "mass" or "simultaneous" or logical terms like "or", has changed its meaning after a scientific revolution.

In "The analytic and the synthetic" (1962), Putnam argues that (i) Quine is wrong in claiming that there just is no analytic-synthetic distinction, but (ii) Quine is right

S. Shapiro (✉)
The Ohio State University, Columbus, OH, USA
e-mail: shapiro.4@osu.edu

G. Hellman and R. T. Cook (eds.), *Hilary Putnam on Logic and Mathematics*, Outstanding Contributions to Logic 9, https://doi.org/10.1007/978-3-319-96274-0_8

in arguing that analyticity plays no significant role in the philosophy or science (except, perhaps, linguistics). According to Putnam, analytic truths are trivial, and easily recognized. They are not the sort of thing that philosophers are specially trained to uncover, via conceptual analysis or any other a priori activity. Uncovering analytic truths requires only competence with the language or, at most, empirical studies on how terms are, in fact, used by competent speakers.

Putnam's most important conclusion here is that Quine is right that analyticity does not explain our knowledge of mathematics or even logic. At most, analyticity may be part of the explanation of our knowledge of some trivial linguistic matters, the classic examples being that all bachelors are unmarried and that vixens are female foxes. Nothing more significant than that. According to Putnam, then, Quine is more right than wrong. One will make far fewer mistakes in following Quine than in following most of his critics.

One group of linguistic items under scrutiny in Putnam (1967) are geometric terms like "point", "plane" and "line". Those are used to set up what Putnam wants to say about other words. General relativity is formulated in the backdrop of a non-Euclidean space-time. Let us assume that this provides the best theory of space-time. Does this mean that expressions like "straight line" have different meanings than they had in older scientific theories that invoke a Euclidean framework? Or were we just wrong in thinking that Euclid's parallel postulate is true? The main focus of Putnam (1968) is on logical terminology. He advocates that quantum mechanics be formulated using a quantum logic, and then raises the question as to whether this involves a change of meaning concerning the logical terms "or", "not", and "and", or, again, whether we were just mistaken in thinking that the distributive principles hold (see also Putnam (1957), where a similar issue is raised concerning excluded middle). He concludes, at the end of (1968, Sect. 6), that "we simply do not posses a notion of 'change of meaning' refined enough to handle" the questions.

In some interesting ways, Putnam's views on these matters connect with those developed in Friedrich Waismann's "Analytic-synthetic", published serially in *Analysis* (1949, 1950, 1951a, b, 1952, 1953), around the same period as Quine's "Two dogmas of empiricism" (1951). Waismann provides a rich and subtle conception of analyticity and meaning, and the role that analyticity and synonymy play in linguistic interpretation (see also Waismann (1945)).[1]

In the third article in the analyticity series, Waismann (1951a, 50) asks whether the sentence "Time is measurable" is analytic. He suggests:

> We are, perhaps, first inclined to answer, yes. What tempts us to do this is that it seems to be part of the meaning of 'time' that time should be measurable. Yet this claim can hardly be substantiated, ... What we could do is, at the most, to point out some of the uses (such as 'timing', 'timepiece', 'What is the right time?', etc.) which seem to indicate that time is measurable. This, however, will lead only to a scarcely enviable position since there is

[1]Each article in the analyticity series, including the last one, ends with "(To be continued)", so it is safe to conclude that the article was never finished. Waismann does not come to a firm conclusion. The only mention of Quine is in the first number (1949), where the main theme of "Truth by convention" (Quine 1936) is endorsed.

> no sharp line which separates those uses which, as one would say, are characteristic of the concept, from those which are not.

Sound familiar?

Waismann invites the reader to consider the situation when people had no precise ways to measure intervals of time ("before sand-glasses, water-clocks, or sun-dials had come into use"), and presents some thought experiments in which there does not seem to be a stable way to measure temporal intervals. Suppose, for example, that time-in-days does not coordinate with time-in-hour-glasses, nor with anything else. In effect, we are asked to suppose that we can find no constant ratios among events that are independent of the mode of measurement. Waismann writes:

> Would you be prepared to say that, in case the world was such that time could not be measured—say, because of the absence of sequences of recurrent events—time would not be what it is now? Here, I suppose, you may be inclined to say that it lies in the nature of time that it can be measured. But what do you mean by the expression 'it lies in the nature of time'? That this is part of the definition of the word 'time'? But as there is no definition to refer to, but only a use, forming a vast maze of lines, as it were, you will feel that this argument loses its point. On what, then, rests your assurance? (Waismann 1951a, 50–51)

Waismann next asks whether it is analytic that pain *cannot* be measured, a perhaps ironic (or prophetic) example in light of later developments in pain science. The upshot is this:

> When we were asking this sort of question, namely, whether the meaning of 'time' or 'pain' changes when a method of measuring is introduced, we were thinking of the meaning of a word as *clear-cut*. What we were not aware of was that there are no *precise rules* governing the use of words like 'time', 'pain', etc., and that consequently to speak of the 'meaning' of a word and to ask whether it has, or has not changed in meaning, is to operate with too blurred an expression. (Waismann 1951a, 53)

The "too blurred expression" here is something like "has the same meaning". This is the same as Putnam's quip that "we simply do not posses a notion of 'change of meaning' refined enough to handle" our questions.

Putnam (1968) suggests that one can assign an "operational meaning" to the terms in question—the geometrical vocabulary and the logical terms—that is preserved across the revolution. In the case of geometry, and relativity, "straight line" can be defined in terms of geodesics, the shortest paths through space (or space-time). It is assumed that light travels along geodesics (provided we think of light as the motion of particles—which we can think of as points via their center of gravities—and not as a wave). So construed, the parallel postulate would be a "synthetic" truth about geodesics in Euclidean spaces.[2]

[2]A similar idea was echoed by Poincaré (1908, 235), with a somewhat ironic prediction (given the accuracy of hindsight):

> In astronomy 'straight line' means simply 'path of a ray of light'. If therefore negative parallaxes were found … two courses would remain open to us; we might either renounce Euclidean geometry, or else modify the laws of optics and suppose that light does not travel rigorously in a straight line. It is needless to add that all the world would regard the latter solution as the most advantageous.

In the case of the logical terms, Putnam proposes "operational meanings" for conjunction and disjunction in terms of some features of combinations of physical systems, together with various scientific laws about them. The basic deductive rules (or the truth-tables) hold for the resulting frameworks, whether they are classical or quantum. The distributive principles hold in some systems (distributive lattices) but not in all. So, again, we get a common "meaning" across the revolution (see also Putnam (1957) for a similar take on excluded middle).

One of Putnam's slogans here is that you can't have a revolution and minimize it, too. The upheaval in scientific thought cannot be dismissed as a mere change in linguistic conventions—what is so revolutionary about that? Conventions, it is supposed, are arbitrary or, at best, pragmatic. To be sure, Putnam is not claiming that his "operational definitions" give us the real meaning of the terms in question, say the thing that linguists and lexicographers are after. That would be to claim that we *do* possess a sufficiently refined notion of "change of meaning". Putnam's claim, I take it, is that sharpening this "too blurry" relation in a certain way helps shed light on the revolutionary nature of the newer theories. One cannot grasp just how far-reaching the advance is if we insist that the theories before and after the revolution are completely incommensurable. In retrospect, at least, we can find enough in common to illustrate and explain the changes. Waismann would agree with this; Quine has no reason to reject it either.

So why not introduce a robust analytic-synthetic distinction, one that will work for explaining certain kinds of theoretical knowledge? In other words, what is Quine right about? Adopting a broadly Wittgensteinian distinction, Putnam (1962) coins an expression:

> In analogy with the notion of a cluster concept, I should like to introduce the notion of a *law-cluster*. Law-cluster concepts are constituted not by a bundle of properties as are the typical general names like 'man' and 'crow', but by a cluster of laws which, as it were, determine the identity of the concept. The concept 'energy' is an excellent example of a law-cluster concept. It enters into a great many laws. It plays a great many roles, and these laws and inference roles constitute its meaning collectively, not individually.

One can re-identify a given law-cluster concept in different theories if the laws that govern it have sufficient overlap. And, as with Putnam's example concerning geometry, we can often illuminate the nature of the change across theories by focusing on the overlap. In the geometric case, the notion of a geodesic figures in the overlap.

However, like cluster concepts and family-resemblance terms generally, no one item in the cluster is essential to the identity of the concept. The notion of "same concept", across theories, is not a sharp one. In principle, I suppose, one can imagine a sorites series of scientific theories—a sort of conceptual ship of Theseus—in which there is a word denoting a law-cluster concept in each theory. Suppose that there is a small change in the cluster from each theory to the next, say that only one law in the cluster is changed. But the cluster in the first theory has preciously little in common with the corresponding cluster in the last.

So the notion of law-cluster cannot play the role that analyticity does in Quine's opponents, at least not concerning scientific languages. There is nothing to focus on as the real meaning of the various theoretical terms, something that remains fixed throughout all uses of the term. It is a pragmatic matter which items in the cluster we focus on, as we look to re-identify the corresponding term after an actual scientific revolution. And we cannot know in advance just which items in the cluster will be preserved in future theories. Presumably, Putnam would say the same of the terms in geometry and even logic, as least in so far as those terms are applied in science—relativity and quantum mechanics in the cases at hand.

For his part, Waismann does not limit the focus to scientific terms. As we saw, he speaks of a "vast maze of lines" in the use of almost any given expression. With Quine and Putnam, he argues that we cannot easily distinguish which of those "lines" go into (or follow from) meaning and those which of them represent common beliefs about the subject, so common that one can hardly imagine things otherwise. Early in the second installment (1950, 25–26) in the analyticity series, Waismann goes after attempts to give clean definitions of key terms[3]:

> What, then is a *definition*? A definition, it seems, is a licence which permits us to replace a word, or a symbol, by the *definiens*, i.e. to translate an expression into a different idiom. When we say this sort of thing, what we have in mind are perhaps *explicit* definitions, … illustrated by such stock examples as "A planet is a heavenly body revolving round the sun". And we are perhaps tempted to think that *every* definition conforms to this archetype. We are apt to forget that definitions of this kind are of use only in comparatively simple and trivial cases. The more interesting concepts such as truth and falsity, meaning and purpose, cause and effect, intelligence, time, number, which fascinate theorists, elude our efforts to pin them down in this way and only mock such clumsy attempts at defining.

The same goes for philosophical efforts to uncover analytic connections between concepts, or at least the complex concepts that concern us. Among the "vast maze of lines", some seem more significant than others, in some situations. But this does nothing to illuminate epistemology, give the meaning of scientific or ordinary statements, or the other jobs traditionally assigned to analyticity. In the aforementioned treatment of the statement that time is measurable, Waismann notes that someone

> may be inclined to say: "Though 'time is measurable' and 'rock salt is cleavable', sound superficially alike, they are very different: the one is accidental, the other is not." And you say that perhaps in the tone of a man who is calling attention to a notorious fact. But in saying this you do not want to object to any of the facts which make it possible to measure time …. What you don't see is that you are *irresistibly urged* to use a certain mode of representation which means a lot to you, in fact *that* mode which enables you to visualize with the greatest ease all sorts of temporal relations. As so often in philosophy, a statement appears so convincing precisely because it is … the obscure expression of a desire to use certain images, or a certain pictorial representation, to satisfy certain needs. (Waismann 1951a, 53)

So much for the agreement between Quine and both Putnam and Waismann. Now, what was Quine wrong about? Here Putnam and Waismann do not agree, although in some ways their views overlap and reinforce each other.

[3]In light of the developments concerning Pluto, not to mention asteroids and planets of other stars, Waismann's example is not a good definition. Maybe something like "a vixen is a female fox" would be better.

Putnam (1962) takes Quine to be arguing that there just is no analytic-synthetic distinction to be had, or at least none that coincides with various traditional conceptions of analyticity—there is, after all, stimulus analyticity. We will go along with this interpretation of Quine here, at least for the sake of argument. If Quine were arguing instead that the notion of analyticity is too thin, vague, obscure, or unstable to bear the philosophical weight sometimes placed on it, then Putnam would happily agree. I presume that Waismann would, too. Indeed, they both argue for that conclusion.

Grice and Strawson (1956) also take Quine to be arguing for the conclusion that there just is no such thing as an analytic-synthetic distinction. They take the fact that theorists agree on wide range of cases to be at least strong prima facie evidence that the distinction exists.

Putnam (1962) agrees:

> [T]he argument used by [Grice and Strawson] to the effect that where there is agreement on the use of the expressions involved with respect to an open class, there must necessarily be some kind of distinction present, seems to me correct and important. Perhaps this argument is the only [argument] of any novelty to have appeared since Quine published his paper.

Grice and Strawson make a similar argument concerning the English expressions "means the same as" and "does not mean the same as":

> Now since [Quine] cannot claim this time that the pair of expressions in question is the special property of philosophers, the strategy ... of countering the presumption in favor of their marking a genuine distinction is not available here (or is at least enormously less plausible). Yet the denial that the distinction ... really exists is extremely paradoxical. It involves saying, for example, that anyone who seriously remarks that "bachelor" means the same as "unmarried man" but that "creature with kidneys" does not mean the same as "creature with a heart" ... *either* is not in fact drawing attention to any distinction at all between the relations between the members of each pair of expressions *or* is making a philosophical mistake about the nature of the distinction between them ... [W]e frequently talk of the presence or absence of synonymy between kinds of expressions—e.g., conjunctions, particles of many kinds, whole sentences—where there does not appear to be any obvious substitute for the ordinary notion of synonymy ... Is all such talk meaningless? (Grice and Strawson (1956, 145–146))

As we saw above, Putnam argues that in the scientific-geometric-logical cases, we simply do not have a sufficiently refined notion of "same meaning", and Waismann finds it to be "too blurred" to do much philosophical work for us. But a vague, unrefined, or otherwise blurry distinction is still a distinction. The important question is what this distinction can do for us in our attempt to understand science, mathematics, and ordinary concepts. What weight will it bear?

Both Quine and Waismann note that we can get a model for analyticity by thinking about stipulations in formal languages. As long as one wants to use the formal language, one must adhere to the stipulation. So, for example, a logician can just define a connective as follows: for any formulas A, B, $(A\&B)$ is equivalent to $\neg(A \rightarrow \neg B)$. The stipulations are constitutive of the formal language.

But, of course, natural languages do not have anything analogous to stipulation. Quine famously insists that even if a word or phrase begins life via some sort of stipulation (or is learned that way), it can then take on a life of its own. In the evolution of natural languages, stipulations do not have the same constitutive character that they do in formal languages developed for specific purposes.

Putnam presents a sort of natural language analogue of a stipulation. The final section of (1962) defines a sentence to be an *analytic definition* if

(1) The statement has the form: 'Something (Someone) is an *A* if and only if it (he, she) is a *B*', where *A* is a single word.
(2) The statement holds without exception, and provides us with a *criterion* for something's being the sort of thing to which the term *A* applies.
(3) The criterion is the only one that is generally accepted and employed in connection with the term.
(4) The term *A* is not a 'law-cluster' word.

For Putnam, a sentence is *analytic* if it is a logical consequence of an analytic definition.

Notice that, on this account, we do not *discover* the analytic definitions and the analytic truths by any kind of philosophical analysis. What matters is how a key term is used by competent speakers of the language (in light of clauses (2) and (3)), and that can only be determined empirically. We have to find out what is "generally accepted" in the linguistic community in question (and not what a philosopher argues ought to be generally accepted).

Note also that there is no direct connection between matters of linguistic or conventional *meaning* and analytic definitions, and thus analyticity itself. It is an open theoretical question to determine just *why* speakers use the criterion the way they do. So Quine has no grounds to reject Putnam's definition, beyond its use of vague phrases like "generally accepted". Of course, a Quinean might wonder what the point of the definition is, whether analyticity is an interesting notion, illuminating anything of substance. Putnam wonders about that, too.

As indicated by Putnam's final clause (4), analyticity does not apply to statements that involve scientific terms essentially. Those terms denote law-cluster concepts. As advertised, then, Putnam's notion of analyticity is limited to rather trivial and obvious cases. The two examples Putnam cites are "Bachelors are unmarried" and "Vixens are foxes".

Putnam gives an amusing thought experiment in which "bachelor" could become a law-cluster term, if we discovered a unique medical condition had mostly by bachelors and hardly anyone else. If that were to happen, then "Bachelors are unmarried" might very well lose its status as analytic—but Putnam rightly points out how far-fetched this scenario is.

It is easy to see the value of analytic truths, or at least of analytic definitions. It is essentially the same value that comes with stipulations in formal languages. Instead of always having to say "female fox", we can say "vixen"'; instead of having to say "unmarried man", we can say "bachelor". Given that these are not law-cluster terms, and probably won't become law-cluster terms, it is safe to assume that the "stipulation" is permanent.

Or is it? As we saw, Waismann (1951a) speaks of a "vast maze of lines" between the common use of words. He did not intend this to be restricted to scientific terms. His example was "time", in its ordinary usage ("in the nick of time", "time-piece").

To focus on Putnam's two examples, "vixen" and "fox" are used for biological classification. The same goes for "female". Terms like those sometimes evolve in light of scientific needs. It is not all that far-fetched to imagine scientific developments in which "vixens are female foxes" would lose its status as analytic. Perhaps "vixen" or "fox" would then become a law-cluster term. Or maybe "female" would.[4]

Putnam (1962) notes that, even with formal languages, speakers can *intend* to make a use permanent and later, in light of developments, change their minds about that—if the proposal loses its utility. Surely something similar can happen with natural language terms. And, in such cases, to wonder if the meaning has thereby changed is to operate with "too blurred an expression".

Putnam's other example is perhaps more interesting. The term "unmarried" invokes the legal/social/religious institution of marriage. That, of course, is not a law-cluster term (at least not concerning *scientific* laws). There was surely a time, not long ago, when a statement that two women cannot be married to each other would meet Putnam's definition for analyticity. As noted, Putnam's analytic definitions apply to statements that hold "without exception", and provide "a criterion is the only one that is generally accepted and employed in connection with the term". There is no mention of what the linguistic community *should do*, nor with what they themselves (or their professional linguists and lexicographers) *take to be* definitional. Putnam's account of analyticity only concerns what the community accepts and employs in connection with the term. Think back to, say, 1940, or 1840, or … Consider a statement that marriage is a (holy?) relationship entered into by a man and a woman that is sanctioned as such by a legal body with relevant jurisdiction. That must have at least come close to meeting the letter of an analytic definition. It held without exception (given the network of statues and religious institutions at the time) and was, in fact, used as a criterion for the application of the term. I would speculate that if a member of one of those linguistic communities were asked about a pair of adult women, they would not bother to inquire as to whether their relationship had the relevant legal sanction. There would have been no need to do so; they would know that, in the case at hand, there just is no legal sanction, and so the women are not married to each other.

For his part, Waismann gives a few hints as to what the role of analyticity is in language. Waismann would certainly agree with Quine and Putnam that analytic truths are not discovered via conceptual analysis or any other a priori activity. Thus, analyticity cannot play the role assigned to it in some of Quine's opponents. But analytic truths are much more robust than Putnam thinks. For Waismann, analyticity is connected with the ways in which speakers interpret each other, and with what one can take for granted in conversation in a given context.

---

[4]There is, of course, a rich literature on the question of whether words like "female" stand for "natural" scientific classifications, whether they are social constructions, etc. See, for example, Haslanger (2015).

At the start of the second number in the series, Waismann (1950, 25) writes:

> it is significant that we do not only "find out" that a given statement *is* analytic; we more often precisify the use of language, chart the logical force of an expression, by *declaring* such-and-such a statement to be analytic. If 'analytic' was as fixed and settled a term as, say, 'tautology' is, this would be hard to understand : can I, e.g., by decree appoint a given statement to the rank of tautology? It is precisely because, in the case of 'analytic', the boundary is left open somewhat that, in a special instance, we may, or may not, recognize a statement as analytic.

As we saw above, in (1951a, 53), he spoke of a speaker being "*irresistibly urged* to use a certain mode of representation", that mode which enables her "to visualize with the greatest ease all sorts of temporal relations".

In the final number in the series (1953, 75), Waismann turns our attention to someone who declares that identity (of length) is transitive[5]:

> He obviously wants to make the inference *independent of experience*, so that he can stick to it whatever may happen to the physical rods. That is to say, he insists on using a language in which '$a=b$, $b=c$, so $a=c$' is an *inference licence*, not an empirical statement, or again, in which this relation is adopted as a *convention*. But as a convention, emptied of content, it does not say anything about the actual world, and in particular it does not help us to infer, or predict, the results of experiments with actual rods. And this makes us see the drawback of this view—namely that it offers no guarantee that the rule adopted will be applicable. Look here, we might say to him, if we were living in a sort of Lewis Carroll world where things expand and shrink unaccountably, what will become of your rule? You may cling to it, yes; and you may insist that any deviation observed must be due to some distorting force, blaming physics for the discrepancy. Yet the fact remains that your rule cannot be relied on. So what is the good of having it? Wouldn't you do better without it?
>
> But these are fancies, it will be said, so why care for them? Even if they are fancies—which is not too sure—it is enough to show that the rule must answer to something in reality, have some empirical backing if it is not to be worthless.

Waismann (1953, 79) emphasizes that "we are not *slaves* of the existing language". Nevertheless, there *are* linguistic connections, inference licenses, that we are (at least for the time being) irresistibly urged to make and maintain. We use those connections to interpret each other, at least until something goes wrong.

Grice and Strawson (1956) provide a now famous thought experiment to undermine what they take to be Quine's overblown claims. It involves two conversations. In one of them, a speaker (X) makes the following claim:

(1) My neighbor's three-year-old child understands Russell's theory of types.

In the second conversation, another speaker (Y) says

(1′) My neighbor's three year old child is an adult.

Grice and Strawson point out, plausibly, that with X we would know what is being said, even if we find it extremely unlikely or perhaps psychologically impossible for

---

[5]The connection between this and Putnam's views on logical and geometrical terms is straightforward. I don't know whether, for these purposes, Putnam would consider identity a logical term (*pace* Quine) or a law-cluster term or something else.

it to be true. In contrast, "we shall be inclined to say that we just don't understand what Y is saying". Indeed, "whatever kind of creature is ultimately produced for our inspection, it will not lead us to say that what Y said was literally true". If "like Pascal, we thought it prudent to prepare against long chances, we should in the first case know what to prepare for; in the second, we should have no idea". Perhaps a more charitable conclusion, still in line with Grice and Strawson's agenda, would be that Y must not be using the word "adult" the same way we do.

Grice and Strawson conclude:

> The distinction in which we ultimately come to rest is that between not believing something and not understanding something; or between incredulity yielding to conviction, and incomprehension yielding to comprehension. (p. 151)

Of course, the distinction Grice and Strawson point to is a real one—and an important one. There is a difference, a difference in kind, between not believing someone, because what they claim is wildly implausible, and not understanding them (or thinking that they are using words differently from the way we do).

Putnam's analytic truths would seem to play a role in this distinction. If someone were to deny one of them, by saying that Sly is a male vixen or that Seymour is a married bachelor, one would probably think they did not know what they were talking about (or did not understand what the words he uses mean).

Waismann also raises the distinction between not believing and not understanding, but he goes on to show how, with a little explanation, one *can* sometimes come to understand, and perhaps even believe, what is being said. In the earlier "Verifiability" paper (1945, 120), he presents a thought experiment much like Grice and Strawson's character Y:

> If, for instance, someone were to tell us that he owned a dog that was able to think, we should at first not quite understand what he was talking about and ask him some further questions. Suppose he described to us in detail the dog's behaviour in certain circumstances, then we should say "Ah, now we understand you, that's what you call thinking".

The same goes for those expressions he called "inference licenses" (1953, 74) and things we are "irresistibly urged" to apply (1951a, 59).

Waismann highlights the dynamic nature of language throughout the "Analytic-synthetic" series. As new situations are encountered, and as new scientific theories (and perhaps social situations and norms) develop, the extensions of predicates change. As new, unexpected cases are encountered, the predicate in question is extended to cover them, one way or the other. When things like this happen, there is often no need to decide—and no point in deciding—whether the application of a given predicate, or its negation, to a novel case represents a change in its meaning or a discovery concerning the term's old meaning—"going on as before", as Wittgenstein might put it. That, we saw, is to "operate with too blurred an expression", namely "means the same as".

In the fourth number of the analyticity series, we find:

> Simply... to refer to "the" ordinary use [of a term] is naive... [The] whole picture is in a state of flux. One must indeed be blind not to see that there is something unsettled about language;

> that it is a living and growing thing, adapting itself to new sorts of situations, groping for new sorts of expression, forever changing. (Waismann 1951b, 122–123)

And in the final published installment:

> What lies at the root of this is something of great significance, the fact, namely, that language is never complete for the expression of all ideas, on the contrary, that it has an essential *openness*" (Waismann 1953, 81–82).

Waismann notes that major advances in science sometimes—indeed usually—demand *revisions* in the accepted use of common terms: "breaking away from the norm is sometimes the *only way* of making oneself understood" (1953, 84).[6]

This dynamic nature of both scientific and ordinary language is common ground between Quine, Putnam, and Waismann. They also agree that the traditional notion of analyticity cannot play its assigned roles in some perhaps overly naive accounts of both scientific and ordinary knowledge. The key items do not quite stay put, nor is it quite correct to say that they are replaced with new terms. Unlike Quine, however, Putnam and Waismann find that there is, nevertheless, an interesting and important notion of analyticity—or at least an interesting and important way to sharpen the intuitive notion—one that does have an important role to play in how we understand language and interpretation, with all of its dynamic elements. I hope I have made a case that it helps to understand and reinforce the two philosophers by placing them side-by-side, looking for connections and differences.

References

Benacerraf, P., & Putnam, H. (1983). *Philosophy of mathematics* (2nd ed.). Cambridge: Cambridge University Press.

Grice, P., & Strawson, P. F. (1956). In defense of a dogma. *Philosophical Review, 65,* 141–158.

Haslanger, S. (2015). *Resisting reality: Social construction and social critique.* Oxford: Oxford University Press.

Poincaré, H. (1908). *Science and method, foundation of science* (H. Poincaré, & G. Halsted, Trans.). New York: The Science Press (1921).

Putnam, Hilary. (1957). Three valued logic. *Philosophical Studies, 8,* 73–80.

Putnam, H. (1962). The analytic and the synthetic. In H. Feigel & G. Maxwell (Eds.), *Scientific explanation, space, and time, Minnesota studies in the philosophy of science* (Vol. 3, pp. 358–397). Minneapolis: University of Minnesota Press (Reprinted in *Readings in the philosophy of language,*

[6]Waismann (1952) illustrates this point in some detail with the evolution of the word "simultaneous". Is there any *semantic* fallout of the theory of relativity? Are we to say that a brand new word, with a new meaning, was coined (even though it has the same spelling as an old word), or should we say instead that we have discovered some new and interesting features of an old word? Did Einstein discover a hidden and previously unnoticed context-sensitivity in the established meaning of the established word "simultaneous" (or its German equivalent), even though he showed no special interest in language as such? Or did Einstein introduce a brand new theoretical term, to replace an old term whose use had scientifically false presuppositions? As above, those are unhelpful questions; they invoke the "too blurred expression" of "same meaning". "Simultaneous" here qualifies as a Putnamian "law-cluster" term.

pp. 94–126, by J. Rosenberg & C. Travis, Eds., 1971, Englewood Cliffs, New Jersey: Prentice-Hall).
Putnam, H. (1967). Mathematics without foundations. *Journal of Philosophy, 64*, 5–22; Benacerraf and Putnam. (1983), 295–311.
Putnam, H. (1968). Is logic empirical?. In R. S. Cohen, & M. W. Wartofsky (Eds.), *Boston studies in the philosophy of science* (Vol. 5, pp. 216–241). Dordrecht: D. Reidel. (Reprinted as "The Logic of Quantum Mechanics" in *Philosophical Papers 1*, pp. 174–197, by H. Putnam, 1975, Cambridge: Cambridge University Press).
Quine, W. V. O. (1936). Truth by convention. In O. H. Lee (Ed.), *Philosophical essays for Alfred North Whitehead* (pp. 90–124). New York: Longmans. (Reprinted in Benacerraf and Putnam 1983, 329–354).
Quine, W. V. O. (1951). Two dogmas of empiricism. *Philosophical Review, 60,* 20–43.
Waismann, F. (1945). Verifiability. In *Proceedings of the Aristotelian Society, Supplementary Volume 19*, 119–150. (Reprinted in *Logic and language*, A. Flew (Ed.) pp. 117–144, 1968, Oxford: Basil Blackwell).
Waismann, F. (1949). Analytic-synthetic I. *Analysis, 10,* 25–40.
Waismann, F. (1950). Analytic-synthetic II. *Analysis, 11,* 25–38.
Waismann, F. (1951a). Analytic-synthetic III. *Analysis, 11,* 49–61.
Waismann, F. (1951b). Analytic-synthetic IV. *Analysis, 11,* 115–124.
Waismann, F. (1952). Analytic-synthetic V. *Analysis, 13,* 1–14.
Waismann, F. (1953). Analytic-synthetic VI. *Analysis, 13,* 73–89.

Stewart Shapiro is the O'Donnell Professor of Philosophy at The Ohio State University, a Distinguished Visiting Professor at the University of Connecticut, and a Professorial Fellow at the University of Oslo. His major works include *Varieties of logic* (Oxford, Oxford University Press, 2014), which articulates and defends a pluralism about logic, *Vaguenessin context* (Oxford, Oxford University Press, 2006), which contains a philosophical account of vague terms in which extensions shift from conversational context to conversational context, and the development of a concomitant model theory; *Philosophy of mathematics: structure and ontology* (Oxford University Press, 1997), a presentation of structuralism; and *Foundations without foundationalism* (Oxford University Press, 1991), an articulation and defense of second-order logic. He has taught courses in logic, philosophy of mathematics, metaphysics, epistemology, philosophy of religion, Jewish philosophy, social and political philosophy, and medical ethics.

Part II
Mathematics, Foundations, and Philosophy of Mathematics

Chapter 9
Putnam on Foundations: Models, Modals, Muddles

John P. Burgess

Abstract Putnam has famously offered a sketch of a mathematics without foundations, existing in two equivalent descriptions, set-theoretic and modal-logical. Here his proposal is critically examined, with attention to difficulties surrounding both the modal-logical description itself and especially the notion of equivalence of descriptions.

9.1 Putnam, Logic, and Foundations

Among the many and varied writings of Hilary Putnam on foundations and philosophy of mathematics, one work holds special prominence. The classic anthology of Benacerraf and Putnam (1964/1983), known to generations of students, was enriched in its second edition by the addition of two papers from each of its editors. But one of Putnam's, "Models and Reality" (1980), though it was a Presidential Address to the Association for Symbolic Logic, and contains theorems on models of set theory, is on its own showing more in the domain of philosophy of language. That leaves the other, "Mathematics without Foundations" (1967b), as perhaps Putnam's most conspicuously displayed publication centrally on philosophy of mathematics. It is that paper, henceforth referred to as MWF, and the issues it raises, that will be my concern here.

MWF has two distinguishable aspects, calling for separate comment. For want of better terms I will call the two the *foreground message* and the *background story*. The foreground message, the ostensible main claim of the paper, vigorously stated in its opening paragraph, is that mathematics is in a safe and flourishing condition; it is not endangered by any foundational crisis, it does not need to be rescued by philosophical heroics. I hope that we can all agree that Putnam is right on this point.

Nonetheless, in the course of developing this message Putnam makes a remark about not taking certain things seriously on which the late Kreisel pounced in a crit-

J. P. Burgess (✉)
Department of Philosophy, Princeton University, Princeton 08544-1006, NJ, USA
e-mail: jburgess@princeton.edu

G. Hellman and R. T. Cook (eds.), *Hilary Putnam on Logic and Mathematics*, Outstanding Contributions to Logic 9, https://doi.org/10.1007/978-3-319-96274-0_9

ical review (1972). Kreisel seems to take Putnam to be insufficiently appreciative of work in mathematical logic and foundations. At the time, there had already been a shift from impassioned advocacy for or against restrictive foundational frameworks such as finitism or constructivism or predicativism to dispassionate attempts to determine the scope and limits of the mathematics that could be developed within one or another such framework if it *were* adopted. More generally the project has become to determine what trade-offs are available in mathematics between, on the one hand, expansion of the power to prove theorems, and on the other hand, contraction of the danger of collapsing in contradiction (as has happened in a few cases: Gottlob Frege's *Grundgesetze*, Alonzo Church's illative λ–calculus, the first edition of W. V. Quine's *Mathematical Logic*, ZFC set theory plus a Reinhardt cardinal). Decades of work has established a detailed scale of "consistency strength", as what might more frankly be labeled "inconsistency risk" is euphemistically called, in which virtually every foundational framework that has ever had advocates has been placed (apart from Quine's "New Foundations", and even that seems to be on the brink of being placed as well).

The best-known part of Putnam's quite extensive early technical work in mathematical logic was closely related to these developments. I mean his work with Martin Davis and Julia Robinson, culminating in their great 1961 paper, pointing the way to the eventual solution of Hilbert's Tenth Problem by Yuri Matijasevich. For what that solution implies is just the following: As one climbs higher and higher up the scale of consistency strength, which near the top means assuming bigger and bigger "large cardinals" in set theory, one continues to get at each step new results in mathematics *of the most down-to-earth kind*, results asserting the nonexistence of solutions to certain Diophantine equations.

It is, I hope and believe, a misreading of Putnam to take him to be disparaging this kind of research, regardless of what he may say about not taking seriously this or that project. There is really nothing in his foreground message that should be offensive even to a logician like Kreisel, provided one carefully distinguishes "foundations" in a traditional, philosophical, "foundationalist" sense, which Putnam holds that mathematics can do without, from "foundations" as an American Mathematical Society subject classification, in which use it is essentially just a synonym for "mathematical logic".

9.2 A View of His Sketch

If Putnam's foreground message ought to be uncontentious, the same can hardly be said of what I am calling his background story, expounded in the section of MWF headed "A Sketch of My View". Here Putnam contrasts two "pictures" of mathematics, one being "mathematics as set theory", which he takes to be familiar, the other being "mathematics as modal logic", his own innovation. He also offers a characterization of the relationship between the two: Borrowing a term from his teacher Hans

Reichenbach's discussion of wave-particle duality in quantum mechanics, Putnam says the two pictures are "equivalent descriptions".

To be accurate, I should not say that this story is "expounded" in MWF, but rather that it is "adumbrated". For the sketch offered begins by saying that it will be "cursory and superficial", and this disclaimer is as accurate as it is disarming. If in discussions in later work Putnam seems to attribute great significance to MWF, one probably should understand the significance as being attributed to *the view of which a sketch is given* rather than to *the sketch that is given of the view*.

The "mathematics as modal logic" picture directly inspired Geoffrey Hellman's book (1989), advocating a "modal structuralist" version of nominalism, and so indirectly inspired a chapter (IIIC) in the book coauthored by myself and Rosen (1997), presenting a "mixed modal" strategy for reconstruing mathematic nominalistically. MWF is mentioned in Burgess and Rosen mainly just as a precursor to Hellman, but to treat the paper this way is to focus on only one half of the background story, "mathematics as modal logic", downplaying the other, "equivalent descriptions"; and of this emphasis Putnam himself cannot approve. His attitude can be seen, among other places, in a pair of twenty-first-century works, one a public lecture (2002), the other a book chapter (2012). In both he suggests that insufficient attention to the what he has written about "equivalent descriptions" in MWF and elsewhere has had unfortunate effects.

A theme of the 2012 chapter is that neglect of MWF has led to misunderstanding of another of Putnam's works, the celebrated booklet *Philosophy of Logic* (1971). Putnam can scarcely deny that the rhetoric about the indispensability of set theory in the booklet is superficially Quine-like, but he does deny that it is appropriate to give this rhetoric any genuinely Quine-like reading, given what he was already on record as saying in MWF. Failure to attend to MWF has led to a conflation of Putnam's with Quine's indispensability argument, creating a chimæra, "the Quine-Putnam indispensability argument". I need not say much about this. That the two "indispensability arguments" are different may be news to many, but it is not news to Burgess and Rosen any more than to Hellman. I think the three of us would all agree that Putnam is right that many commentators have got him wrong.

I will just say, however, that ignorance of MWF may not be the sole cause at work. Even a reader well-acquainted with the position of Putnam in MWF might still have attributed an incompatible, Quine-like position to the Putnam of the later booklet, simply because the Putnam of those years was as famous or infamous as Bertrand Russell for frequent radical changes of views, philosophical and other, and because there is little in the booklet to indicate that it does not just confront us with another of the tergiversations for which its author had become notorious, comparable to the flip-flop over whether the Continuum Hypothesis has a determinate truth value between his paper on Russell (1967a) and MWF later the same year. So far as I can see, there is in the booklet just one late, brief mention of "equivalent descriptions" as a topic not being gone into pro or con.

A theme of the 2002 lecture is that neglect of the lessons of MWF has resulted in consequences so grave as to suggest that analytic philosophy as such may simply be a bad thing; and the book of Burgess and Rosen is cited as the best example of what is wrong with it. As one of the coauthors, I am naturally not happy to see the book thus negatively evaluated by an eminent senior figure; but as Doctor Johnson said, if the reader of a book has not been pleased, it is no use to tell him why he ought to have been pleased. I can at least console myself that I am *only* a target in the lecture. My poor coauthor Rosen is a *double* target, being not just criticized with me in the public lecture for our joint work, but also further criticized in the book chapter for later solo work. But Rosen is perfectly capable of speaking for himself, if he sees any point in doing so, and I will speak only for myself in what follows.

I will not, however, be speaking for either of us right away. My immediate aim will be to address some of difficulties that arise when one tries to fill in the gaps in the sketch of a view in MWF. These emerge with both components of Putnam's background story, the "mathematics as modal logic" picture, which I will consider first, and the "equivalent descriptions" claim, which I will turn to later. It is in connection with the latter that occasion will arise to look at Putnam's criticisms of Burgess and Rosen, and say a few words in self-defense.

9.3 Modality and Mathematics

In the 2002 lecture there is early on a section entitled "The modal-logical interpretation of mathematics" in which Putnam enunciates his view thus:

(A) In mathematics the assertion that objects or structures 'exist' is completely fungible with the assertion that their existence is *possible*.

Formulation (A), considered by itself, is most naturally read as suggesting the following:

(B) In pure mathematics the assertion than an object exists is equivalent to the assertion that it possibly exists.
(C) In pure mathematics the assertion than a structure exists is equivalent to the assertion that it possibly exists.

These formulations are suboptimal. For (C) is redundant, since it is already implied by (B), mathematical structures being after all just one kind of mathematical object, capable of appearing as elements in higher structures such as categories. And (B) is too restricted, applying only to existence assertions what it seems could be applied to *all* purely mathematical assertions, giving us among other principles the following:

(D) In pure mathematics the assertion that an object has a certain property is equivalent to the assertion that it possibly has that property.
(E) In pure mathematics the assertion that two objects stand in a certain relation is equivalent to the assertion that they possibly stand in that relation.

The appropriate generalization of (D), (E), and so on, would seem to be this:

(F) In pure mathematics, adding "Possibly" or "It could have been that" in front of an assertion makes no difference.

Given the equivalence of *necessarily* to *not possible not* we get from (F) the following:

(G) In pure mathematics, adding "Necessarily" or "It could not have failed to be that" in front of an assertion makes no difference.

Note that (F) and (G) do *not* imply that there is anything special about the *logic* of "exists" in mathematics, that the introduction and elimination rules for the quantifiers somehow differ between mathematical and extra mathematical discourse. What (F) and (G) *do* say is that there is something very special about the logic of "possibly" and "necessarily" in mathematics: It trivializes.

I am tempted to suggest that the real principle is after all not (F) and (G) but something more radical, namely this:

(H) In pure mathematics, we just don't add things like "Possibly" or "Necessarily" in front of assertions.

But I must hasten to add a clarification that perhaps should have been made earlier: In all the above formulations it is to be tacitly understood that we are setting aside modals used in a merely *epistemic* sense. Perhaps the best evidence in favor of the hypothesis that a rule like (H), understood as banning *non-epistemic* modals, is at work in our mathematical language is that when modals *are* used, as indeed they sometimes are, we automatically and spontaneously interpret them epistemically. Surely we take "Possibly Euler's constant is rational, and possibly it is irrational," to mean something like, "Euler's constant isn't known to be irrational, and it isn't known to be rational," rather than anything like "There is an alternate universe where Euler's constant is rational and an alternate universe where it is irrational."

Principle (H) is essentially a negative grammatical rule: The usual distinctions of grammatical mood have no application in pure mathematics. If we try to transpose from "the formal mode" to "the material mode", we get something like the following:

(I) Purely mathematical facts are necessary.

But such a formulation, even if it can be understood as merely a rewording of (G), is dangerously misleading. It makes a humble, negative, grammatical observation look like a grand, positive, metaphysical principle. Start talking this way and philosophers are bound to be reminded of Plato. Remind them of Plato, and many will want to deny what you say; and unfortunately, experience shows that some will think they can only deny what you say if they also deny a lot of perfectly ordinary non-philosophical assertions. These include perfectly ordinary mathematical existence theorems, and thus some philosophers fall into nominalism.

All this was in the air when Rosen and I were writing our book. Some of it is hinted at there (article I.A.1.b). I claim no originality for any of it. To illustrate, it will be useful to put down some examples of specific purely mathematical assertions. Though I have claimed that existential assertions have no special status as compared with ascriptions of properties or relations, and that structures are included among objects and hardly require separate mention, I will take an existential assertion as my first example and an assertion about structures as my second, since Putnam puts emphasis on these special cases.

Consider, then, the following:

(1) There is a counterexample to Polya's conjecture.
(2) There is a standard model in which "There is a counterexample to Polya's conjecture" holds.
(3) There could have been a counterexample to Polya's conjecture.
(4) There could have been a standard model in which "There is a counterexample to Polya's conjecture" held.

The conjecture in question is that most of the numbers less than any given number have an odd number of prime factors; the smallest counterexample is 906,150,257. Never mind, for the moment, just what is meant by a standard model, apart from its being a kind of structure. Never mind, for the moment, exactly how to understand the modal "could have", beyond recognizing that it is not supposed to be merely *epistemic*.

What does our earlier discussion tell us about (1)–(4)? Well, (B) already implies that (1) is equivalent to (3), besides implying (C), which then implies that (2) is equivalent to (4). When the notion of standard model is properly worked out, model theory will presumably tell us that (1) is equivalent to (2); and when the notion of possibility at issue is properly pinned down, modal logic will presumably then tell us further that (3) is equivalent to (4). And so the whole quartet of formulations will be equivalent, at any rate for those who accept the pertinent model theory, which means accepting the background set theory within which it is developed, along with the pertinent modal logic.

9.4 Mathematics as Modal Logic

This is a pretty picture, but unfortunately our starting point, Putnam's formulation (A), does not accurately represent his actual position, as one can see by attending to what is said beyond the enunciation of (A) in the lecture, or by reading the corresponding material in MWF. A more accurate representation would be this:

(J) In mathematics the existence of a given *object* is completely fungible with the possible existence of a certain *structure*.

Here there is no mention of the possible existence of objects or of the actual existence of structures. Putnam's claim is that (1) is equivalent to (4) directly, and not via intermediates such as (2) or (3), of which he makes no mention.

All this makes the issue of the proper formulation of (4) an absolutely crucial one on Putnam's approach, which it was not in my discussion above. One must thus address the questions I postponed earlier, about the proper understanding of the notions of standard model and of the possibility modality. Putnam's hints about these matters in MWF are few and brief. The Hellman book examines, and the pertinent chapter of the Burgess and Rosen book re-examines, these matters at some length. Here I can do no more than mention a few questions telegraphically.

Structuresor models. The modal-logical formulation is supposed to contrast with a set-theoretic formulation, yet it speaks of the possibility of certain structures or models, and structures or models are usually explained set-theoretically. Thus it seems that the modal-logical formulation requires the possible existence of certain sets, which we have just been seeing is arguably the same as the actual existence of those sets. The problem could be avoided if talk of structures or models could be construed in something other than a set-theoretic way. Much of the labor in Hellman and in Burgess and Rosen, involving a good deal of auxiliary apparatus, is directed at just this problem.

Mathematical possibility. Linguists writing on modality (as in Palmer 2001) distinguish three kinds: *deontic* and *epistemic* and *dynamic*. The deontically possible is what may permissibly be done, a notion irrelevant to mathematics and to the present discussion. The epistemically possible is what for all we know may actually be, a notion already dismissed from consideration. The dynamically possible is what (actually is or actually isn't but potentially) could have been, which philosophers perhaps misleadingly call the "metaphysically" possible; and this is the notion I had in mind in my earlier discussion of possibility in mathematics. Putnam insists that *his* interest is in a specifically *mathematical* modality, but he tells us next to nothing about it. There's no denying that we do in ordinary language sometimes speak of "mathematical possibility". For instance, when the prophetess claims to have had a vision of 144,000 saints arranged in a perfect square, we may say, "That's mathematically impossible!" But what we mean here seems to be just the following. First, it's dynamically or "metaphysically" impossible: She couldn't have had such a vision. Second, purely mathematical considerations alone are enough to tell us as much, notably the consideration that the number 144,000 isn't a perfect square.

De re modality. There is a problem, however, about "modalities" such as "purely mathematical considerations alone are enough to tell us that..." The problem is that while this modality seems to make sense *de dicto*, meaning in application to a whole sentence, it seems not to make sense *de re*, meaning in application to a predicate with respect to an object. For instance, it seems mathematical considerations alone can tell us that the following is true: "The set {Hesperus, Hesperus} is a singleton," since any unordered pair $\{a, a\}$ of an element with itself is equal to the singleton $\{a\}$. And it seems mathematical considerations alone cannot tell us that the following is true: "The set {Hesperus, Phosphorus} is a singleton." For *that*, we need the nonmathematical, astronomical information that Hesperus is Phosphorus. But of course, what the astronomical information tells us is that it is the same set in both cases, differently described. What then of the predicate "...is a singleton"? Are mathematical considerations alone enough to tell us that it applies to *that set*, as it is in itself, so to speak, *independently* of how or whether it is described? This kind of difficulty is familiar from critical discussions by Quine. It matters because "quantifying in", or making assertions of the form "There is an x such that it is possible that x..." only makes sense where the kind of "possibility" involved is one for which *de re* modality makes sense.

Applications of mathematics. The project of developing a modal-logical version of mathematics seems to depend not only on our being able to "quantify into" contexts of the relevant kind of possibility, but on our being able to make cross-comparisons between how things there actually are are, and how merely possible things would have been. We do in ordinary language make such comparisons with dynamic or "metaphysical modality", as when we say that the building that has been put up on a given site is shorter than the one would have been put up there instead if the architect's original plans had been followed. Standard formalisms of modal logic don't accommodate such cross-comparison well, but that is a defect of the standard formalisms, an issue that gets a fair amount of attention in Burgess and Rosen. We need cross-comparison to provide modal-logical versions of *mixed* as opposed to *pure* mathematical assertions, ones that mention physical as well as mathematical objects in the same sentence; and these in turn are needed to provide modal-logical versions of *applied* as well as *pure* mathematics, as we must if the modal-logical approach is going to provide a genuine alternative.

There are a lot of interconnected issues here. So far as I know, only three writers, of whom Putnam is not one, have struggled at length in print with the complexities involved. And I doubt that any of us (Hellman, Burgess, Rosen) would claim to have gotten to the bottom of all the issues. Nonetheless, I won't press further here issues about "mathematics as modal logic", since the deeper issues seem to me to be those surrounding "equivalent descriptions", the other part of the background story in MWF.

9.5 Equivalent Descriptions

As the Council of Nicæa declared that the Father and the Son are somehow the same and yet somehow different, so Putnam declares the "mathematics as set theory" and "mathematics as modal logic" pictures, represented in my discussion by (1) and (4), are somehow the same and somehow different. I find the Nicene Creed easier to understand than Putnam's notion of equivalent descriptions.

Well, I *think* I may after all understand his claim of *difference*. Putnam explicitly says that formulations like (1) and (4) are not *synonymous*. In the jargon of Burgess and Rosen this means that no *hermeneutic* claim is being made, no claim to the effect that the modal-logical formulation reveals what, despite misleading surface appearances, the set-theoretic formulation has deep down meant all along. I also understand that no *revolutionary* claim is being made either, no claim that the set-theoretic picture should be suppressed and superseded by the modal-logical picture. But I don't understand the claim of *sameness*. What consubstantiality is being asserted?

Even here there is one claim that I believe everyone can accept, namely, that there is a mapping back and forth between sentences of the set-theoretic and modal-logical kinds about which one can say the following: *If* one accepts all pertinent set-theoretic apparatus needed to develop model theory, and *if* one accepts all appropriate modal-logical apparatus, *then* in a grand combined background modal set theory one should be able to deduce that the mapping in question is truth-preserving: One should be able to deduce biconditionals on the order of "(1) if and only if (4)".

Putnam wants to say that the two pictures "cognitively equivalent", but whose cognition is at issue here? Presumably not that of the nominalist, who rejects set theory, or that of the extensionalist, who rejects modal logic. *They* cannot accept the grand comprehensive background theory, or the proof that the mapping is truth-preserving. Hence that mapping provides *them* with no bridge that would permit the cognizer to pass back and forth between the two pictures. And I'm afraid ordinary mathematicians are in about the same position as the extensionalists, not because they *reject* modal logic, but because they know little or nothing about it.

What amounts to Putnam's first attempt to clarify what he means by "equivalent descriptions" consists simply in comparing the relationship between two pictures of mathematics to the relationship between the wave and particle pictures in quantum mechanics. But I'm afraid I see here only a *dis*analogy between the case Putnam wants to consider and the case of quantum mechanics as Reichenbach viewed it. In the latter, *the physicists themselves* were supposed to be aware of the particle and the wave pictures, and used to moving back and forth between them. But mathematicians know little about model theory and less about modal logic. Among them one does not have two pictures in use, but just one, the set-theoretic, worked out in great deal in many books, covering all branches of mathematics. The other picture, the modal-logical, receives a "cursory and superficial" sketch in one paper, MWF, and some elaboration in one book, and along variant lines in one chapter of another, all of which material remains wholly unknown to the mainstream mathematical community.

Putnam also sometimes says that a pair like my (1) and (4) express "the same fact". I find this, too, unhelpful. Such a formulation requires a metaphysics of "coarse-grained" facts at variance with ordinary, unselfconscious, non-philosophical fact-talk. In the *ordinary* usage of "fact", when someone is aware that *p* and unaware that *q*, the someone in question may always be said, pleonastically, to be aware *of the fact* that *p* but not *of the fact* that *q*. Now mathematicians are aware that (1) holds but, not knowing much about model theory or modal logic, unaware that (4) holds. Hence according to ordinary ways of speaking, (1) is a fact of which they are aware, (4) a fact of which they are not aware. That makes them two different facts, and so any claim they are the same requires what P. F. Strawson would have called a *revisionary* metaphysics, as opposed to a *descriptive* metaphysics that merely follows ordinary usage. And *all* claims of revisionary metaphysics are highly contentious.

9.6 Objections and Replies

Hints as to what Putnam may mean by "equivalent descriptions" are implicit in his 2002 critique of Burgess and Rosen, so let me turn to that barrage of criticisms, indicating my response to the four objections that seem to me most important, the last of which will bring the issue of "equivalent descriptions" back front and center. I will state the complaints in my own terms, as if made by an anonymous objector, and the responses in the first-person plural, though really I am speaking only for myself.

First Objection. Burgess and Rosen claim at one early point to be describing the naive, pre-philosophical view of mathematical existence assertions, but in so doing they speak of numbers as "things", and to do that *is already to import a contentious philosophical position.*

Reply. It is not to import anything, since naive, pre-philosophical subjects themselves *already* speak of numbers as "things". Perhaps the objector means something extraordinary by "thing", but we do not: There is nothing extra in *our* use of "things" beyond the ordinary use of expressions on the order of "such things as numbers" and "things like numbers", and these phrases are found in unselfconscious, non-philosophical writing. To be sure, one cannot demonstrate this by Googling on those exact phrases for examples, because doing so unfortunately brings up mainly links to discussions of nominalism versus platonism, useless for illustrating anything about unselfconscious, non-philosophical usage. However, Googling on "such things as X" and "things like X" does turn up examples of unselfconscious, non-philosophical usage for all the following values of X: Euler's constant, quaternions, triangular numbers, Mersenne primes, measurable cardinals, perfect numbers, the golden ratio, Gaussian integers, Bernoulli numbers, algebraic integers, transfinite ordinals, and p-adic numbers. If one doesn't have to be a philosopher with a hidden agenda to speak of all these kinds of numbers as "things", then surely one doesn't have to be a philosopher with a hidden agenda to speak of other kinds of numbers as "things", either.

Second Objection. Burgess and Rosen, in the same discussion in which they call numbers "things", go on to imply, what is wildly implausible, that the plain sense of such an assertion as

(5) There are prime numbers greater than a million.

amounts to something like

(6) There are causally inert things that are not in space and time and which are prime numbers greater than a million.

Reply. We do not claim that the plain sense of (5) amounts to anything like (6), any more than we would claim that the plain sense of

(7) There is a cat on the mat.

amounts to something like

(8) There is a furry, four-legged, whiskered creature that is a cat on the mat.

The plain senses of (5) and (7) are (5) and (7), and if you want to say something whose plain sense is (6) or (8), you will have to say (6) or (8). What we *do* claim is that, just as it is implicit in the ordinary understanding of what cats are that they are furry, four-legged, whiskered creatures, so it is implicit in the ordinary understanding of what numbers are that they are not to be described as, for instance, exerting a gravitational attraction on the Milky Way, or being located at the center of the Andromeda galaxy. Philosophers would summarize such negatives using such expressions as "causally inert" and "not in space and time", just as linguists would characterize how English speakers form plurals using such expressions as "sibilant" and "voiced". In neither case is familiarity with the technical jargon of specialists being ascribed to the ordinary people whose thought or practice is being summarized or characterized. Our claim about what is implicit in the ordinary understanding of what sorts of things numbers are is open to challenge, but the objection as stated is not a challenge but a caricature.

The chief evidence for our claim is simply that people don't ordinarily go around describing numbers as exerting gravitational attraction on the Milky Way, or being located at the center of the Andromeda galaxy, or whatever. Admittedly, our claim does predict something more than this, namely, that if exceptionally some eccentric *did* describe numbers in one of these ways, ordinary people would boggle. Admittedly, it is difficult to test this further prediction, simply because it is difficult to find anyone quite so eccentric as to go around making astrophysical assertions that bring in numbers as alleged *participants* in astrophysical processes. One has to resort to a hypothetical example. Let me do so now.

There is a famous problem in cosmology called that of "missing mass" or "dark matter". Observed gravitational effects suggest that there is something massive at the centers of galaxies that we cannot see because it does not emit light. One hypothesis that has been floated is that neutrinos, originally assumed to be massless, on the contrary have some nonzero mass, and that there are lots of them concentrated in the centers of galaxies. I take it this proposal is still under discussion, but let us imagine it comes to be recognized that it doesn't solve the problem. And suppose some bright young physics graduate student then comes forward with a novel hypothesis: It is not neutrinos but *numbers* that have some nonzero mass, and not neutrinos but *numbers* that are concentrated at the centers of galaxies. The proposal may even include conjectured formulas for the mass $\mu(n)$ and spatial coordinates $x(n)$, $y(n)$, $z(n)$ of the number n. What if you were in the position of the student's dissertation supervisor, hearing this proposal for the first time?

I am pretty sure that my own reaction would run much as follows: "Huh? Nonsense! [or an unprintable synonym] Is it April 1st? Or is this student crazy? Should I call the university counseling services? I hope I won't need to call campus security." Readers can judge for themselves, but those who imagine that they, too, would boggle should not find unreasonable our assumptions about what is implicit in the ordinary understanding of what sorts of things numbers are.

Third Objection. Burgess and Rosen, in their defense of the supposed ordinary belief in causally inert things outside space and time, raise an objection on grounds of complexity against proposed modal-logical alternatives to set-theoretic formulations. But objections on grounds of complexity would rule out virtually every analysis offered by philosophers from Frege onwards.

Reply. This objection overlooks the distinction we make between two kinds of nominalist. In contemporary linguistics it is common to associate with quite short sentences fairly elaborate structures that are supposed to be "psychologically real", though unconscious. Complexity in itself is no objection to analyses in linguistics, nor is it any objection to analyses in philosophy. But if a philosopher puts forward an alternative to a set-theoretic formulation as an *analysis*, then we call that philosopher a *hermeneutic* nominalist, and our objections against *that* kind of nominalist are not primarily based on complaints about complexity, but on complaints about the lack of *evidence* in favor of nominalistic analyses comparable to kind the evidence linguists advance for their hypotheses. Complexity is relevant only secondarily, insofar as more complex hypotheses require more compelling evidence. We advance complexity as the *primary* objection against a *different* kind of nominalism, *revolutionary* as contrasted with hermeneutic, the kind of nominalism that proposes to *replace* current formulations with novel ones. And what could be more legitimate, when faced with such a proposal, than to protest that the novel formulations are very elaborate and awkward, if they are so?

Fourth Objection. Burgess and Rosen are not alive to the possibility of equivalent descriptions. The assumption that

(K) Reality determines just *one* privileged language that we are to use when our interests are theoretical.

is so deeply ingrained in them, as in other analytic philosophers, that they are not even aware that they are making a potentially contestable assumption here.

Reply. Disambiguation is called for. The formulation (K), whose wording is taken *verbatim* from the 2002 lecture, admits of two readings:

(L) Reality picks a language for us and requires us to use this one language in all theoretical contexts.

(M) Reality requires us to pick a language for ourselves and use this one language in all theoretical contexts.

Here (L) is, nearly enough, the denial of the view of Quine, according to which theory forms a web on which reality impinges only at the boundary, through the evidence sensory experience, leaving indeterminate how the web should be woven in the middle, both as to content and as to language, if indeed one can separate the two. No one but a dunce could read our approving quotations from Quine in the conclusion to our book and imagine that we subscribe to (L). Since the objector is not a dunce, we must assume that (K) in the objection is understood in the sense of (M).

I thus take it that the suggestion is that we do not accept the modal-logical formulation because we (rightly think that it cannot be accepted as an analysis and) wrongly think that if we did accept it we could only accept it as supplanting rather than supplementing the set-theoretic formulation. The charge seems to be that we overlook the option of accepting the modal-logical formulation while *keeping* the set-theoretic formulation, varying which we use from occasion to occasion, letting two if not a hundred flowers bloom, filling the world with pluralistic or at least dualistic sweetness and light. And all this tells me something about the content of the doctrine of equivalent descriptions, which has otherwise eluded my efforts to grasp it: Whatever exactly it amounts to, it is a doctrine subscription to which would make the foregoing seem a cogent objection. But *is* it a cogent objection?

9.7 Contrasting Pictures

Do we overlook the option of just accepting the modal-logical picture *alongside* the set-theoretic? No, I say, because there is not really any such option for us to overlook. To articulate this more fully, let me contrast my view with a view that is "Putnamian" at least in the sense that a reader might well come away with it from reading MWF, whether or not it is what the author of that suggestive but cryptic discussion had in mind, a matter about which I make no exegetical claims. But before presenting the rival views or pictures of mathematics, let me offer my own analogy with physics, and begin with two rival views or pictures of that science.

A first picture of physics is as follows: *Our best theory of gravitation is general relativity, and our best theory of the other fundamental forces is quantum field theory. These theories and their languages are incompatible or incommensurable. Yet we do not discard one and adhere to the other. Rather, we make use of both, the one in one set of contexts, the other in another set of contexts.*

This picture is false, and the true situation is as follows: Our best theory of gravitation is a body of expert opinion *about* general relativity, roughly to the following effect, that it provides the best available formalism for dealing with gravity, but needs a quantum correction no one at present quite knows how to give it, though fortunately the error is small, and the formalism remains usable in a wide range of applications. Similarly our best theory of the other fundamental forces is a body of expert opinion *about* quantum field theory. Our best theories are not incompatible or incommensurable, nor are they formulated in different languages. They are formulated in one comprehensive language in which we can both develop various formalisms and also express various opinions about them.

A first picture of mathematics is as follows: *There is no need to fix on a* single *language for thinking about mathematical reality. There are two available, set-theoretic and modal-logical, and a translation between them that lets one go back and forth to suit the occasion. If listening to nominalist rhetoric has raised qualms, one can tell oneself, "Oh, don't worry about the apparent ontological commitments of set theory. We could put everything modal-logically." If listening to extensionalist rhetoric has raised qualms, one can tell oneself, "Oh, don't worry about the apparent ideological commitments of modal logic. We could put everything set-theoretically." The key is to think of the two theories as complementary and not as in competition.*

This picture is false, and the true situation is as follows: On the one hand, as already in effect indicated in my earlier objections to talk about "cognitive equivalence", if one just has the two languages and the mapping of sentences of one to sentences of the other, with no guarantee that this mapping is truth-preserving, then one has nothing that should allay anyone's qualms about anything. On the other hand, the only place to go for a proof that the mapping is truth-preserving seems to be a grand comprehensive background theory, combining all the commitments of set theory and modal logic. And if you go *there,* you no longer in any serious sense have a dualism of two languages. You have a monism of one great, big one.

As for what we accept, we accept set-theoretic mathematics, and reject modal-logical mathematics, as a *nominalist* might offer it, as an analysis of or replacement for set-theoretic mathematics. But we might be persuaded to accept modal-logical apparatus *in addition to* what we now accept. (We do at places in the book at least suspend disbelief in it for the sake of argument.) In that case, we would be accepting a grand composite background theory, a modal set theory, and *it* would be our mathematics. The set-theoretic and modal-logical pictures would be merely components, together making up less than the whole, as synthetic geometry and pure algebra are merely components, together making up less than the whole, of coordinate geometry. But I cannot see us accepting set-theoretic mathematics and modal-logical mathematics *just side by side* without modal set theory in the background to provide a bridge between them.

The long and the short of it is this. Given two descriptions, if we have a comprehensive background incorporating both, we may be able to see that they are equivalent, but we do not then have in the two descriptions two languages or theories, but only two fragments of one comprehensive language and theory, together making up less than the whole. By contrast, if we have no such background, it is unclear that or how or in what sense we would be able to see that the two descriptions are equivalent. If Putnam has some clever way to evade this dilemma, it has escaped me.

Postscript.

Though the frequency of major reversals in Putnam's views has lessened with the passing decades, he still shows a remarkable intellectual flexibility. A few weeks after the deadline for contributions to this volume, and the submission of the present paper, one of the editors has written to me to point out a couple of posts on Putnam's blog (putnamphil.blogspot.com) from late 2014 (dated 12 and 13 December) in which, replying to an as yet unpublished criticism by Steven Wagner (unavailable to me at the time of this writing), he retracts his doctrine, upheld for well over 40 years, that the set-theoretic and modal-logical pictures of mathematics are "equivalent descriptions". In particular, Putnam now acknowledges that the comparison with wave and particle pictures in quantum mechanics is not apt for reasons including one of those noted above, that whereas physicists are aware of *both* alternative pictures, mathematicians are only aware of *one*. He now prefers to describe the modal-logical picture as a "rational reconstruction" of the set-theoretic picture, which I suppose implies that the set-theoretic picture is, if not wholly irrational, anyhow inferior in point of rationality to the modal-logic, and not its equal or equivalent from a rational point of view. (I freely confess that I have never pretended to understand the notion of "rational reconstruction" in Carnap or anyone else.) Putnam would now compare the relationship of the set-theoretic to the modal-logical pictures to the relationship of confused seventeenth and eighteenth century ideas about imaginary numbers to the improved understanding that came in during the nineteenth century (complex numbers as simply pairs of real numbers).

There are, nonetheless, several reasons to hope that the foregoing discussion of Putnam's former view may remain of some interest. In order of increasing importance I would mention the following: First, so far as I know, the retraction of the doctrine of equivalent descriptions has not been accompanied by any retraction of the criticisms of Burgess and Rosen, or of analytic philosophy generally, whose defects *A Subject with No Object* was supposed to exemplify. Second, Putnam has not abandoned the doctrine of equivalent descriptions generally, but only in application to the mathematical case, whereas some at least of the above criticisms of the doctrine (such as the dilemma in my closing paragraph) seem to me to apply beyond that case. Third, the retraction has appeared so far only in the blogosphere and not in the print literature, and to that extent it may be said that it is not yet "official". Fourth, with Putnam one must always allow for the possibility that he will return at a later date to some version of a former view apparently definitively abandoned. Fifth, any view maintained by a philosopher of Putnam's stature for the better part of a half-century must remain worthy of discussion even if its author eventually moves beyond it. For all these reasons, and because I lack the time and energy to keep trying to hit a moving target, I am allowing the above text to stand unamended.

References

Benacerraf, P., & Putnam, H. (Eds.). (1964). *Philosophy of mathematics: Selected readings* (1st ed.). Englewood Cliffs: Prentice-Hall.

Benacerraf, P., & Putnam, H. (Eds.). (1983). *Philosophy of mathematics: Selected readings* (2nd ed.). Cambridge: Cambridge University Press.

Burgess, J. P., & Rosen, G. (1997). *A subject with no object*. Princeton: Princeton University Press.

Davis, M., Putnam, H., & Robinson, J. (1961). The decision problem for exponential diophantine equations. *Annals of Mathematics, 74*, 425–436.

Hellman, G. (1989). *Mathematics without numbers*. Oxford: Oxford University Press.

Kreisel, G. (1972). Review of Putnam 1967b. *Journal of Symbolic Logic, 37*, 402–404.

Palmer, F. R. (2001). Mood and modality (2nd ed.). In *Cambridge textbooks in linguistics*. Cambridge: Cambridge University Press.

Putnam, H. (1967a). The thesis that mathematics is logic. In R. Schoenman (Ed.), *Bertrand Russell: Philosopher of the century* (pp. 273–303). London: Allen and Unwin.

Putnam, H. (1967b). Mathematics without foundations. *Journal of Philosophy, 64*, 1–22. (Reprinted in Benacerraf and Putnam, 295–313, 1983).

Putnam, H. (1971). *Philosophy of logic*. New York: Harper and Row.

Putnam, H. (1980). Models and reality. *Journal of Symbolic Logic, 45*, 464–482. (Reprinted in Benacerraf and Putnam, 421–445, 1983).

Putnam, H. (2002). Is analytic philosophy a good thing? Why I am ambivalent. Lecture at the Einstein Forum: *Die Zukunft der Analytischen Philosophie* [*The Future of Analytic Philosophy*], Potsdam.

Putnam, H. (2012). Indispensability arguments in the philosophy of mathematics. In M. De Caro & D. Macarthur (Eds.), *Chapter 9 of philosophy in an age of science: Physics, mathematics, and skepticism* (pp. 181–201). Cambridge: Harvard University Press.

John P. Burgess John P. Burgess is the John N. Woodhill Professor of Philosophy at Princeton University, where he has taught since 1975, serving as director of undergraduate studies for much of that time. He has written or co-written eight books and a hundred or so papers and reviews in logic, philosophy of mathematics, and adjoining fields.

Chapter 10
Pragmatic Platonism

Mathematics and the Infinite

Martin Davis

Abstract It is argued that to a greater or lesser extent, all mathematical knowledge is empirical.

Although[1] I have never thought of myself as a philosopher, Harvey Friedman has told me that I am "an extreme Platonist". Well, extremism in defense of truth may be no vice, but I do feel the need to defend myself from that description.

10.1 Gödel's Platonism

When one thinks of Platonism in mathematics, one naturally thinks of Gödel. In a letter to Gotthard Günther in 1954, he wrote:

> When I say that one can ...develop a theory of classes as objectively existing entities, I do indeed mean by that existence in the sense of ontological metaphysics, by which, however, I do not want to say that abstract entities are present in nature. They seem rather to form a second plane of reality, which confronts us just as objectively and independently of our thinking as nature.[2]

If indeed that's extreme Platonism, it's not what I believe. I don't find myself confronted by such a "second plane of reality".

---

[1]This essay is based on a paper with the same title (but without the subtitle) read at a conference celebrating Harvey Friedman's 60th birthday. Much of the text is taken verbatim from that paper.
[2]Feferman et al. (2003), vol. IV, pp. 502–505.

M. Davis (✉)
Courant Institute of Mathematical Sciences, New York University,
New York, NY, USA
e-mail: martin@eipye.com

M. Davis
3360 Dwight Way, Berkeley, CA 94704, USA

G. Hellman and R. T. Cook (eds.), *Hilary Putnam on Logic and Mathematics*,
Outstanding Contributions to Logic 9, https://doi.org/10.1007/978-3-319-96274-0_10

In his Gibbs lecture of 1951, Gödel made it clear that he rejected any mechanistic account of mind, claiming (with no citations) that

> ...some of the leading men in brain and nerve physiology ...very decidedly deny the possibility of a purely mechanistic explanation of psychical and nervous processes.[3]

In a 1974 letter evidently meant to help comfort Abraham Robinson who was dying of cancer, he was even more emphatic:

> The assertion that our ego consists of protein molecules seems to me one of the most ridiculous ever made.[4]

Alas, I'm stuck with precisely this ridiculous belief. Although I wouldn't mind at all having the transcendental mind Gödel suggests, I'm aware of no evidence that our mental activity is anything but the work of our physical brains.

In his Gibbs lecture Gödel suggests another possibility:

> If mathematics describes an objective world just like physics, there is no reason why inductive methods should not be applied in mathematics just the same as in physics. The fact is that in mathematics we still have the same attitude today that in former times one had toward all science, namely we try to derive everything by cogent proofs from the definitions (that is, in ontological terminology, from the essences of things). Perhaps this method, if it claims monopoly, is as wrong in mathematics as it was in physics.[5]

I will claim that mathematicians have been using inductive methods, appropriately understood, all along. There is a simplistic view that induction simply means the acceptance of a general proposition on the basis of its having been verified in a large number of cases, so that for example we should regard the Riemann Hypothesis as having been established on the basis of the numerical evidence that has been obtained. But this is unacceptable: no matter how much computation has been carried out, it will have verified only an infinitesimal portion of the infinitude of the cases that need to be considered. But inductive methods (even those used in physics) need to be understood in a much more comprehensive sense.

10.2 Gödel Incompleteness and the Metaphysics of Arithmetic

Gödel has claimed that it was his philosophical stance that made his revolutionary discoveries possible and that his Platonism had begun in his youth. However, an examination of the record shows something quite different, namely a gradual and initially reluctant embrace of Platonism as Gödel considered the philosophical implications of his mathematical work (Davis 2005). It is at least as true that Gödel's philosophy was the result of his mathematics as that the latter derived from the former.

---

[3]Feferman et al. (2003), vol. III, p. 312.

[4]Feferman et al. (2003), vol. V, p. 204.

[5]Feferman et al. (2003), vol. III, p. 313.

In 1887, in an article surveying transfinite numbers from mathematical, philosophical, and theological viewpoints, Georg Cantor made a point of attacking a little pamphlet on counting and measuring written by the great scientist Hermann von Helmholtz. Cantor complained that the pamphlet expressed an "extreme empirical-psychological point of view with a dogmatism one would not have thought possible ..." He continued:

> Thus, in today's Germany we see, as a reaction against the overblown Kant-Fichte-Hegel-Schelling Idealism, an *academic-positivistic skepticism* that powerfully dominates the scene. This skepticism has inevitably extended its reach even to *arithmetic*, in which domain it has led to its most fateful conclusions. Ultimately, this may turn out most damaging to this positivistic skepticism itself.

In reviewing a collection of Cantor's papers dealing with the transfinite, Frege chose to emphasize the remark just quoted, writing (Frege 1892):

> Yes indeed! This is the very reef on which this doctrine will founder. For ultimately, the role of the infinite in arithmetic is not to be denied; yet, on the other hand, there is no way it can coexist with this epistemological tendency. Thus we can foresee that this issue will provide the setting for a momentous and decisive battle.

In a 1933 lecture, Gödel, considering the consequences of his incompleteness theorems, and perhaps not having entirely shaken off the positivism of the Vienna Circle, showed that the "battle" Frege had predicted was taking place in his own mind:

> The result of our previous discussion is that our axioms, if interpreted as meaningful statements, necessarily presuppose a kind of Platonism, which cannot satisfy any critical mind and which does not even produce the conviction that they are consistent.[6]

The axioms to which Gödel referred were an unending sequence produced by permitting variables for ever higher "types" (in contemporary terminology, sets of ever higher rank) and including axioms appropriate to each level. He pointed out that to each of these levels there corresponds an assertion of a particularly simple arithmetic form, what we now would call a Π_1^0 sentence, which is not provable from the axioms of that level, but which becomes provable at the next level.

It may be worth digressing to emphasize that later work has shown how to express these Π_1^0 sentences in a particularly simple form. Π_1^0 statements have the form $(\forall x) x \in S$ where x ranges over the natural numbers and where S is a computable set, meaning that there is an algorithm that enables one for each given input a to determine whether or not $a \in S$. In 1959 Hilary Putnam and I, basing ourselves on results and methods from my dissertation and from Julia Robinson's (1996), succeeded in proving that for any such computable set S, there is an expression $p(a, x_1, x_2, \ldots, x_n)$ constructed from variables raging over natural numbers as well as particular natural numbers using addition, subtraction, multiplication, and exponentiation such that

$$a \in S \leftrightarrow (\exists x_1, x_2, \ldots, x_n)[p(a, x_1, x_2, \ldots, x_n) = 0]$$

[6]Feferman et al. (2003), vol. III, p. 50.

Our proof made use of the fact that there are arbitrarily long sequences of prime numbers in arithmetic progression something that had not yet been proved.[7] Julia Robinson then showed how the proof could be modified to avoid depending on what was then an unproved (but generally believed) conjecture (Davis et al. 1996). Later Yuri Matiyasevich showed that exponentiation was not needed so that p was just a polynomial.[8] In the light of this work, a Π_1^0 sentence can be seen as simply asserting that some particular equation

$$p(x_1, x_2, \ldots, x_n) = 0,$$

where p is a polynomial with integer coefficients, has no solutions in natural numbers. To say that such a proposition is *true* is just to say that for each choice of natural number values $a_1, a_2, \ldots, a_n$ for the unknowns,

$$p(a_1, a_2, \ldots, a_n) \neq 0.$$

Moreover a proof for each such special case consists of nothing more than the sequence of additions and multiplications needed to compute the value of the polynomial together with the observation that that value is not 0. So in the situation to which Gödel is calling attention, at a given level there is no single proof that subsumes this infinite collection of special cases, while at the next level there is such a proof.

This powerful way of expressing Gödel incompleteness is not available to one who holds to a purely formalist foundation for mathematics. For a formalist, there is no "truth" above and beyond provability in a particular formal system. Post had reacted to this situation by insisting that Gödel's work requires "at least a partial reversal of the entire axiomatic trend of the late nineteenth and early twentieth centuries, with a return to meaning and truth as being of the essence of mathematics".[9] Frege's reference to the "role of the infinite in arithmetic" is very much to the point here. It is the infinitude of the natural numbers, the infinitude of the sequence of formal systems, and finally, the infinitude of the special cases implied by a Π_1^0 proposition that point to some form of Platonism.

10.3 Infinity in the Seventeenth Century

Hilbert saw the problem of the infinite as central to resolving foundational issues. Perhaps succumbing a bit to hyperbole, he said:

---

[7]A proof did not appear until the next millennium! See Green and Tao (2008).

[8]This was a celebrated result because the unsolvability of Hilbert's tenth problem was an immediate corollary (Davis et al. 1996).

[9]Post (1994) p. 295.

> The infinite has always stirred the emotions of mankind more deeply than any other question; the infinite has stimulated and fertilized reason as few other ideas have; but also the infinite, more than any other notion is in need of clarification.[10]

People have pronounced and speculated about what is and isn't true about infinity since they began thinking abstractly. Aristotle's views on the subject in particular had a great influence. A discovery made by the Italian mathematician Toricelli in 1641 provides a very revealing example.[11] He found that the volume of a certain solid of infinite extent is finite. The solid in question is obtained by rotating about an axis a certain plane figure with infinite area. Specifically, in modern terminology, it is the figure bounded by the hyperbola whose equation is $y = 1/x$, the line $x = 1$, and the horizontal asymptote of the hyperbola, namely the X-axis. Toricelli's solid is formed by rotating this figure about the X-axis. Although showing that this solid of revolution has a finite volume is a routine "homework" problem in a beginning calculus course,

$$\pi \int_1^\infty \frac{1}{x^2} dx = \pi,$$

at the time it created a sensation because it contradicted prevalent views about the infinite. Toricelli himself remarked "…if one proposes to consider a solid, or a plane figure, infinitely extended, everybody immediately thinks that such a figure must be of infinite size." In 1649, Petri Gassendi wrote,

> Mathematicians …weave those famous demonstrations, some so extraordinary that they even exceed credibility, like what …Torricelli showed of a certain…solid infinitely long which nevertheless is equal to a finite cylinder.

Writing in 1666, Isaac Barrow found Torricelli's result contradicting what Aristotle had taught. He referred to Aristotle's dictum, "there is no proportion between the finite and the infinite":

> The truth of which statement, a very usual and well known axiom, has been in part broken by …modern geometricians [who] demonstrate …equality of …solids protracted to infinity with other finite …solids which prodigy …Torricelli exhibited first.

Much can be learned from this example about the way in which mathematicians expand the applicability of existing methods to new problems and with how they deal with the philosophical problems that may arise. Toricelli used a technique called the method of *indivisibles* , a method pioneered by Cavalieri that provided a short-cut for solving area and volume problems. Toricelli used this technique to prove that his infinite body had the same volume as a certain finite cylinder. The method conceived of each of the two bodies being compared as constituted of a continuum of plane figures. Although there was no rigorous foundation for this, Cavalieri and later Toricelli showed how effective it could be in easily obtaining interesting results. They were well aware of the Eudoxus-Archimedes method of exhaustion (which they

---

[10] van Heijenoort (1967) p. 371.

[11] This discussion, including the quotations, is based on Paolo Mancosu's wonderful monograph (Mancosu 1996).

called "the method of the ancients"), and used it to confirm their results and/or to convince skeptics.[12] But, Toricelli insisted on the validity of the new method.

What can we say about Toricelli's methodology? He was certainly not seeking to obtain results by "cogent proofs from the definitions" or "in ontological terms, from the essences of things". He was *experimenting* with a mathematical technique that he had learned, and was attempting to see whether it would work in an uncharted realm. In the process, something new about the infinite was discovered. I insist that this was induction from a body of mathematical experience.

10.4 Putnam's Defense of "Realism"

When I wrote an earlier version of this essay, I was not familiar with Putnam (1975) in which Hilary Putnam argues that "mathematics should be interpreted realistically and objectively". I believe that the position Putnam advocates in that paper is essentially the same as my view expressed in this essay. Indeed Putnam even writes of the use of "quasi-empirical", that is *inductive* methods in mathematics. "Why not," he writes "use *both* deductive proof *and* confirmation by mathematical 'experiment' in the search for truth?"[13] It isn't entirely clear what is to be understood by the term "mathematical experiment". The crude understanding might be that when an assertion like Goldbach's Conjecture or the Riemann Hypothesis has been confirmed up to some enormous upper bound, we should simply accept it, at least tentatively. as proven. Indeed some suggest accepting the Riemann Hypothesis as an axiom. The reply to the "why not" taken in this sense is: because it may not be true. The integers up to this "enormous" bound is but the tiniest fragment of the infinitude of the natural numbers. A famous example from number theory will bring this out. The wonderful prime number theorem relates $\pi(x)$ defined as *the number of primes* $< x$[14] and

$$\mathsf{Li(x)} = \int_2^\infty \frac{1}{\log(\mathsf{x})}.$$

Namely, it states that they are asymptotic, that is,

$$\lim_{x\to\infty} \frac{\pi(x)}{\mathsf{Li}(x)} = 1.$$

An equivalent version is

$$\lim_{x\to\infty} \frac{\pi(x)}{x/\log x} = 1.$$

[12]The method of exhaustion typically required one to have the answer at hand, whereas with indivisibles the answer could be computed.

[13]Emphasis is in the original.

[14]Of course this conventional use of the Greek letter π has nothing to do with the number $\pi = 3.14159\ldots$.

Now, say for convenience for $x > 10$, all numerical computation has provided values such that $\pi(x) < \mathsf{Li}(x)$. Should that justify accepting this inequality as being true? Although no numerical example violating this inequality is known, Littlewood famously proved that the inequality sign reverses from $<$ to $>$ and back again, *infinitely many times*! Littlewood's student Skewes was able to show that the first time the inequality sign changes to $>$ is a number no larger than

$$10^{10^{10^{963}}}.$$

This is a number larger than the presumed number of quarks in the observable universe!

It is the infinitude of the objects of mathematics that lies behind the call for some kind of metaphysics. I think this can be seen clearly by considering chess problems. To be definite consider problems of the form: Achieve a check-mate in three moves. Solving one of these problems successfully requires proving a little theorem. We must prove that for our chosen first move M1, for whatever legal reply the opponent chooses, we have found a move M2, such that however the opponent responds, we have a move M3 which places the opponent in check-mate. This is all quite abstract. The physical representation of the pieces in wood or plastic is as irrelevant as the representation of integers as tally marks, decimal numbers, binary numbers, or von Neumann, or Zermelo sets, is to the properties of prime numbers. Yet would anyone claim there is a philosophical issue as to whether knights or rooks are "real", or whether the inability of a bishop to move to a different colored square or of a knight to move to the same color is an a priori truth? I think it is clear that the reason such issues do not arise in this case is simply that it is all finite.

In fact the examples that Putnam chooses of "quasi-empirical" methods in mathematics are far more sophisticated and may be worth more discussion than the brief mention in his article. The first is Euler's method of obtaining the sum of the infinite series $\sum_{n-1}^{\infty} 1/n^2$. This method depending on assuming that infinite power series would have certain of the crucial properties of an ordinary polynomial involving the relationships between its zeros and its coefficients. For example the polynomial $x^3 - 8x^2 + 19x - 12$ has the zeros 1, 3, and 4. Therefore

$$x^3 - 8x^2 + 19x - 12 = (x - 1)(x - 3)(x - 4).$$

If we wish to verify this by multiplying the three factors we can see that the $-8x^2$ term is obtained by the addition $(-1x^2) + (-3x^2) + (-4x^2)$. So the coefficient of $-x^2$ is the *sum* of the zeros. Euler's idea was to somehow use this idea in summing $1/n^2$. His "polynomial" was

$$\frac{\sin x}{x} - 1 \quad \frac{x^2}{6} + \frac{x^4}{24} - \frac{x^6}{144} + \cdots$$

whose zeros are $\{\pm\pi n \mid n = 1, 2, 3, \ldots\}$. But this can't work: the exponents in our example are decreasing from 3 to 0, whereas in the power series for $\sin x/x$ the exponents are increasing from 0 in infinity in jumps of 2. To see what to do, let's rewrite our polynomial as

$$x^3 - 8x^2 + 19x - 12 = -12\left(1 - \frac{19}{12}x + \frac{8}{12}x^2 - \frac{1}{12}x^3\right).$$

Of course the zeros are still 1, 3, and 4, but the factorization now looks like this:

$$1 - \frac{19}{12}x + \frac{8}{12}x^2 - \frac{1}{12}x^3 = \left(1 - \frac{x}{1}\right)\left(1 - \frac{x}{3}\right)\left(1 - \frac{x}{4}\right),$$

and the sum $1 + 1/3 + 1/4 = 19/12$ is the coefficient of $-x$. So boldly (if hesitantly) following Euler, remembering not to neglect the negative zeros, we write[15]:

$$\begin{aligned}1 - \frac{x^2}{6} + \frac{x^4}{24} - \frac{x^6}{144} + \cdots &= \left(1 - \frac{x}{\pi}\right)\left(1 + \frac{x}{\pi}\right)\left(1 - \frac{x}{2\pi}\right)\left(1 + \frac{x}{2\pi}\right)\left(1 - \frac{x}{3\pi^2}\right)\left(1 + \frac{x}{3\pi^2}\right)\cdots \\ &= \left(1 - \frac{x^2}{\pi^2}\right)\left(1 - \frac{x^2}{4\pi^2}\right)\left(1 - \frac{x^2}{9\pi^2}\right)\cdots\end{aligned}$$

So the coefficient of $-x^2$ which is 1/6 should be

$$\frac{1}{\pi^2} + \frac{1}{4\pi^2} + \frac{1}{9\pi^2} + \cdots = \frac{1}{\pi^2}[1 + \frac{1}{4} + \frac{1}{9} + \cdots]$$

Thus we get Euler's result:

$$1 + \frac{1}{4} + \frac{1}{9} + \cdots = \frac{\pi^2}{6}.$$

The story goes that Euler verified this result by what might well be called a "mathematical experiment". It was easy for Euler to compute a pretty good approximation to $\pi^2/6$. He could also calculate $\sum_{k=1}^{n} 1/k^2$ for some suitably large value of n and then compare the two numbers. He was known to be a prodigious calculator but that infinite series converges very slowly. From Wikipedia we have

$$\frac{\pi^2}{6} = 1.644934\ldots.$$

I wrote and ran a simple Pascal program on my PC to compute $1 + 1/4 + 1/9 + \cdots + 1/n^2$ for various values of n; the results are shown in a table:

---

[15] Although Euler obtained this infinite product by using a doubtful analogy, Weierstrass eventually obtained it quite rigorously as a special case of a general theorem about complex analytic functions.

n	$1 + 1/4 + 1/9 + \cdots + 1/n^2$
100	1.63498390
1,000	1.64439346
30,000	1.64490073

From this data it is clear that Euler would not have been able to obtain what he needed by direct calculation. Even adding 30,000 terms, far beyond what what is humanly possible by direct calculation, would only have yielded four correct decimal digits. Instead he used a formula, found by Euler himself and, independently, by the British mathematician MacLaurin, that approximates sums by integrals. He used this "Euler-MacLaurin formula" to calculate the sum of the series correct to 20 places. The agreement with approximations to $\pi^2/6$ gave Euler the confidence to announce the result to the mathematical public. However, using the Euler-MacLaurin formula required considerable care. It required truncating a divergent (so-called *asymptotic*) series at just the right point when the sum still provided a useful approximation before the series marches off to infinity. In Euler's time the rigorous treatment of such series had still not been developed, so in effect his "experimental" method for verify his heuristic method for summing $\sum_1^\infty 1/n^2$ was to use another heuristic method!

Though Euler was never able to make this argument rigorous (the necessary basis for such was yet to be developed), he did make sure to develop another fully convincing proof that the sum in question really is $\pi^2/6$. Wasn't the heuristic argument fully convincing? Well no: the computational evidence was perfectly consistent with, say,

$$1 + \frac{1}{4} + \frac{1}{9} + \cdots = \frac{\pi^2}{6} + \gamma,$$

where γ is some real number with $0 < \gamma < 10^{-10^{100}}$.

Putnam's other example of heuristic reasoning brings probabilistic reasoning to bear on properties of primes and divisibility. On the face of it this may appear strange, because these matters are totally determined, and probabilistic methods seem inappropriate. Nevertheless they have proved quite useful.

Although not the example that Hilary cites, I can't resist mention the probabilistic use of Euler's summation we have just been discussing. It is the problem of computing the probability of two natural numbers having no prime factor in common, i.e., of being relatively prime. The probability of a specific prime p being a divisor of a randomly chosen number is $1/p$ because it is every p-th number that is divisible by p. So the probability that p is a divisor of each of a pair of randomly, but independently, chosen numbers is $1/p^2$ and the probability that it fails to be a divisor of at least one of them is $1 - 1/p^2$. So the probability of such a pair being relatively prime is the infinite product of $1 - 1/p^2$ over all primes. Since every natural number (except 1) is the product of primes, this infinite product is equal to

$$\frac{1}{1 + \frac{1}{4} + \frac{1}{9} + \frac{1}{16} + \cdots} = \frac{6}{\pi^2}$$

The example that Hilary mentions has to do with so called *twin primes*. This are pairs of prime numbers like 11 and 13 that differ by exactly 2. It is an old, and still unproved, conjecture that there are infinitely many of these twin primes.[16] Here is a heuristic argument for the infinitude of twin primes: The Prime Number Theorem, already referred to above can be interpreted probabilistically to say that the probability of a number $<x$ being prime is $1/\log x$. Since we can regard x and $x+2$ being prime as "independent" of one another, we can say that the probability of a pair of numbers $<x$ being twin primes is $1/(\log x)^2$.[17] Although this is hardly a decisive argument it does have the corollary that there are infinitely many twin primes.

10.5 Robustness of Formalism

An interesting example is provided by the development of complex numbers. The fact that the square of any non-zero real number is positive had been generally accepted as implying that there could be no number whose square is negative. Sixteenth century algebra brought this into question. The quadratic formula, essentially known since antiquity, did seem to lead to solutions which involve square roots of negative quantities. But those were simply regarded as impossible. But the analogous formula for cubic equations, discovered by Tartaglia and published in Cardano's book of 1545, forced a rethinking of the matter. In the case of a cubic equation with real coefficients and three real roots, the formula led to square roots of negative numbers as intermediary steps in the computation. Bombelli discussed this in his book of 1572. In particular, he noted that although the equation $x^3 - 15x - 4 = 0$ had the three roots $4, -2+\sqrt{3}, -2-\sqrt{3}$, the Tartaglia formula forced one to consider $\sqrt{-109}$. Soon mathematicians were working freely with complex numbers without questioning whether they really exist in some "second plane of reality". What this experience illustrates is the robustness of mathematical formalisms. These formalisms often point the way to expansions of the subject matter of mathematics before any kind of convincing justification can be supplied. This is again a case of induction in mathematical practice.

Leibniz referred to this very experience when asked to justify the use of infinitesimals. As Mancosu explains

---

[16]That there are infinitely many primes has been proved in many different ways. The first proof is already in Euclid. It is very simple and elegant. To show that given any finite collection of primes, $p_1, p_2, \ldots, p_n$, there is another prime not in that collection, one multiplies together all the primes in that collection and adds 1:

$$N = p_1 p_2 \cdots p_n + 1$$

Since none of $p_1, p_2, \ldots, p_n$ are divisors of N, either N is itself a prime different from any of them or it is divisible by a prime different from them.

[17]A similar conclusion is obtained heuristically in Hardy and Wright (1960) pp. 371–372 without probabilistic considerations.

> ...the problem for Leibniz was not, Do infinitely small quantities exist? but, Is the use of infinitely small quantities in calculus reliable?[18]

In justifying his use of infinitesimals in calculus, Leibniz compared this with the use of complex numbers which had become generally accepted although at the time, there was no rigorous justification.

In another example, the rules of algebra, including the manipulation of infinite series was applied to operators with scant justification. This can be seen in Boole's (1865) massive tract on differential equations in which marvelous manipulative dexterity is deployed with not a theorem in sight.

10.6 The Ontology of Mathematics

If the objects of mathematics are not in nature and not in a "second plane of reality," then where are they? Perhaps we can learn something from the physicists. Consider for example, the discussion of the "Anthropic Principle" (Barrow and Tipler 1986). The advocates of this principle note that the values of certain critical constants are finely tuned to our very existence. Given even minor deviations, the consequence would be: no human race. It is not relevant here whether this principle is regarded as profound or merely tautological. What I find interesting in this discussion of alternate universes whose properties exclude the existence of us, is that no one worries about their ontology. There is simply a blithe confidence that the same reasoning faculty that serves physicists so well in studying the world that we actually do inhabit, will work just as well in deducing the properties of a somewhat different hypothetical world. A more mundane example is the ubiquitous use of idealization. When Newton calculated the motions of the planets assuming that each of the heavenly bodies is a perfect sphere of uniform density or even a mass particle, no one complained that the ontology of his idealized worlds was obscure. The evidence that our minds are up to the challenge of discovering the properties of alternative worlds is simply that we have successfully done so. Induction indeed! This reassurance is not at all absolute. Like all empirical knowledge it comes without a guarantee that it is certain.

My claim is that what mathematicians do is very much the same. We explore simple austere worlds that differ from the one we inhabit both by their stark simplicity and by their openness to the infinite. It is simply an empirical fact that we are able to obtain apparently reliable and objective information about such worlds. And, because of this, any illusion that this knowledge is certain must be abandoned. If, on a neo-Humean morning, I were to awaken to the skies splitting open, hearing a loud voice bellowing, "This ends Phase 1; Phase 2 now begins," I would of course be astonished. But I will not say that I *know* that this will not happen. If presented with a proof that Peano Arithmetic is inconsistent or even that some huge natural number is not the sum of four squares, I would be very very skeptical. But I will not say that I *know* that such a proof must be wrong.

---

[18]Mancosu (1996) p. 172.

10.7 Infinity Today

Mathematical practice obtains information about what it would be like if there were infinitely many things. It is not at all evident a priori that we can do that. But mathematicians have shown us that we can. Our steps are tentative, but as confidence is acquired we move forward. Our theorems are proved in many different ways, and the results are always the same. Our formalisms are robust and yield information beyond the original intent. To doubt the significance of the concrete evidence for the objectivity of mathematical knowledge is like anti-evolutionists doubting the evidence of paleontology by suggesting that those fossils were part of creation. As was discussed above, Gödel's work has left us with a transfinite sequence of formal systems involving larger and larger sets. Models of these systems can be obtained from initial segments of the famous hierarchy obtained by iterating transfinitely the power set operation $\mathcal{P}$:

$$V_0 = \emptyset; \quad V_{\alpha+1} = \mathcal{P}V_\alpha; \quad V_\lambda = \bigcup_{\alpha<\lambda} V_\alpha, \ \ \lambda \text{ a limit ordinal}$$

Thus, $V_{\omega 2}$ is a model of the original Zermelo axioms. To obtain a model of the more comprehensive Zermelo–Fraenkel (ZF) axioms, no ordinal whose existence is provable in ZF will do.[19] To continue the transfinite sequence of formal systems, it is necessary to enter the realm of large cardinals in which there has been intensive research. Workers in this realm are pioneers on dangerous ground: although we know that no proof of the consistency with ZF of the existence of these enormous sets is possible, it is always conceivable that a proof in ZF of the inconsistency of one of them will emerge thereby destroying a huge body of work. But the empirical evidence is encouraging. Although the defining characteristics of the various large cardinal types that have been studied seem quite disparate, they line themselves up neatly in order of increasing consistency strength. Moreover, they have shown themselves to be the correct tool for resolving open questions in descriptive set theory.

So far Gödel incompleteness has had only a negligible effect on mathematical practice. Cantor's continuum hypothesis remains a challenge: although the Gödel-Cohen results prove its undecidability from ZF, if the iterative hierarchy is taken seriously, it does have a truth value whether we can ever find it or not. In the realm of arithmetic many important unsolved problems, including the Riemann Hypothesis and the Goldbach Conjecture, are equivalent to Π^0_1 sentences. However, so far no undecidable Π^0_1 sentences have been found that are provably equivalent to questions previously posed (as has been done for uncomputability). However, Harvey Friedman has produced a remarkable collection of Π^0_1 and Π^0_2 arithmetic sentences with clear combinatorial content that can only be resolved in the context of large cardinals.

[19] Because otherwise the consistency of ZF would be provable in ZF contradicting Gödel's second incompleteness theorem. For that matter the set $V_{\omega 2}$ cannot be proved to exist from the Zermelo axioms alone; in ZF its existence follows using Replacement.

10.8 Hilary Punam's "Depressing Survey"

In his Putman (1995) Putnam laments about philosophy of mathematics that "nothing works". This turn of phrase suggests that a point of view regarding the nature of mathematics be thought of as being like a machine that has failed to carry out its intended function, a toaster that doesn't make toast. This intended function, he seems to tell us, is "explaining the phenomenon of mathematical knowledge". In this essay I have claimed that what is mysterious about mathematical knowledge is that so much of it involves the infinite although mathematicians manage to obtain it using their finite brains. How it is that human brains can do this is perhaps a question for neuroscientists. How it is that this knowledge is reliable, if indeed it is, is another matter. I have suggested that certainty is not to be had in the realm of mathematics any more than in empirical science. Indeed as we ascend the hierarchy of ever stronger formalisms to which Gödel incompleteness pushes us, we face ever less assurance that we will not encounter an outright inconsistency.

Mark Twain suggested the lovely notion of a "Sunday truth": something fervently believed in church on Sunday but having no effect on behavior in the rest of the week. Many mathematicians will profess a belief in formalism when foundational matters are discussed. But in their day-to-day work as mathematicians, they remain thoroughgoing Platonists. The "crisis" in foundations from the turn of the 20th century to the 1920s has quietly dissipated. Set theory as a foundation is evident in the initial chapter of many graduate-level textbooks. The obligation to always point out a use of the axiom of choice is a thing of the past. I haven't heard of anyone calling the proof of Fermat's Last Theorem into question because of the large infinities implicit in Grothendieck universes.[20] So foundational issues are being discussed mainly by philosophers.

It is striking that logicism, formalism, and constructivism, three of the philosophical approaches Putnam discusses, were each developed by working mathematicians in terms of programs that could be developed and examined by mathematical methods, enhanced by rigorous formalisms. While none of these programs provided any guarantee of the certainty of mathematical knowledge, they led to new insights, some of them surprising and quite unexpected.

Frege and then Russell didn't content themselves with a verbal defense of logicism, they built systems. Frege had to first develop his formalism for Russell to be able to use his famous paradox to prove its inconsistency. (That Paradox, by the way, was distilled by Russell from Cantor's proof that no set can be in a one-one correspondence with its power set.) Russell didn't content himself with a paper advocating type theory as a means of avoiding that paradox, he convinced Whitehead to join him in the massive endeavor that resulted in the three-volume *Principia Mathematica*. As Putnam points out, whatever defects these systems had, they showed that the project of formalizing mathematics was a feasible undertaking.

[20] Number theorists seem to regard the use of Grothendiek universes as a mere convenience. See McLarty (2010) for a careful discussion.

Hilbert was attracted to the Whitehead-Russell formalism and thought at first of basing his metamathematical investigations on it. But repelled by the awkwardness of that system, he was led to propose the much simpler first order logic as the real logic of mathematics, and to the formalist program. This was the context in which Gödel produced his incompleteness theorems which were so devastating for both logicism and formalism.

Hilbert and the logicists wanted to provide an unimpeachable formal foundation for the entire corpus of mathematics including Cantor's transfinite which Hilbert had called "the most admirable flower of the mathematical intellect and in general one of the highest achievements of purely rational human activity". But Hilbert's best student, Hermann Weyl, was attracted by the heretical views of L.E.J. Brouwer and his constructive style of mathematics that rejected this and much more of twentieth century mathematics. "Brouwer, that is the revolution!" he exclaimed. Incensed, Hilbert declared

> ...Weyl and Brouwer ...seek to provide a foundation for mathematics by pitching overboard whatever discomforts them and declaring an embargo ...No! Brouwer's [program] is not as Weyl thinks, the revolution, but only a repetition of an attempted putsch with old methods, that in its day was undertaken with greater verve yet failed utterly.

Many years later, writing Hilbert's obituary, Weyl was much less enthusiastic. While he still insisted that Brouwer had forced the discourse on mathematical foundations to a higher level, he found "an almost unbearable awkwardness" in Brouwer's actual intuitionistic mathematics (Weyl 1944). While intuitionistic mathematics found few followers, the formalization of intuitionistic logic by Brouwer's student Arend Heyting provided a rich source of important work by researchers in mathematical foundations and led to unexpected insights. Perhaps the most surprising of these is that, at least in the realm of first order number theory, intuitionism leads to no restrictions: In "classical" (that is, ordinary) logic the operations $\vee$ $\exists$ are definable in terms of the remaining operations $\wedge$ $\neg$ $\forall$. However the provable assertions of first order number theory written using only these last operations are the same whether classical or intuitionistic logic is used.

There still are those who wish to draw a line between safe and unsafe proof methods. The line is drawn by some who insist on some variety of constructivity. Others demand predicativity. Contemporary foundational research makes such notions precise and obtains theorems on the relative strengths of different methods. But any attempt to restrict mathematicians will be pointless. History suggests that they will use whatever methods work, including the higher realms of the infinite.

References

Barrow, J. D., & Tipler, F. J. (1986). *The anthropic cosmological principle*. Oxford: Oxford University Press.

Boole, G. (1865). *A treatise on differential equations*. London: Macmillan and Co.

Davis, M. (2005). What did Gödel believe and when did he believe It? *Bulletin of Symbolic Logic, 11*, 194–206.
Davis, M., Putnam, H., & Robinson, J. (1961). The decision problem for exponential diophantine equations. *Annals of Mathematics, 74*, 425–436. Reprinted in S. Feferman (Ed.). (1996). *The collected works of Julia Robinson* (pp. 77–88). American Mathematical Society.
Davis, M., Matiyasevich, Y., & Robinson, J. (1976). Hilbert's tenth problem. Diophantine equations: positive aspects of a negative solution. In *Proceedings of Symposia in Pure Mathematics, vol XXVIII: Positive Aspects of a Negative Solution* (pp. 323–378). Reprinted in S. Feferman (Ed.). (1996). *The collected works of Julia Robinson* (pp. 269–378). American Mathematical Society.
Feferman, S., et al. (1986–2003). *Kurt Gödel collected works, volume I–V*. Oxford: Oxford University Press.
Frege, G. (1892). Rezension von: Georg Cantor. Zum Lehre vom Transfiniten. *Zeitschrift fr Philosophie und philosophische Kritik*, new series, *100*, 269–272.
Green, B., & Tao, T. (2008). The primes contain arbitrarily long arithmetic progressions. *Annals of Mathematics, 167*, 481–547.
Hardy, G. H., & Wright, E. M. (1960). *An introduction to the theory of numbers* (Fourth Edition ed.). Oxford: Clarendon Press.
Mancosu, P. (1996). *Philosophy of mathematics and mathematical practice in the seventeenth century*. Oxford: Oxford University Press.
McLarty, C. (2010). What does it take to prove Fermat's Last Theorem? Grothendiek and the logic of number theory. *Bulletin of Symbolic Logic, 16*, 359–377.
Post, E. L. (1944). Recursively enumerable sets of positive integers and their decision problems. *Bulletin of the American Mathematical Society, 50*, 284–316. Reprinted: Davis, M. (1965, 2004). *The undecidable*. New York: Raven Press; New York: Dover. Reprinted: E. L. Post, & M. Martin Davis (Ed.). (1994). *Solvability, provability, definability: The collected works*. Birkhäuser.
Putnam, H. (1975). What is mathematical truth? *Historia Mathematica, 2*, 529–533.
Putman, H. (1995). Philosophy of mathematics: why nothing works. *Works and life* (pp. 499–511). Cambridge: Harvard University Press.
Robinson, J. (1952). Existential definability in arithmetic. In: *Transactions of the American Mathematical Society* (Vol. 72, pp. 437–449). Reprinted in S. Feferman (Ed.). (1996). *The collected works of Julia Robinson* (pp. 47–59). American Mathematical Society.
van Heijenoort, J. (Ed.). (1967). *From Frege to Gödel: A source book in mathematical logic, 1879–1931*. Cambridge: Harvard University Press.
Weyl, H. (1944). David Hilbert and his mathematical work. *Bulletin of the American Mathematical Society, 50*, 612–654.

Martin Davis was born in New York City in 1928. He was a student of Emil L. Post at City College and his doctorate at Princeton in 1950 was under the supervision of Alonzo Church. Davis's book "Computability and Unsolvability" (1958) has been called "one of the few real classics in computer science." He is best known for his pioneering work in automated deduction and for his contributions to the solution of Hilbert's tenth problem. Among other awards, he has been a Guggenheim Foundation Fellow. Davis has been on the faculty of the Courant Institute of Mathematical Sciences of New York University since 1965 and is now Professor Emeritus.

Chapter 11
Abstraction, Axiomatization and Rigor: Pasch and Hilbert

Michael Detlefsen

To proceed axiomatically means nothing other than to think with awareness (mit Bewußtsein denken)

Hilbert (1922), 201

Abstract In the late nineteenth century, Pasch made a well known statement concerning the conditions of attaining rigor in geometrical proof. The criterion he offered called not only for the elimination of appeals to geometrical figures, but of appeals to meanings of geometrical terms more generally. Not long after Pasch, Hilbert (and others) proposed an alternative standard of rigor. My aim in this paper is to clarify the relationship between Pasch's and Hilbert's standards of rigor. There are, I believe, fundamental differences between them.

Keywords Rigor · Proof · Pasch · Hilbert · Lambert · Freudenthal, premisory surreption · Abstraction from meaning · Semantic abstraction · Abstraction condition · Axiomatic method · Axiomatizaton · Formalization

AMS Categories 00A30; 03A05

11.1 Introduction

In his 1882 lectures on geometry, Moritz Pasch described and endorsed a standard of rigor for geometrical proof.

> [I]f geometry is to be genuinely deductive, the process of inferring (Process des Folgerns) must be everywhere independent of (unabhängig sein vom) the *sense* (*Sinn*) of geometrical concepts just as it must be independent of figures. It is only *relations* between geometrical concepts that should be taken into account in the propositions and definitions that are dealt with. In the course of the deduction, it is certainly legitimate (statthaft) and useful (nützlich),

M. Detlefsen (✉)
University of Notre Dame, Notre Dame, IN, USA
e-mail: mdetlef1@nd.edu

G. Hellman and R. T. Cook (eds.), *Hilary Putnam on Logic and Mathematics*, Outstanding Contributions to Logic 9, https://doi.org/10.1007/978-3-319-96274-0_11

> though *by no means necessary* (*keineswegs nöthig*), to think of the meaning (Bedeutung) of the geometrical concepts involved. In fact, if it is necessary so to think, the gappiness (Lückenhaftigkeit) of the deduction and the insufficiency (Unzulänglichkeit) of the means of proof is thereby revealed, unless it is possible to remove the gaps (Lücke) by modifying the reasoning used.
>
> Pasch (1882), 98 (emphases in text)

The part of this statement that will most concern me here is that in which Pasch states that the "process of inferring" in proper geometrical proof, whatever that might most reasonably be taken to be, must be "independent of" the meanings of geometrical terms. I will offer a view of what the notion of independence referred to here comes to. I will further consider what I take to be most significantly at stake in the enforcement of such a condition of independence.

The understanding of Pasch's standard on which I will focus sees it as a restriction on the justification of judgments of deductive inferential validity in geometrical proofs. By implication, it therefore also sees it as a constraint on proper judgements of deductive validity in mathematical proofs more generally.

If we let $\mathcal{C}$ be a sentence and $\mathcal{P}$ a set of sentences in a given mathematical language, we may roughly state this constraint as follows:

> Abstraction Condition: Justification of a judgment that an inference from $\mathcal{P}$ to $\mathcal{C}$ is deductively valid ought not to be based on any judgment whose contents concern the meanings or contents of non-logical expressions that occur in $\mathcal{P}$ or $\mathcal{C}$.[1]

Pasch presented this standard as a standard for rigor, where he seems to have seen this as centering on the attainment of proper justification for our judgments of validity. The featured element of propriety, moreover, was the avoidance of surreptious fillings of deductive "gaps" in inferences judged to be deductively valid. Given that such fillings are paradigmatic cases of failure of rigor, it seems appropriate to refer to Pasch's standard as a standard of rigor.

It also seems right to call it an "abstractionist" standard in as much as the justificative prescission it calls for amounts to a type of abstraction. Pasch's condition requires that the justification of a judgment of deductive validity for a mathematical proof (or for an inference in a mathematical proof) should *abstract away from*—more clearly, perhaps, should prescind from—all justificative appeal to the senses or meanings of non-logical expressions.

Pasch suggested that failure to observe this condition (or another condition to like effect) incurs a non-negligible and avoidable risk of misjudgment of validity—particularly, misjudgment owing to misidentification of the constitutive elements of the inference(s) judged to be valid.[2]

---

[1]Roughly speaking, an expression E may be said to *occur* in a class of expressions $\mathcal{K}$ if (i) E is an element of $\mathcal{K}$ or (ii) E is an expression upon whose meaning the meaning of an element of $\mathcal{K}$ depends.

[2]There have also been important alternative conceptions of rigor concerning which failure of rigor is not conceived as it is conceived here. One such conception is what I have elsewhere referred to as *probative* rigor. This is rigor which, roughly speaking, concerns the extent to which everything

Failure of rigor of this type, though it concerns validity and judgements of validity, concerns more as well. Pasch in fact focused most tightly not on misjudgments of validity per se, but on certain reasons for making such misjudgments—namely, those based on misidentifications of the elements (e.g., the premises and/or conclusion) of an inference or proof judged to be valid. To say more exactly what such misidentifications consist in, and what Pasch took them to consist in, are central tasks of this paper.

In this connection, let me begin by noting that Pasch's special concern seems to have been misjudgment of validity due to misidentification of premises. More particularly still, he focused on misjudgments of validity based on failure to recognize the use of illicitly imported premises.[3]

Following tradition, I'll refer to this type of misidentification of premises as *premisory surreption*. For a given inference Inf_{Id} and a given inferring agent R,[4] I take its key elements to be as follows:

(i) Judgment by R that Inf_{Id} is valid.
(ii) Failure of R to recognize that her judgment that Inf_{Id} is valid is based on her taking it to include a premise(s) which, properly speaking,[5] it does not (or ought not to be taken to) include.[6]

Pasch's proposed antidote for the failure mentioned in (ii) was application of the Abstraction Condition. In what follows, this proposal will be my main preoccupation. I will consider, in particular, some similarities and differences between it and another proposal of roughly the same period—namely, the so-called *axiomatic method* of Hilbert and others. I will argue that despite their having important similarities, these two proposals, and their underlying conceptions of rigor, are also importantly different.

Before turning to these matters, I'll present some points of historical background that I hope serve to clarify Pasch's and Hilbert's proposals by putting them into

adverted to in a proof that is in some sense capable of being proved is in fact proved. Both Bolzano and Dedekind, as I read them, advocated probative conceptions of rigor, although the particulars of their conceptions were different. For more on this and related matters, see Detlefsen (2010, 2011).

[3]There are distinctions between different types of inferential failures that should be borne in mind here. Among these is failure to recognize *that* premisory importation has occurred when it has occurred. Related to, though also distinct from, this type of failure is failure properly and correctly to identify the premise(s) imported. These seem to be distinct types of failures though their differences will not feature in what follows.

[4]Conceived as I conceive them, inferring agents include not only those who may devise a given piece of reasoning, but those who, though they may not devise it, nonetheless judge it to be valid.

[5]Proper, that is, for purposes of judging the validity of Inf_{Id}.

[6]What is fundamentally wrong, then, with judging a premisorily surreptious argument to be valid is not that, taken to include its surreptious premises, it is not valid. Rather, it is that premises sufficient to warrant a judgement of validity have not been properly identified or registered as premises.

clearer historical and logical perspective. These include, though they do not center on, a challenge to an influential claim(s) concerning Pasch's priority as an advocate of an abstractionist standard of rigor.

11.2 Background

The Abstraction Condition represents a significant departure from older standards of rigor that importantly influenced the thinking of eighteenth and nineteenth century mathematicians. Chief among these was a standard I have elsewhere referred to as the *presentist* standard.[7]

Like the standard of rigor based on the Abstraction Condition, the presentist standard rested on a conception of rigor which sees it as attainment of a type of gaplessness in reasoning. The gaplessness of concern to the presentist, though, was different in character from that pursued by supporters of semantic abstraction as a standard of rigor.

On the presentist conception, mathematical reasoning—particularly, proof—was regarded as having a *subject* of some type (e.g. a geometrical figure). A proof (or, perhaps better, a proving) was judged to be rigorous to the extent that its subject was gaplessly retained before a prover's mind throughout the course of a proof (or proving) as the *subject of* the various judgments whose deductive arrangement makes up the proof.[8]

Poncelet expressed the core idea of such a view as follows:

> In ordinary geometry, which one often calls synthetic ...the figure is described, one never loses sight of it (jamais on ne la perd de vue), one always reasons with quantities and forms that are real (réelles) and existing (existantes), and one never draws consequences which

---

[7]Cf. Detlefsen (2005), 237, 264–66 and Detlefsen (2010), 176.

[8]There are at least two different ways to understand gapless retention of subject. One is to emphasize a notion of awareness, and to take gapless retention of subject to consist in some type of continuity of the objects of *awareness* of a prover throughout the course of a proof.

Gapless retention of subject might also be conceived along more logical lines. On such a view, proofs would be seen as characteristically having parts—in particular, constituent judgments and inferences. Each of these parts would itself have a subject, and gapless retention of subject throughout the course of a proof would consist in the subjects of the relevant parts of a proof standing in a certain relationship to each other (e.g. being *identical to* or in a relevant sense *continuous with* each other) and to the overall subject of the proof.

Of these two broad understandings of gapless retention of subject, the latter might seem the more appealing. On the surface, at least, it would appear to allow that gapless retention of subject be an objective matter. This may be deceiving, though, in that it is possible that any satisfactory understanding of the central notion of a proof's having a *subject* would have to make use of a subjective element, perhaps in the form of an appeal to a prover's awareness. There may ultimately be no other way to make sense of the idea of a proof's being *about* something that a prover, in order properly to be a prover, must associate with it as its subject.

> cannot be depicted in the imagination (à l'imagination) or before one's eyes (à la vue) by sensible objects (objets sensibles).
>
> Poncelet (1822), xxj[9,10]

There seems to be a tension between abstractionism and presentism. Presentism sees proofs as characteristically having contentual subjects (e.g. geometrical figures), and it takes rigor to consist in some type of *constancy* or *continuity* concerning the subject-bearing parts of a proof throughout its course.[11,12]

The abstractionist reasoner, on the other hand, seeks detachment from rather than continuous contact with or immersion in the contentual subjects of proofs. More particularly, she requires that no judgment concerning the validity of an inference or proof should depend for its justification on judgments whose contents are even partially constituted by contents of non-logical expressions. Indeed, Pasch and other abstractionists sometimes went farther and advocated practical measures whose intent seems to have been to reduce, at least in particular contexts, the role of geometrical contents in geometrical reasoning. Only in this way, they believed, could the dangers posed to rigor by contentual associations be reasonably managed.

---

[9]Despite what this passage may suggest, Poncelet's endorsement of traditional synthetic procedure was qualified. He seems particularly to have had reservations concerning its laboriousness, which he saw as being primarily due to a perceived need for the prover to take things back to rudimentary constructions—or, as he put it, "to reproduce the entire series of primitive arguments from the moment where a line and a point have passed from the right to the left of one another, etc." (*ibid.*).

[10]Presentist standards of rigor seem to have been familiar to writers well before Poncelet's time. My reason for mentioning him is to indicate the influence that such ideas still had on nineteenth century mathematicians.

An older description of presentism, and (some of) its supposed virtues, can be found in Berkeley.

> It hath been an old remark that Geometry is an excellent Logic. And it must be owned, that ...when from the distinct Contemplation and Comparison of Figures, their Properties are derived, *by a perpetual well-connected chain of Consequences, the Objects being still kept in view, and the attention ever fixed upon them*; there is acquired a habit of reasoning, close and exact and methodical: which habit strengthens and sharpens the Mind ...
>
> Berkeley (1734), sec. 2, emphasis added

It should be noted that though Berkeley described a "presentist" conception of rigor in this remark, he did not generally subscribe to such a conception.

[11]By *constancy* of the subject-bearing parts of a proof, I mean constancy or identity of subjects throughout the subject bearing parts of a proof (or, more exactly, throughout the series of judgments and inferences which together make up a proof).

In speaking of subjectively *continuous* proof, I mean roughly proof in which the subjects of the subject-bearing parts of a proof are in some sense *continuous with* each other and with the overall subject of the proof, even though they may not be constant. Roughly speaking, continuity in this sense assumes that though the subjects of the subject-bearing parts of a proof may be distinct, the transitions from one to another are in some important way(s) conservative. No clearer formulation of these ideas is necessary for my purposes here.

[12]In mentioning the "course" of a proof here, I am assuming that proofs are characteristically divided, or at least divisible, into stages or steps. Nothing I propose here, though, depends on a particular working out of this idea.

It has been suggested that Pasch had some type of priority as a defender of such a view and standard of rigor. Hans Freudenthal, for example, referred to him as "the father of rigor in geometry" (cf. Freudenthal 1962, 619). And some fifty years before Freudenthal's statement, J.W. Young remarked Pasch's abstractionist emphasis. "The abstract formulation of mathematics", he wrote, "seems to date back to the German mathematician Moritz Pasch." (cf. Young et al. 1911, 51). Later in the same essay he noted the link Pasch saw between abstraction and rigor: namely, that "to be rigorous ...an argument must be abstract" (op. cit., 218).

It seems clear, however, that there were clear expressions of such ideas well before Pasch and his writings. An example is J. H. Lambert, who wrote:

> [It] can and must be required that one nowhere in a proof call on the thing itself (auf die Sache selbst berufe) but that the proof should be carried forward symbolically throughout (durchaus symbolisch vortrage)—if this is possible. In this aspect Euclid's postulates are the same as so many algebraic equations which one has before oneself, and from which x, y, z &c will be brought out (herausgebracht) without one's looking back at the thing itself (ohne daß man auf die Sache selbst zurücke sehe).
>
> Lambert (1786), 149–150

Lambert made this remark in the context of discussing the question of the derivability of the parallel postulate from the other Euclidean axioms and postulates. He took this to be the question whether the parallel postulate can be "properly derived" (in richtige Folge hergeleitet werden könne) from the other Euclidean postulates, taken in conjunction with what might be other commonly recognized basic propositions (übrigen Grundsätze) Lambert (1786), 149 of Euclidean geometry.[13,14]

Lambert claimed that proper derivation of the parallel postulate from the other basic Euclidean propositions would require derivation which "abstracts" (abstrahiert)

---

[13]This suggests that Lambert may have seen properly rigorous proof as allowing not only inclusion of axioms among the legitimate ultimate premises of a proof, but inclusion of other propositions as well—specifically, propositions which were commonly recognized as having a basicness appropriate for use in proofs of the propositions being proved. This suggests a view of proof in which the basic qualification for premises is that they be appropriately more basic than the theorems they're used to prove.

Lambert didn't say in a precise way what he took the salience of such relative basicness to be. It seems sensible enough, though, to allow for the possibility that there be propositions which are axiom-like in certain respects (e.g. their relative evidentness, or their relative evidensory primitivity), but not in others (e.g. their deductive power, or their simplicity). It is also sensible enough to hold that the basic aim of proof is to justify the seemingly less basic by the seemingly more basic to the fullest extent feasible or practicable. On such a view of the aim of proof, a proof which used relatively more basic propositions to justify relatively less basic propositions could be seen as making progress even if the progress made were not that of justificative reduction to the most basic propositions.

[14]Lambert raised a related question as well, namely, whether, supposing the parallel postulate to not be so derivable, it might nonetheless become derivable by adding to the basic Euclidean propositions other propositions which have "the same evidentness" (die gleicher Evidenz hätten) (loc. cit.) as them (i.e. the basic Euclidean propositions). This, however, seems to have been more a comment concerning how to think about the independence of the parallel postulate and its significance than a comment concerning premisory rigor per se.

(loc. cit.) from all "representation and conceivability of the things talked about" (von der Vorstellung und der Gedenkbarkeit der Sache die Rede ist) (Lambert 1786, 155) and which thus proceeds by the application of what are essentially symbolical rules.[15]

In this way, and perhaps only in this way, Lambert suggested, can one adequately guard against surreptitious importation of information (ein *Vitium subreptionis*, Lambert 1786, 156) into—hence failure of rigor of—geometrical proof.

If this is right, Pasch was not the first to propose semantic abstraction as an effective, perhaps even a necessary means of securing rigor in geometrical proof. My purpose in noting this, however, is not to diminish Pasch's importance as an advocate of abstractionist approaches to rigor.

He was neither the first[16] nor the last,[17] even of his day, to express concerns regarding the rigor of Euclid's proofs. This notwithstanding, his discovery of what has come to be known as Pasch's axiom[18] added materially to the perceived urgency of these concerns. In addition, his insistence that rigor requires the avoidance not only of appeals to diagrams, or diagrammatically conveyed contents, but to geometrical contents however conveyed, both strengthened and clarified the place of the Abstraction Condition as a constraint on geometrical proof.

These points having been noted, let me turn now to the questions identified earlier—namely, how, if at all, application of the Abstraction Condition might reasonably be taken to advance rigor, and how such application compares to and differs from application of the so-called *axiomatic* method of Hilbert and others.

11.3 Semantic Abstraction and Premisory Surreption

How is it, exactly, that application of the Abstraction Condition should provide protection against premisory surreption? An historically sensible answer would be: "By mitigating the effects of unrecognized semantically borne psychological association in inference."

According to associationist views, successional form of thinking such as proof (and reasoning more generally) are subject to influences of psychological association. Generally speaking, repeated association of one idea with another, or one proposition with another, increases the likelihood of their co-application (e.g. their being "thought" together, their being affirmed together, etc.), independently of whether such co-application is logically warranted or whether the reasoner is aware of it.

Experiences, thinkings, etc. have contents, and patterns of succession among such mental events not uncommonly induce corresponding associations among their

[15]There are indications that Lambert took Euclid to have been trying to develop a means of arguing which left no room for thought or judgment concerning things-in-themselves in geometrical reasoning. He saw the axioms as functioning symbolically, not semantically.

[16]Cf. Todhunter (1869).

[17]Cf. Smith and Bryant (1901) and Russell (1902).

[18]On one variation, Pasch's axiom states that, in a plane, if a line that does not pass through a vertex of a triangle intersects one side of it internally (i.e., at a point between vertices of the triangle), it then internally intersects another side and externally intersects the third.

contents. These associations may in turn give rise to affirmations, hypothetizations and such other propositional attitude-takings as may generally be suited for use as premises in proofs.

Tendencies to associate contents, however, generally expose reasoners to premisory surreption in proof by dint of provers' unrecognized co-application of associated premises with recognized premises. The seriousness of such exposure was widely recognized as regards the use of geometrical figures in geometrical reasoning.[19]

Pasch seems to have been concerned with threats to rigor that are posed by the forces of psychological association. He saw these forces as posing a threat to non-diagrammatically presented as well as diagrammatically presented contentual appeals. Accordingly, he proposed a standard of rigor which called for judgments of inferential validity to be "independent" not only of uses of diagrams, but of uses of all appeals to semantical contents of geometrical terms.

My reading of the remark by Pasch quoted in the introductory section thus sees it as supporting not only a broadly logical but a psycho-criteriological understanding of this "independence."[20] On this understanding, in order to certify a putative inference as valid it should not only be logically unnecessary for a reasoner to know or even to be aware of the senses or referents of the non-logical terms that occur in the inference, it should be psychologically unnecessary as well. This at any rate is what I take Pasch's statement of the desired independence of geometrical inference from the meanings of geometrical terms to suggest.

Pasch didn't give specific directions for the practical achievement of such independence, but he seems to have believed that it *is* practically achievable. Somehow and in some sense, he suggested, we schematize inferences in axiomatic reasoning by treating their non-logical terms as "variables" rather than as constants. That is, we treat them as terms which range over or admit of different contents and not as terms that have fixed (or relatively fixed) particular contents.

At the same time, Pasch believed, we come to realize that (i) deductive validity depends only on *relationships between* non-logical terms (and not on the contents of those terms themselves), and that (ii) judgments concerning deductive validity ought only to appeal to such relationships.[21] Therefore, by whatever practical means we may achieve abstraction from the meanings of non-logical terms in our judgments of validity, our doing so is key, in Pasch's view, to minimizing the threat of premisory surreption.

---

[19]Here by the "use" of a geometrical figure I mean a justificative appeal to a judgment(s) concerning the properties of said diagram or of the figure(s) it may be taken to represent. The justification of such a judgment is presumably based on some type of "diagrammatic" grasp or examination of the figure involved.

[20]By a broadly logical understanding of this independence, I mean a view according to which to know that a proposition follows deductively from other propositions, it is not (broadly) logically necessary to know or to be in any way aware of senses, referents or images commonly associated with non-logical terms these propositions contain.

[21]This is a way of affirming the traditional idea that, properly speaking, deductive validity ought only to depend on the (logical) *forms* of the premises and conclusion of an inference.

As Pasch saw it, then, practical achievement of rigor requires a kind of psychological discipline in geometrical proof—a discipline aimed at psychological separation of the inferences in geometrical proofs from considerations of the contents of geometrical terms. Pasch seems to have seen the exercise of such discipline as a practically effective means of mitigating the rigor-compromising risks of contentual association.

In light of this, it is perhaps the more remarkable that Pasch did not propose the elimination of *all* appeals to contents in judgments of validity, including, specifically, appeals to the meanings of *logical* terms.

To my mind, that he did not represents an asymmetry in his views concerning the relationship between (attainment of) rigor and appeals to contents in the justification of validity judgements. On the one hand, he took the threat of premisory surreption posed by appeals to the contents of geometrical terms as substantial. On the other hand, he seems to have treated the threat of premisory surreption arising from contentual uses of logical terms as (at least relatively) insubstantial.

This requires explanation, and it suggests that Pasch may have held some such view as the following

> Asymmetry: Contentual use of an expression or figure that is peculiar to or distinctive of reasoning in a given topic- or subject-area τ[22] poses a greater risk of premisory surreption in a proof belonging to τ than does contentual use of an expression or figure that is not peculiar to or distinctive of reasoning in τ.

Supposing that Pasch did hold Asymmetry, or something like it, certain additional premises would also be necessary for justification of his views on rigor and contentual discipline in reasoning. Prominent among these would be a second type of asymmetry claim intended to help articulate what is meant by saying that an expression is "peculiar to" or "distinctive of" reasoning in a given subject-area. Part of the thinking here would presumably be that appeal to logical terms is necessary for reasoning generally and that it does not therefore apply in any peculiar or asymmetic way to any particular area reasoning such as geometry.

My purpose here, however, is not to evaluate or even to analyze Asymmetry. Rather, it is to call attention to a way not taken in the further modern development of the rigor concept and of standards for its attainment. The remarks just made concerning Asymmetry might lead one to expect that post-Paschian development of standards of rigor would follow increasingly fine-grained analyses of surreptive contentual association and means of avoiding it.

This does not seem to describe the post-Paschian development, though, nor even the development from Pasch to Hilbert. Instead of finer analysis of contentual association and of possible means of managing it, there seems rather to have been a basic

---

[22]Here, by contentual use of an expression or figure $\mathcal{E}$, I mean, roughly, use of a judgment $\mathcal{J}$ to justify belief in the validity of an inference or proof where the (propositional) content of $\mathcal{J}$ is in part determined by the content of $\mathcal{E}$.

I am also supposing that, to count as being *proper* to a theory or subject-area τ, a use of $\mathcal{E}$ must be thought to apply in some special or distinctive—some asymmetric—way, or to some asymmetric extent, to reasoning belonging to τ.

if also largely unremarked change in the conception of rigor and even the underlying conception of reasoning to which it has been attached. This at any rate is how I propose we approach understanding of Hilbert's mature post-Paschian writings concerning mathematical proof and rigor.

11.4 Axiomatic Reasoning and Rigor in Hilbert

Pasch emphasized abstraction away from non-logical contents together with reliance on judgments of logical form as means of achieving rigor in mathematical proof. Somewhat more accurately, he saw mathematics as having two parts. One was a more "rigid", properly mathematical part exclusively concerned with deduction. The other was a more "pliable" not-properly-mathematical part concerned with provision of material (i.e. basic starting propositions) for deductions.[23]

In Pasch's view, the proofs of properly mathematical geometry were exclusively concerned with deductive relationships between geometrical propositions and not with their truth or evidentness. Rigor seems similarly to have been conceived as avoidance of surreption in judgments of logical or deductive connection.

Hilbert's views of proof and rigor were different. Weyl called attention to what he saw as a chief such difference in his comments on Hilbert's 1927 address to the Hamburg Mathematical Seminar.[24]

He particularly stressed what he took to be a pivotal difference between Hilbert's and Brouwer's views as regards adherence to the traditional contentual conception of proof.[25]

> Before Hilbert constructed his proof theory everyone thought of mathematics as a system of contentual (inhaltliche), meaningful (sinnerfüllte), and evident (einsichtige) truths; this point of view was the common platform of all discussions. ...Brouwer, like everyone else, required of mathematics that its theorems be (in Hilbert's terminology) "real propositions", meaningful truths.
>
> Weyl (1928), 22[26]

---

[23]Cf. "Mathematics is a system with two parts that should be distinguished. The first, properly mathematical, part, is focused exclusively on deduction. The second makes deduction possible by introducing and elucidating a series of insights that are to serve as material for deduction."

Pasch (1918), 228

For a useful discussion of this and related ideas of Pasch's see Pollard (2010).

[24]The text of this address was published as Hilbert (1928).

[25]According to this view, a proof is a finite sequence of judgments whose propositional contents are judged to stand in an appropriate deductive relationship to one another. This traditional view of proof, however, was something that Brouwer shared with many a non-intuitionist. Thus, though Hilbert directed his criticism towards Brouwer, it might just as justifiably have been aimed at Frege (cf. Frege 1906, 387), or any of a number of other thinkers of the late nineteenth and early twentieth centuries.

[26]Cf. Weyl (1944), 640, Brouwer (1923), 336 and Brouwer (1928), 490–492 for related statements concerning the traditional view of proof.

Hilbert rejected what Weyl here described as the traditional view as representing a distortion of traditional mathematical practice. Its chief inaccuracy, he believed, was its under-estimation of the importance of non-contentual reasoning to traditional mathematics.

Hilbert did not deny the importance of contentual judgement and proof to traditional mathematics. He insisted only that non-contentual methods also figured importantly in making traditional mathematics the successful science it evidently was.

For present purposes, the salient difference between contentual and non-contentual proof is that the latter, unlike the former, does not require the logical or deductive connection of (the propositional contents of) conclusory judgements with (the propositional contents of) premisory judgements. Rather, at least generally speaking, it requires only the formal or symbolic connection of formulae.[27] Hilbert thus offered the following general description of a new view of mathematical reasoning.

> [I]n mathematics the objects of our thinking are concrete signs (konkreten Zeichen) themselves, whose shapes (Gestalt), according to the conception adopted, are immediately clear and re-cognizable (unmittelbar deutlich und wiedererkennbar). ...The propositions (Aussagen) which constitute mathematics are replaced (umgesetzed) by formulae, so that, mathematics proper (die eigentliche Mathematik), becomes a stock of formulae (Bestande an Formeln). ...A proof becomes an array of formulas given as such to our perceptual intuition.
>
> ...
>
> [I]n my theory [of proof, MD] contentual inference (das inhaltliche Schließen) is replaced by outwardly manifest manipulation of signs according to rules (äußeres Handeln nach Regeln). In this way the axiomatic method attains that reliability (Sicherheit) and perfection that it can and must reach if it is to become the basic instrument of all theoretical research.
>
> Hilbert (1928), 2, 4[28,29]

Unlike real or contentual proof, then, with its deductive connection of genuine (i.e., contentual) propositions, Hilbert's ideal proofs featured symbolic expressions connected by applications of rules stated in terms of their (i.e., the expressions') outward appearances.

---

[27]Hilbert referred to such processes of reasoning as "formaler Denkprozesse" in later writings (cf. Hilbert 1930, 380).

[28]To be more exact, the view described here was taken to apply to what Hilbert and Bernays later referred to as formal (formale) axiomatic reasoning, a type of axiomatic reasoning they distinguished (cf. Hilbert and Bernays 1934, §1) from contentual (inhaltliche) axiomatic reasoning. "[I]n contentual axiomatics (inhaltlichen Axiomatik)", they said, "the basic relations are taken to be something found in experience or in intuitive conception (anschaulicher Vorstellung), and thus something contentually determined, about which the sentences of the theory make assertions (Behauptungen)." (Hilbert and Bernays 1934, 6.)

In formal axiomatization (formale Axiomatik), on the other hand, "the basic relations are not taken as having already been determined contentually. Rather, they are determined implicitly by the axioms from the very start. And in all thinking with an axiomatic theory only those basic relations are used that are expressly formulated in the axioms." (op. cit., 7).

In his proof-theoretic writings Hilbert sometimes wrote 'axiomatic' where, more strictly speaking, he meant 'formal axiomatic'.

[29]Cf. Hilbert (1926), 177 for one of a number a similar statements by Hilbert.

In Hilbert's view, the history of mathematics had amply illustrated the benefits of using such non-contentual (or ideal) methods in mathematics. He considered these benefits to be, broadly speaking, benefits of simplicity or, perhaps better, efficiency, and he considered them to be considerable enough to warrant development of a general plan for their systematic justification.

Allowing the use of ideal methods in mathematical proof presents a problem concerning rigor, though, and a problem that seems to go quite deep. Rigor, as Pasch conceived of it, was a property taken to apply to contentual inference. More accurately, it was a property taken to apply in the first instance to judgments of validity concerning contentual inferences. A contentual inference was to qualify as rigorous just in case the justification of its validity avoided all premisory (and other relevant types of) surreption.

Such a conception of rigor does not apply even in principle to ideal proofs. The "premises" and "conclusions" of inferences in ideal proofs, generally speaking, are not and do not express propositions. Nor are they intended to.[30]

As a consequence, they do not admit of genuine logical connection or failure of genuine logical connection. Rather, they are formulae whose use in our reasoning consists in their being manipulated according to rules stated in terms of the outward appearances of the expressions to which they are intended to apply.

What becomes of rigor when contentual inference and proof is replaced by formal manipulation of the type just described? Is there a meaningful and important conception of rigor that remains and is capable of serving as an ideal of formal reasoning in something like the way that avoidance of premisory surreption (or, more generally, of logical gaps of all types) serves as an ideal of contentual deductive inference?

I believe there is. In saying this, though, I do not intend to deny that the differences between genuinely logical reasoning and symbolic reasoning are considerable and that they dictate a change in the very conception of rigor. On the new conception, the aim of rigor will no longer be avoidance of premisory surreption and other validity-nullifying gaps in reasoning. Rather, it will be avoidance of deficiencies of explicitness or transparency in formal reasoning.

In Hilbert's view, axiomatic reasoning[31] was intended to avoid just such deficiencies. Formal axiomatic proofs were taken to be concrete objects that are distinguished from each other, and from non-proofs, by outwardly manifest characteristics.

The axioms of a formal axiomatic system, in particular, were supposed to be syntactically rather than semantically specified. In the end, this meant that they were to be elements of reasoning whose use involves (indeed, substantially consists in) their being exhibited. That is, they are elements of reasoning whose use is to be made

[30] I put 'premises' and 'conclusions' in scare-quotes because, in the present case, they are formulae, and not what premises and conclusions have traditionally been taken to be, namely, propositions or propositional attitude-takings (e.g., judgement or hypothesis).

[31] More specifically, what he and Bernays called *formal* axiomatic reasoning.

manifest by their being displayed or exhibited, and whose contributions to reasoning are a function of their use in various formal-manipulatory procedures or activities.[32]

They are not elements of reasoning that are to be identified by giving a formula that *expresses* them, and whose contributions to reasoning are essentially a function of contents (e.g., propositions, or propositional-functions) they express.

In Hilbert's view, since it is formulae rather than propositions that are capable of being exhibited, it is formal rather than contentual proofs that are capable of being fully explicit and, so, fully rigorous, according to a conception of rigor in which rigor is taken to consist in transparency or explicitness of usage. In my view, it is this or something like this explicitness of formal axiomatic reasoning that Hilbert intended to emphasize when, as in the epigraph at the top of this paper, he described axiomatic thinking as thinking with awareness or consciousness.

Proper rigor in our reasoning should guarantee avoidance of logical surreption when our reasoning is of such a type as to include genuine logical inference. Not all our reasoning is reasoning of this type, however. That it is not suggests the need for an adjustment in our understanding of rigor—one which sees it as applying not only to contentual reasoning but to formal or symbolic reasoning as well.

Extended in this way, rigor consists in a type of explicitness—explicitness in which every element of a piece of reasoning, as well as its use within that reasoning, is outwardly manifest. Hilbert believed formalization to be the key to attaining such rigor, regardless of whether the reasoning in question was what Hilbert and Bernays referred to as *contentual* (inhaltlich) axiomatic reasoning or what they generally termed *formal* (formale) axiomatic reasoning. In each case, it was formalization that was supposed to provide for that explicitness or transparency of use on which rigor was taken to fundamentally depend.[33]

---

[32] Roughly speaking, exhibition in the current sense consists in the presentation (whatever, exactly, that might mean) of a particular concrete expression as an exemplar for other concrete expressions—specifically, expressions whose external features are sufficiently similar to those of the exemplar to qualify them as tokens of the same type as it.

[33] In Hilbert's view, formal axiomatic thinking was not only the "basic instrument of all theoretical research" (Hilbert 1928, 4), it was also a general and pervasive form of human thought.

> In our theoretical sciences we are accustomed to the use of formal thought processes (formaler Denkprozesse) and abstract methods ...[But] already in everyday life (täglichen Leben) one uses methods and concept-constructions (Begriffsbildungen) which require a high degree of abstraction and which only become plain through unconscious application of the axiomatic method (nur durch unbewußte Anwendung der axiomatischen Methoden verständlich sind). Examples include the general process of negation and, especially, the concept of infinity.
>
> Hilbert (1930), 380

The last sentence of this remark raises questions concerning how Hilbert might have understood "unconscious" applications of the formal axiomatic method. Would it be possible to unconsciously apply a method of reasoning whose essence is consciousness of its own elements? As a matter of strict logical possibility, the answer would seem to be 'yes.' Whether this represents some other type of incoherence, though, is more difficult to say and something I lack space to consider further here.

11.5 Conclusion

Pasch believed that achievement of rigor in mathematical inference required practical separation of judgments of inferential validity from the mathematical subject-matters with which the premises and conclusions of inferences might be concerned. To achieve this separation, he suggested, every judgment of inferential validity should be justified in complete abstraction from the meanings of the mathematical terms that occur in it. Only by applying such abstraction, he believed, could a reasoner be properly assured that there is no deductive gap between the premises and conclusion of a given piece of reasoning.

Hilbert too adopted a view according to which attainment of rigor in mathematical reasoning requires a type of separation of that reasoning from the contents of mathematical terms that occur in it. If I am not mistaken, though, the separation he envisioned was quite different both in character and in intended purpose from that which Pasch had in mind.

Pasch generally conceived of mathematical theories as what, since the late nineteenth century, have been called *abstract* sciences. On this conception, the axioms of mathematical theories are not taken to be propositions that are intended to characterize determinate pre-axiomatically given classes of objects (e.g., traditionally conceived points or lines) and relations between them. Rather, they are regarded as propositional-schemata (or perhaps propositional functions)[34] which, though perhaps applying to pre-axiomatically foreseen domains, nonetheless characteristically apply to unforeseen domains as well.

Conceived in this schematic way, the axioms of abstract sciences were taken to have only their schematic forms to contribute to proofs in which they occurred. Specifically, they had no propositional contents to contribute to them. If, then, as Pasch suggested, the justification of a judgment that a proof or an inference in a proof is valid were to appear to appeal to the content of a mathematical term, there would be reason to view it (i.e., the justification) with suspicion.

Pasch's commitment to the Abstraction Condition, and to the type of separation from contents that it brought to geometrical reasoning, reflected his view that axiomatic geometries are generally best seen as abstract sciences whose axioms are propositional schemata rather than propositions. This meant in turn that geometrical proofs were characteristically to be seen as finite sequences of items (viz. propositional schemata) whose contributions to the proofs in which they occurred were their schematic forms.

Hilbert, too, particularly in his writings around the turn of the twentieth century (cf. Hilbert 1899, 1900), stressed a conception of axiomatic reasoning according to which the axioms of an axiomatic system are not intended to describe or capture some pre-axiomatically given content, but, rather, to give "an exact (genaue) and, for mathematical purposes, complete (vollständige) specification (Beschreibung)" (Hilbert 1899), ch. 1, §1; (Hilbert 1900, 181) of those elements which may rightly be used without proof in an axiomatic proof.

[34]Cf. Whitehead (1906), 2, Huntington (1911), §20.

Hilbert didn't expand further in Hilbert (1899, 1900) on the exactness and completeness mentioned in this last remark. In the fuller course of his work (e.g. in Hilbert 1922, 1928), though, he did. The remarks referred to earlier in which he proposed replacing contentual inference (inhaltliche Schließen) with operations on concretely exhibited (konkret ausweisbaren)[35] symbolic expressions according to explicit rules illustrates the development in his understanding of the axiomatic method.

This development is in my view most plausibly seen as the adoption of a new view concerning the nature of rigor—a view which focuses on explicitness or transparency. Judged from this vantage, failures of rigor are fundamentally failures to recognize or to identify elements that ought properly to be seen as belonging to a piece of reasoning. Such failures, in turn, are generally taken to be due to deficiencies of explicitness or transparency in our reasoning.

Axiomatic proof, as Hilbert conceived of it, was intended to protect against such deficiencies by offering the ultimate in explicitness. Axioms were to be identified by their outward shapes, and these shapes, in turn, were to be given by their being exhibited. The idea, if I am right, is that full explicitness in proof can be achieved only through such exhibition. It cannot be achieved by semantical expression. In other words, rigor, or full explicitness in proof, can only be achieved by axiomatization if the axioms of the system are themselves objects which can be exhibited or displayed, and not merely, as with Hilbert's predecessors, propositions, propositional-schemata or other contents taken to be *semantically expressed by* exhibitable objects.

Rather, it requires that axioms be given by being exhibited—that is, by presenting a concrete expression[36] that is identifiable by its outwardly manifest characteristics. An expression given in this way is to serve as an exemplar of similarly shaped concrete expressions that are taken to belong to a given syntactical category of a given formal language.[37]

Hilbert's "decontentualization" of proof—his proposed replacement of propositions and other contentual items which figure centrally in the traditional conception of proof by the formal objects of (his formal conception of) axiomatic proof was thus in his view a transformation that is necessary if the legitimate demands of rigor are generally to be met.

His view is complicated by the fact that, in addition to urging a place for the above-described conception of axiomatic proof and its accompanying conception of rigor, Hilbert continued to see a place in mathematics (and metamathematics) for contentual proof as well. How he may have conceived of rigor for such proof, to what extent he may have taken rigor so conceived to be achievable for this type of proof and how his views on these matters may have compared to and/or contrasted with Pasch's views are matters I will leave for another occasion.

---

[35]Cf. Hilbert (1928), 1.

[36]By 'expression' here, I mean simply a string of characters in a language. I do not mean that this string serves to express a semantical content of some type.

[37]The need to bring syntactical categories into the picture is necessary in order to distinguish between similarly shaped syntactical objects that belong to different syntactical categories (e.g., a formula considered as a line of a proof versus a similarly shaped object which is taken to constitute a one line proof).

Postscript

In 1990–1991, I coordinated one of the Notre Dame philosophy department's *Perspectives in Philosophy* lecture series. Hilary was one of the speakers I invited. His lectures primarily concerned various of Wittgenstein's ideas on proof.

This led at one point to a discussion concerning "Hilbert's Thesis" and its possible bearing on and/or presuppositions concerning matters of rigor. In his 1984 paper "Proof and experience", Hilary presented Hilbert's Thesis as the claim that "derivability in quantification theory[38] captures the intuitive mathematical notion of *deduction*, just as recursiveness captures the intuitive mathematical notion of *computability*" (cf. Putnam (1984), 32, emphases in text).

Hilary was of the view that the completeness of quantification theory provides strong evidence for Hilbert's Thesis.

> [C]ompleteness is easily explained: a sentence which cannot be derived from given axioms by means of quantification theory doesn't, in fact, *follow* from those axioms. "Doesn't follow" in the very intuitive sense that there is, in fact, a possible structure which can be used to interpret the language in such a way that the axioms come out *true* while the sentence that wasn't derivable comes out *false*. This is very strong evidence for ..."Hilbert's Thesis" ...
>
> Putnam (1984), 31–32 (emphases in text)

For such a view to be plausible, I think, one must adopt a semantical or contentual understanding of deduction or deducibility—that is, an understanding according to which a sentence ϕ is properly said to be deducible from a set of sentences Γ only if the propositions expressed by the elements of Γ logically imply the proposition expressed by ϕ.

I have tried to indicate reasons for doubting that Hilbert held such a view of deduction or deducibility. More accurately, I have tried to indicate why I think Hilbert did not take deducibility to consist in or to be constituted by logical implication. (This does not suggest, of course, that Hilbert would have denied their extensional coincidence.)

Relatedly, I have argued that Hilbert's mature view of rigor was one which took it to consist not in logical gaplessness per se, but in full explicitness as regards the constituent elements of a piece of reasoning. To put it another and, I think, not very surprising way, whether Hilbert would have accepted Hilbert's Thesis—the claim that "derivability in quantification theory captures the intuitive mathematical notion of *deduction*" (loc. cit.)—depends crucially on how one understands the notion of *capture* that figures here.

[38] Hilary glossed the term "quantification theory" as "first-order logic" on p. 31 of Putnam (1984).

References

Berkeley, G. (1734). The analyst; or, a discourse addressed to an infidel mathematician. Wherein it is examined whether the object, principles, and inferences of the modern analysis are more distinctly conceived, or more evidently deduced, than religious mysteries and points of faith, Printed for J. Jonson, London.

Bour, P.-E., Rebuschi, M., & Rollet, L. (Eds.). (2010). *Construction*. London: College Publications.

Brouwer, L. E. J. (1923). Über die Bedeutung des Satzes vom ausgescholessenen Dritten in der Mathematik, insbesondere in der Funktiontheorie. *Journal für die reine und angewandte Mathematik, 154*, 1–7. English translation in van Heijenoort (1967), pp. 334–345. Page references are to this translation.

Brouwer, L. E. J. (1928). Intuitionistische Betrachtungen über den Formalismus. *Koninklijke Akademie van wetenschappen de Amsterdam, Proceedings of the Section of Sciences, 31*, 324–329. English translation with introduction in van Heijenoort (1967), pp. 490–492. Page references are to this translation.

Cellucci, C., Grosholz, E., & Ippoliti, E. (Eds.). (2011). *Logic and knowledge*. Newcastle upon Tyne: Cambridge Scholars Publishing.

Detlefsen, M. (2005). Formalism. In Shapiro (2005), pp. 236–317.

Detlefsen, M. (2010). Rigor, Re-proof and Bolzano's Critical Program. In Bour et al. (2010), pp. 171–184.

Detlefsen, M. (2011). Dedekind against intuition: Rigor, scope and the motives of his logicism. In Cellucci et al. (2011), pp. 273–288.

Engel, F., & Stäckel, P. (1895). Die Theorie der Parallelinien von Euklid bis auf Gauss. Leipzig: Teubner.

Frege, G. (1906). Über die Grundlagen der Geometrie, *Jahresbericht der Deutschen Mathematiker-Vereinigung, 15*, 293–309, 377–403, 423–430. English translation in Frege (1971).

Frege, G. (1971). *Gottlob Frege: On the foundations of geometry and formal theories of arithmetic*, trans. and ed., with an introduction, by Eike-Henner W. Kluge. New Haven: Yale University Press.

Freudenthal, H. (1962). The main trends in the foundations of geometry in the 19th century. In Suppes et al. (1962), pp. 613–621.

Hilbert, D. (1899). *Grundlagen der Geometrie*, Published in *Festschrift zur Feier der Enthüllung des Gauss-Weber Denkmals in Göttingen*. Leipzig: Teubner.

Hilbert, D. (1900). Über den Zahlbegriff. *Jahresbericht der Deutschen Mathematiker-Vereinigung, 8*, 180–194.

Hilbert, D. (1922). Neubegründung der Mathematik. Erste Mitteilung. *Abhandlungen aus dem Mathematischen Seminar der Hamburgischen Universität, 1*, 157–77. Page references are to the reprinting in Hilbert (1935), Vol. III.

Hilbert, D. (1926). Über das Unendliche. *Mathematische Annalen, 95*, 161–190.

Hilbert, D. (1928). Die Grundlagen der Mathematik. *Abhandlungen aus dem mathematischen Seminar der Hamburgischen Universität, 6*, 65–85. Reprinted in *Hamburger Einzelschriften, 5*, pp. 1–21. Leipzig: Teubner. Page references are to the latter.

Hilbert, D. (1930). Naturerkennen und Logik. *Die Naturwissenschaften, 18*, 959–963. Printed in Hilbert (1935), Vol. III. Page references are to this printing.

Hilbert, D., & Bernays, P. (1934). *Grundlagen der Mathematik* (Vol. I). Berlin: Springer. Page references are to Hilbert and Bernays (1968), the second edition of this text.

Hilbert, D. (1932–1935). *Gesammelte Abhandlungen*, three volumes. Bronx: Chelsea Pub. Co.

Hilbert, D., & Bernays, P. (1968). *Grundlagen der Mathematik* (Vol. I, 2nd ed.). Berlin: Springer.

Huntington, E. V. (1911). *The fundamental propositions of algebra*. Long Island University, Brooklyn: Galois Institute Press.

Lambert, J. H. (1786). Theory der Parallelinien I. *Leipziger Magazin für reine und angewandte Mathematik, 1*(2), 137–164. Reprinted in Engel and Stäckel (1995), pp. 152–176.

Lambert, J. H. (1786). Theory der Parallelinien II. *Leipziger Magazin für reine und angewandte Mathematik, 1*(3), 325–358. Reprinted in Engel and Stäckel (1995), pp. 176–207.
Pasch, M. (1882). *Vorlesungen über neuere Geometrie*. Leipzig: Teubner.
Pasch, M. (1918). Die Forderung der Entscheidbarkeit. *Jahresbericht der Deutschen Mathematiker-Vereinigung, 27*, 228–232.
Pollard, S. (2010). 'As if' reasoning in Vaihinger and Pasch. *Erkenntnis, 73*, 83–95.
Poncelet, J.-V. (1822). *Traité des propriétés projectives de figures*. Paris: Bachelier.
Putnam, H. (1984). Proof and experience. *Proceedings of the American Philosophical Society, 128*, 31–34.
Russell, B. (1902). The teaching of Euclid. *The Mathematical Gazette*, 2(33), 165–167.
Shapiro, S. (Ed.). (2005). *Oxford handbook of the philosophy of mathematics and logic*. Oxford: Oxford University Press.
Smith, C., & Bryant, S. (1901). *Euclid's elements of geometry: Books I–IV, VI and XI*. London: Macmillan.
Suppes, P., Nagel, E., & Tarski, A. (Eds.) (1962). *Logic, methodology, and philosophy of science: Proceedings of the 1960 International Congress*. Stanford: Stanford U Press.
Todhunter, I. (1869). *The elements of Euclid for the use of schools and colleges*. London: Macmillan.
van Heijenoort, J. (1967). *From Frege to Gödel: A source book in mathematical logic, 1879–1931*. Cambridge: Harvard University Press.
Weyl, H. (1928). Diskussionsbemerkungen zu dem zweiten Hilbertschen Vortrag über die Grundlagen der Mathematik. *Abhandlungen aus dem mathematischen Seminar der Hamburgischen Universität, 6*, 86–88. Reprinted in *Hamburger Einzelschriften, 5*, pp. 22–24, Teubner, Leipzig, 1928. English translation in van Heijenoort (1967).
Weyl, H. (1944). David Hilbert and his mathematical work. *Bulletin of the American Mathematical Society, 50*, 612–654.
Whitehead, A. N. (1906). *The axioms of projective geometry*. Cambridge: University Press.
Young, J. W., Denton, W. W., & Mitchell, U. G. (1911). *Lectures on fundamental concepts of algebra and geometry*. New York: Macmillan Co.

Michael Detlefsen is McMahon–Hank professor of philosophy at the University of Notre Dame, where he also serves as the editor of the Notre Dame Journal of Formal Logic. His current projects include a study of the different conceptions of rigor and of the relationship of rigor to other ideals of proof in the history of mathematics. Recent publications include: "Gentzen's Anti-formalist Ideas", in *Gentzen's Centenary: The Quest for Consistency*, M. Rathjen and R. Kahle (eds.); "On the Motives for Proof Theory", in *Dag Prawitz on Proofs and Meaning*, H. Wansing (ed.), *Outstanding Contributions to Logic series*; "Duality, Epistemic Efficiency and Consistency", in *Formalism & Beyond*, G. Link (ed.).

Chapter 12
Concrete Mathematical Incompleteness: Basic Emulation Theory

Harvey M. Friedman

Abstract By the modern form of Gödel's First Incompleteness Theorem, we know that there are sentences (in the language of ZFC) that are neither provable nor refutable from the usual ZFC axioms for mathematics (assuming, as is generally believed, that ZFC is free of contradiction). Yet it is clear that the usual examples are radically different from normal mathematical statements in several glaring ways such as the mathematically remote subject matter and the essential involvement of uncharacteristically intangible objects. Starting in 1967, we embarked on the Concrete Mathematical Incompleteness program with the principal aim of developing readily accessible thematic mathematical research areas with familiar mathematical subject matter replete with examples of such incompleteness involving only characteristically tangible objects. The many examples developed over the years represent Concrete Mathematical Incompleteness ranging from weak fragments of finite set theory through ZFC and beyond. The program has reached a mature stage with the development of Emulation Theory. Emulation Theory, in its present basic developed form, involves finite length tuples of rational numbers. Only the usual ordering of rationals is used, and there is no use of even addition or multiplication. The basics are fully accessible to early undergraduate mathematics majors and gifted high school mathematics students, who will be able to engage with some simple nontrivial examples in two and three dimensions, with illustrations. In this paper, we develop the positive side of the theory, using various levels of set theory for systematic development. Some of these levels lie beyond ZFC and include familiar large cardinal hypotheses. The necessity of the various levels of set theory will be established in a forthcoming book (Concrete Mathematical Incompleteness in preparation).

H. M. Friedman (✉)
The Ohio State University, Columbus, OH 43210, USA
URL: https://u.osu.edu/friedman.8/

G. Hellman and R. T. Cook (eds.), *Hilary Putnam on Logic and Mathematics*,
Outstanding Contributions to Logic 9, https://doi.org/10.1007/978-3-319-96274-0_12

12.1 Introduction

According to the modern form of Gödel's First Incompleteness Theorem (Gödel 1931, 1940), there are sentences in the language of ZFC that are neither provable nor refutable from the usual ZFC axioms for mathematics, assuming that ZFC is consistent (going back to Concrete Mathematical Incompleteness in preparation). This seminal result already puts an end to one important facet of the multifaceted Hilbert's Program. Since then, attention has naturally and inevitably focused on the nature of the examples of this incompleteness from ZFC.

The first clear specific example of incompleteness from ZFC is already given by the modern form of Gödel's Second Incompleteness Theorem: the consistency of ZFC, Con(ZFC), is neither provable nor refutable from ZFC (going back to Concrete Mathematical Incompleteness in preparation). The second clear specific example is Cantor's continuum hypothesis, CH, that every uncountable set of real numbers can be mapped onto all real numbers. That CH is not refutable in ZFC is by Gödel (1940), and that CH is not provable in ZFC is by Cohen (1963).

Attention naturally focuses on the nature of the examples of incompleteness from ZFC - and in particular, their subject matter. This move to the consideration of the underlying subject matter in examples of incompleteness is entirely natural and inevitable. Both of these examples are easily recognized to be profoundly different from normal mathematical propositions in vividly important ways that are instantly recognized by the general mathematical community. Con(ZFC) is a statement about provability in a certain formal system, and thus involves mathematically (but not philosophically) remote subject matter. CH is a statement in abstract set theory, involving uncontrolled sets of real numbers, which are immediately recognized as uncharacteristically intangible mathematical objects.

Starting in 1967, we embarked on the Concrete Mathematical Incompleteness program with the principal aim of developing readily accessible mathematical research areas with familiar mathematical subject matter replete with examples of such incompleteness involving only characteristically tangible objects. The many examples developed over the years represent Concrete Mathematical Incompleteness ranging from weak fragments of finite set theory through ZFC and beyond. A detailed discussion and presentation of the major results in Concrete Mathematical Incompleteness before this Emulation Theory can be found in Embedded maximal cliques and incompleteness 2013; Friedman 2011; Crangle et al. (2014), Introduction.

Concrete Mathematical Incompleteness has reached a mature stage with the development of Emulation Theory. Emulation Theory, in its present basic developed form, involves finite length tuples of rational numbers. We use the usual ordering of rational numbers, but not addition, subtraction, or multiplication. The basics are fully accessible to early undergraduate mathematics majors and gifted high school mathematics students, who will be able to engage with some simple nontrivial examples in two and three dimensions, with illustrations. In this paper, we develop the positive side of the theory, using various levels of set theory for systematic development. Some of these levels lie beyond ZFC and include familiar large cardinal hypotheses.

The large cardinal hypotheses used here are given by the SRP hierarchy (stationary Ramsey property), which are beyond strongly inaccessible, strongly Mahlo, weakly compact, and indescribable cardinals, are intertwined with the subtle and ineffable cardinal hierarchy, which lives well below $\kappa \rightarrow \omega$, and thus is compatible with ZFC + V = L (see Kanamori 1994). Emulation Theory has also been extended involving necessary uses of the HUGE cardinal hierarchy, which are stronger than measurable cardinals, supercompact cardinals, and Vopenka's Principle, but lie below nontrivial $j{:}V(\kappa) \rightarrow V(\kappa)$. See Kanamori (1994). Use of the HUGE cardinal hierarchy and the necessity of the various levels of set theory throughout Emulation Theory will appear in a forthcoming book.

In Sect. 12.1.1, we start with the most general form of Basic Emulation Theory presented relative to any given relational structure in the usual sense of elementary logic. We successively lower this generality in several steps down to the particular relational structure $M = (Q[0, 1], <)$ that drives Sect. 12.3.

In Sect. 12.1.2 we give a systematic account of the major results in the paper.

Throughout the paper, we follow the convention that any free variables are implicitly universally quantified in front. Until Sect. 12.3.5, we mostly see implicit free variables for dimension k, for a relational structure M, for a subset E of M^k, for a subset E of $Q[0, 1]^k$, and for a relation $R \subseteq Q[0, 1]^k \times Q[0, 1]^k$. In Sect. 12.3.5, we also see the free variable r used for r-emulations.

In this paper we focus on the use of certain large cardinal hypotheses to develop Emulation Theory. In fact, we only use the consistency of these large cardinal hypotheses, and in particular, Con(SRP). For the presentation of SRP, see Appendices A, B. We have shown that the basic results of Emulation Theory proved here from Con(SRP) are in fact provably equivalent to Con(SRP) over WKL_0 (see Appendix B). These so called reversals will appear in Concrete Mathematical Incompleteness (in preparation).

The reversal asserting that RCA_0, or even ZFC proves $A \rightarrow$ Con(SRP), guarantees that A cannot be proved in SRP (unless SRP is inconsistent). Also, ZFC proves $A \rightarrow$ Con(ZFC) tells us that A cannot be proved in ZFC (unless ZFC is inconsistent). The reason for this is Gödel's Second Incompleteness Theorem, that no reasonable system can prove its own consistency (unless it be inconsistent).

Already in the currently available (Concrete Mathematical Incompleteness in preparation), Boolean Relation Theory, reversals of combinatorial statements are fully worked out (in that case, 1-Con(MAH) is used with the somewhat weaker MAH). In both cases, the general method is the same. We start with the statement A (from Boolean Relation Theory or from Emulation Theory). We then build a series of structures, rather explicitly, which become more and more like models of set theory with large cardinals, until finally they really are. However, as can be seen with Boolean Relation Theory, (Concrete Mathematical Incompleteness in preparation), there are many obstacles that have to be overcome while traveling down that long path.

12.1.1 Basic Emulation Theory

Basic Emulation Theory starts with a given relational structure M in the usual sense of elementary logic. Thus M = (D, ...), where the domain D is a nonempty set, and ... are the components, as in Definition 2.1.

It is convenient to use M^k for D^k with the understanding that the working space is the relational structure M. Tuples x, y are M equivalent if and only if x, y obey the same unnested atomic formulas. See Definition 2.2.

Now comes the crucial definition which we first give in its most compact form.

Maximal Emulation Definition/1. ME/DEF/1. S is a maximal emulator of $E \subseteq M^k$ if and only if $S \subseteq M^k$ and every element of S^2 is M equivalent to an element of E^2, where this is false if S is replaced by any proper superset of S.

Here E^2, S^2 are viewed as sets of 2k-tuples. The where clause is equivalent to saying that if we add a new point x ∈ M to S then this ruins S^2 having all of its elements M equivalent to an element of E^2.

We can equivalently break this definition into the following two parts.

Maximal Emulation Definition/2. ME/DEF/2. S is an emulator of $E \subseteq M^k$ if and only if $S \subseteq M^k$ and every element of S^2 is M equivalent to an element of E^2. S is a maximal emulator of $E \subseteq M^k$ if and only if S is an emulator of $E \subseteq M^k$ which is not a proper subset of an emulator of $E \subseteq M^k$.

In Basic Emulation Theory, we investigate basic properties of maximal emulators of $E \subseteq M^k$.

Maximal Emulation/1. ME/1. Every $E \subseteq M^k$ has a maximal emulator.

There is an important sharper version.

Maximal Emulation/2. ME/2. Every $E \subseteq M^k$ has a maximal emulator containing any given emulator.

In Sect. 12.2, we prove that ME/1, 2 are equivalent to the full axiom of choice over ZF.

We also show that if M is countable (i.e., D = dom(M) is countable) with finitely many components, then ME/1 is provable in RCA_0 and ME/2 is provable in ACA_0 by ordinary recursion along the nonnegative integers. In fact, for such M, the sharper form is provably equivalent to ACA_0 over RCA_0.

We now start with the most general formulation of Basic Emulation Theory as a Template. We then take the generality down five steps to BETA/6, which is where the current development of Basic Emulation Theory resides.

Maximal Emulation Use Definition. MEU/DEF. $R \subseteq M^k \times M^k$ is ME usable if and only if for all subsets of M^k, some maximal emulator contains its R image.

The R image of S is the forward image $\{y: (\exists x \in S)(R(x, y))\}$.

This is a convenient place to make a purely expositional point. The reader might question the wisdom of the particular English construction we have used to present MEU/DEF. In particular, it might appear more natural, grammatically, to write if and only if every subset of M^k has a maximal emulator containing its R image.

However, this introduces an ambiguity. Might this be an emulator that, among the emulators containing its R image, is maximal? That is quite different, and in fact easily seen to be true on general grounds. What we of course mean is that we have a maximal emulator that just happens to also contain its R image. Note how the formulation of MEU/DEF completely avoids this ambiguity.

Basic Emulation Theory Aim/1. BETA/1. Investigate the ME usable $R \subseteq M^k \times M^k$.

The following necessary condition for ME usability is used throughout the paper.

Maximal Emulation/4. ME/4. If $R \subseteq M^k \times M^k$ is ME usable then R is M preserving in the sense that $(\forall x, y)(R(x, y) \rightarrow x, y$ are M equivalent).

We have yet to place any restrictions on the relation R or the subset E of M^k. We will naturally want to focus on reasonably well behaved R and subsets of M^k. The M elementary sets are the subsets of M^k that are defined by a quantifier free formula over M, with parameters allowed. E.g., the M elementary sets, where M is the ordered field of real numbers, are the semi algebraic subsets of $\Re^k$.

If M has finitely many components, then in the Maximal Emulation Use Definition, we can replace "subsets of M^k" with "finite subsets of M^k" and have an equivalent definition.

Basic Emulation Theory Aim/2. BETA/2. Investigate the ME usable $R \subseteq M^k \times M^k$, where M is a relational structure with finitely many components and $R \subseteq M^k \times M^k$ is elementary.

It is also natural to focus on well behaved M. We propose three main rich sources of such M. For uncountable M, there are the structures that are definable over the ordered field of real numbers. For countable M, there are the structures that are definable over the ordered field of real algebraic numbers or over the structure (Q, Z, <, +).

Basic Emulation Theory Aim/3. BETA/3. Investigate the ME usable $R \subseteq M^k \times M^k$ where M has finitely many components and is definable over the ordered field of real numbers, and $R \subseteq M^k \times M^k$ is M elementary.

Basic Emulation Theory Aim/4. BETA/4. Investigate the ME usable $R \subseteq M^k \times M^k$ where M has finitely many components and is definable over the ordered field of real algebraic numbers, and $R \subseteq M^k \times M^k$ is M elementary.

Basic Emulation Theory Aim/5. BETA/5. Investigate the ME usable $R \subseteq M^k \times M^k$ where M has finitely many components and is definable over (Q, Z, <, +), and $R \subseteq M^k \times M^k$ is M elementary.

We now arrive at the particular case in Basic Emulation Theory that is discussed in this paper, which uses only $M = (Q[0, 1], <)$. Here $Q[0, 1] = Q \cap [0, 1]$, where Q is the set of all rational numbers with the usual numerical <.

Basic Emulation Theory Aim/6. BETA/6. Investigate the ME usable $R \subseteq M^k \times M^k$, where $M = (Q[0, 1], <)$ and R is order theoretic.

Here the order theoretic R are exactly the elementary R, and, in fact, the definable R by the well known quantifier elimination for dense linear orderings.

For the rest of the paper, we focus on BETA/6 for $(Q[0, 1], <)$. We use Q[0, 1] as an abbreviation for $(Q[0, 1], <)$, as the $<$ is tacitly understood.

12.1.2 ON Q[0, 1]

Section 12.2 presents proofs of the results discussed in Sect. 12.1.1. Section 12.3 is the heart of the paper, and treats Basic Emulation Theory on Q[0, 1]. We use ME/DEF/1, 2 and MEU/DEF with $M = (Q[0, 1], <)$.

In Sect. 12.3.1, we present eight illustrative examples of $E \subseteq Q[0, 1]^2$ and their emulators and maximal emulators to familiarize the reader with these notions.

We show that the ME usability of any given order theoretic $R \subseteq Q[0, 1]^k \times Q[0, 1]^k$ forms a sentence φ which is implicitly Π^0_1 over WKL_0 (Corollary 3.1.7). I.e., is provably equivalent to a Π^0_1 sentence over WKL_0. This is shown by use of the Gödel Completeness Theorem. A consequence of φ being implicitly Π^0_1 over WKL_0 is that φ is WKL_0 falsifiable, in the sense of Definition 3.1.6. Falsifiability resonates with the idea in physical science that in order for a statement to be physically meaningful, it must be refutable by experimentation. According to Theorem 3.1.4, implicitly Π^0_1 and falsifiable are nearly the same notions.

Thus the usability of a given order theoretic R forms a statement that is, at least implicitly, of the most concrete level of complexity for mathematical statements involving infinitely many objects. The highlight of this paper is how this leads to independence from the usual ZFC axioms for mathematics. These statements still involve the use of an infinite object, namely the maximal emulator. This is a feature that is much stronger than merely having infinitely many objects. So although these statements are implicitly Π^0_1, they are not explicitly Π^0_1 or even explicitly finite. Thus it would be of great interest to have similarly interesting and strategic mathematical examples of independence from ZFC that are explicitly finite or even explicitly Π^0_1. We have arguably achieved this in a particularly satisfying way, and this work will appear elsewhere in Concrete Mathematical Incompleteness (in preparation).

Recall our basic necessary condition for ME usability in ME/4. For $M = (Q[0, 1], <)$, we call this necessary condition order preserving. I.e., $R \subseteq Q[0, 1]^k \times Q[0, 1]^k$ is order preserving if and only if $(\forall x, y)(R(x, y) \rightarrow x, y$ are order equivalent). (E.g., (0.5, 0.7, 0.6) and (0.2, 1, 0.9) are order equivalent).

We illustrate the depth of ME usability through two simple examples.

Maximal Emulation Example/1. MEX/1. For finite subsets of $Q[0, 1]^2$, some maximal emulator is equivalent at (1/2, 1/3), (1/3, 1/4).

Maximal Emulation Example/2. MEX/2. For finite subsets of $Q[0, 1]^2$, some maximal emulator is equivalent at (1, 1/2), (1/2, 1/3).

I.e., we demand, for some maximal emulator S, that (1/2, 1/3) ∈ S ↔ (1/3, 1/4) ∈ S in MEX/1 and (1, 1/2) ∈ S ↔ (1/2, 1/3) ∈ S in MEX/2. Thus MEX/1 asserts the ME usability of the symmetric R = {((1/2, 1/3), (1/3, 1/4)), ((1/3, 1/4), (1/2, 1/3))} of cardinality 2 in dimension 2, and analogously for MEX/2.

We show that MEX/1 is rather superficial in that it is merely a consequence of the fact that isomorphic copies of maximal emulators are maximal emulators. MEX/2 cannot be proved this way because of the use of the right endpoint 1, and so there is something deeper going on here.

In Sect. 12.3.2, we address the problem of finding a necessary and sufficient condition for finite R to be ME usable. Two key notions are "appearance" and "alteration" in Definition 3.2.1, which we repeat here for the reader's convenience.

Definition 3.2.1 Let (x, y) ∈ $Q[0, 1]^k \times Q[0, 1]^k$ and R ⊆ $Q[0, 1]^k \times Q[0, 1]^k$. p is present in (x, y) if and only if p is a coordinate of x or y. p is altered in (x, y) if and only if there exists i such that $p = x_i \neq y_i$ or $p = y_i \neq x_i$. p is present in R if and only if there exists i such that $p = x_i$ or $p = y_i$. p is altered by R if and only if p is altered in some element of R. We also write "p appears in (x, y)", "(x, y) alters p", "p appears in R", and "R alters p".

Example In ((0.5, 0.7, 0.5), (0.6, 0.7, 0.5)), 0 is not present, 0.7 is present but not altered, 0.5, 0.6 are present and altered.

We start with a particularly easy result, which generalizes MEX/1.

Maximal Emulation Finite Use/1. MEFU/1. Any finite order preserving R ⊆ $Q[0, 1]^k \times Q[0, 1]^k$ in which neither 0 nor 1 appear, is ME usable.

More difficult is

Maximal Emulation Finite Use/2. MEFU/2. Any finite order preserving R ⊆ $Q[0, 1]^k \times Q[0, 1]^k$ in which not both 0, 1 appear, is ME usable.

which generalizes MEX/2. The full result along these lines reads

Maximal Emulation Finite Use/3. MEFU/3. Any finite order preserving R ⊆ $Q[0, 1]^k \times Q[0, 1]^k$ that does not alter both 0 and 1, is ME usable.

which is the form that we prove. It immediately implies MEX/1, 2, MEFU/1, 2. The proof of MEFU/3 is given in ACA′, and we don't have a proof in ACA_0. We conjecture that ACA_0 does not suffice to prove MEU/2, 3, although RCA_0 suffices for MEFU/1.

If finite R alters both 0, 1, then open issues arise. However, we do have the following.

Maximal Emulation Finite Use/4. MEFU/4. Any order preserving R ⊆ $Q[0, 1]^k \times Q[0, 1]^k$ of cardinality 1 is ME usable.

We show that MEFU/4 is false with cardinality 2, even for R = {((0, 1/2), (1/2, 1)), ((1/2, 1), (0, 1/2))}, which is order preserving and symmetric of cardinality 2 in dimension k = 2.

We present a complete determination of the symmetric ME usable R of cardinality 2 in dimension $k \leq 2$. However, we do not have a complete determination of the symmetric ME usable R of cardinality 2 in any dimension $k \geq 3$.

In Sects. 12.3.3, and 12.3.4, we address ME usability for infinite order theoretic $R \subseteq Q[0, 1]^k \times Q[0, 1]^k$. In Sect. 12.3.3 we focus on the "large" R where infinitely many numbers are altered, and in Sect. 12.3.4 we focus on the "small" R where finitely many numbers are altered.

We show that the large $Q(0, 1)^{2<} \times Q(0, 1)^{2<}$ is not ME usable. However, we show that the large relation $Q[1/2, 1)^{2<} \times Q[1/2, 1)^{2<}$ is ME usable. On the other hand, we show that for dimension $k \geq 3$, the large relation $Q[1/3, 1/2]^{k<} \times Q[1/3, 1/2]^{k<}$ is not ME usable. There are many issues left open for the ME usability of large order theoretic relations R, even in dimension 2. So far, the results about large relations have not even come close to challenging ZFC.

The highlight of the paper is in Sect. 12.3.4 where we focus on the ME usability for small order theoretic $R \subseteq Q[0, 1]^k \times Q[0, 1]^k$. It is here that it is necessary and sufficient to go well beyond the usual ZFC axioms in order to obtain basic information.

We start with the obvious parameterization of finite $R \subseteq Q[0, 1]^k \times Q[0, 1]^k$ obtained by merely adding a new dimension given by $R'(x, y)$ if and only if $R((x_1, \ldots, x_k), (y_1, \ldots, y_k))$ and $x_{k+1} = y_{k+1}$. This is a crude false start because generally this R' is not even order preserving – the basic necessary condition for ME usability.

But we can easily and very naturally recover by going to lower parameterizations. Here we use $R'(x, y)$ if and only if $R((x_1, \ldots, x_k), (y_1, \ldots, y_k))$ and $x_{k+1} = y_{k+1} < x_1, \ldots, x_k, y_1, \ldots, y_k$. This is promising because the lower parameterization of an order preserving $R \subseteq Q[0, 1]^k \times Q[0, 1]^k$ is order preserving.

Maximal Emulation Small Use/1. MESU/1. The lower parameterization of any order preserving finite $R \subseteq Q[0, 1]^k \times Q[0, 1]^k$ is ME usable.

This lower parameterization idea is familiar from indiscernibles in set theory. Let $R \subseteq \lambda^k$, where λ is a suitable large cardinal. A strong kind of SOI (set of indiscernibles) often considered is an $I \subseteq \lambda$ such that for all $\alpha_1 < \cdots < \alpha_{k-1}$ and $\beta_1 < \cdots < \beta_{k-1}$ from I, and for all $\gamma < \min(\alpha_1, \beta_1)$, we have $(\gamma, \alpha_1, \ldots, \alpha_{k-1}) \in R \leftrightarrow (\gamma, \beta_1, \ldots, \beta_{k-1}) \in R$. Various forms of this lower parameterization idea requires that λ be a large cardinal in what is called the SRP hierarchy, Appendix A. Thus from the results of this paper, we can view Emulation Theory as a particularly natural discrete form of the SRP hierarchy.

MESU/1 is the first of the paper's two most immediately transparent statements independent of ZFC. MED/1 is more specific but more specialized. We say that S is drop equivalent at x, y if and only if $x, y \in Q[0, 1]^k \wedge x_k = y_k \wedge (\forall p \in [0, x_k))(S(x_1, \ldots, x_{k-1}, p) \leftrightarrow S(y_1, \ldots, y_{k-1}, p))$.

We think of x, y as raindrops in the space $Q[0, 1]^k$, at the same height $x_k = y_k$ over the ground. As they fall to the ground in tandem, they generally go in and out of a given set $S \subseteq Q[0, 1]^k$. Drop equivalence says that as they fall in tandem, one is in S if and only if the other is in S.

Maximal Emulation Drop/1. MED/1. For finite subsets of $Q[0, 1]^k$, some maximal emulator is drop equivalent at (1, 1/2, …, 1/k), (1/2, …, 1/k, 1/k).

We derive MED/1 easily from MESU/1. We also present two strengthenings MESU/2, 3 of MESU/1, and two strengthenings MED/2, 3 of MED/1.

MESU/2 is based on the equivalence relations $R_k(A)$, on $Q[0, 1]^k$, associated with each $A \subseteq Q[0, 1]^k$ with the finiteness condition, not altering 0, is ME usable.

MED/2 strengthens MED/1 by giving a necessary and sufficient condition (droppable) for a pair of k-tuples to work for MED/1. MED/3 strengthens MED/2 by asserting that we can simultaneously use any finite list of pairs of k-tuples for MED/1 if and only if we can use any one of the pairs for MED/1. The necessity of these conditions is provable in RCA_0.

We derive all six statement. MESU/2 asserts that for finite $A \subseteq Q(0, 1]$, $R_k(A)$ is ME usable. MESU/3 is based on a natural finiteness condition on $R \subseteq Q[0, 1]^k \times Q[0, 1]^k$, and asserts that every order preserving $R \subseteq Q[0, 1]^k \times Q[0, 1]$s from MESU/2 and derive MED/1 from all six statements. We have shown that MED/1 implies Con(SRP) over RCA_0. This reversal will appear elsewhere in Concrete Mathematical Incompleteness (in preparation).

In Sect. 12.3.5, we derive MESU/2 in dimension $k = 2$ using a transfinite construction of uncountable length. This puts MESU/2 in dimension $k = 2$ well within ZFC, and, with some modification, even in Z and even Z_3. We then derive full MESU/2 in $WKL_0 + \text{Con(SRP)}$. Thus, with the help of the reversal to appear elsewhere (Concrete Mathematical Incompleteness in preparation), we have established that MESU/1, 2, 3, MED/1, 2, 3 are all provably equivalent to Con(SRP) over WKL_0. It follows that MESU/1, 2, 3, MED/1, 2, 3 are all independent of ZFC, assuming SRP is consistent.

We conjecture that MESU/2 for dimension $k = 2$ is provable already in RCA_0. However, we also conjecture that such a proof would be much more difficult than the proof given in Sect. 12.3.5 using the transfinite construction of uncountable length. It would require essentially a complete analysis of the sets of maximal emulators of finite subsets of $Q[0, 1]^2$. We know that there are finitely many such sets of maximal emulators in any dimension k, so an exhaustive analysis is theoretically possible. But see below for a sharpened form of MESU/2 for dimension $k = 2$.

In Sect. 12.3.6, we introduce r-emulators, where the emulators are the 2-emulators. Most of the earlier results go through without serious modification for r-emulators, with noted exceptions. We conjecture that MESU/1 in dimension $k = 1$ and MESU/2, 3, MED/1, 2, 3 in dimension $k = 2$, sharpened with r-emulators, are not provable in ZFC\P, or equivalently, not in Z_2.

We also conjecture that MESU/1 in dimension $k = 2$ and MESU/2, 3, MED/1, 2, 3 in dimension $k = 3$, sharpened with r-emulators, are not provable in ZFC (assuming ZFC is consistent). In fact, we conjecture that they are provably equivalent to Con(ZFC + "there exists a subtle cardinal") over WKL_0. Note that MED/1 has the "raindrops falling in tandem" interpretation, which is particularly vivid in 3 dimensions.

In Sect. 12.4, we discuss a number of General Conjectures. These do not specifically pertain to the statements discussed in Sect. 12.3. The first of these strategic conjectures is

General Conjecture 1. GC1. There is an algorithm for determining whether a given order theoretic $R \subseteq Q[0, 1]^k \times Q[0, 1]^k$ is usable. For inputs, use a standardly digitized form of quantifier free formulas over $(Q[0, 1], <)$ with parameters.

about which we know essentially nothing. However we establish that its sharpening.

General Conjecture 2. GC2. There is a Turing machine with at most $2^{2\hat{}1000}$ states/symbols each, for determining whether a given order theoretic $R \subseteq Q[0, 1]^k \times Q[0, 1]^k$ is usable. For inputs, use a standardly digitized form of quantifier free formulas over $(Q[0, 1], <)$ with parameters.

is not provable in ZFC, assuming SRP is consistent.

12.2 General Maximal Emulation

Here we work in the most general context of relational structures M.

Definition 2.1 A relational structure is a system $M = (D, \ldots)$, where the domain D is a nonempty set, and … are the components, consisting of named constants from D, named relations on D of finite arity, and named functions from and into D of finite arity. In full generality, the number of components is arbitrary, although most commonly there are finitely many components. Equality is considered implicitly present, and does not have to be a component. M is countable if and only if its domain and number of components is countable.

We use M^k for the set D^k with the understanding that the environment is M.

Definition 2.2 $x, y \in M^k$ are M equivalent if and only if x, y obey the same unnested atomic formulas in the sense that

$$x_i = x_j \leftrightarrow y_i = y_j$$
$$x_i = c \leftrightarrow y_i = c$$
$$R(x_{i1}, \ldots, x_{in}) \leftrightarrow R(y_{i1}, \ldots, y_{in})$$
$$F(x_{j1}, \ldots, x_{jm}) = x_b \leftrightarrow F(y_{j1}, \ldots, y_{jm}) = y_b$$

where $1 \leq i, j, i_1, \ldots, i_n, j_1, \ldots, j_m, b \leq k$, c is a constant, R is an n-ary relation, and F is an m-ary function of M.

$EQR(M, k) \subseteq M^k \times M^k = M^{2k}$ is the equivalence relation of M equivalence on M^k.

Maximal Emulation Definition/1. ME/DEF/1. S is a maximal emulator of $E \subseteq M^k$ if and only if $S \subseteq M^k$ and every element of S^2 is M equivalent to an element of E^2, where this conjunction is false if S is replaced by any proper superset of S.

Here ME is read "maximal emulator". Also S^2, E^2 are viewed as sets of 2k-tuples.

We rely on presenting E as a subset of M^k in order to specify the environment in which we are operating. Thus only supersets of S that are subsets of M^k are relevant.

We can equivalently break this definition into the following two parts.

Maximal Emulation Definition/2. ME/DEF/2. S is an emulator of $E \subseteq M^k$ if and only if $S \subseteq M^k$ and every element of S^2 is M equivalent to an element of E^2. S is a maximal emulator of $E \subseteq M^k$ if and only if S is an emulator of $E \subseteq M^k$ which is not a proper subset of an emulator of $E \subseteq M^k$.

Maximal Emulation/1. ME/1. Every $E \subseteq M^k$ has a maximal emulator.

There is an important sharper form.

Maximal Emulation/2. ME/2. Every $E \subseteq M^k$ has a maximal emulator containing any given emulator.

This general situation is nicely clarified as follows.

Theorem 2.1 *(Z) The following are equivalent.*

i. The axiom of choice.
ii. ME/2.
iii. ME/1.
iv. Every finite $E \subseteq M^2$ has a maximal emulator, where M is an equivalence relation.

Proof For i → ii, let $E \subseteq M^k$ and S be an emulator of $E \subseteq M^k$. Let $<_D$ be a well ordering of dom(M) = D. Perform the usual greedy transfinite algorithm where S′ is unique such that $S' = \{x: (S' \cap \{y: y <_D x\}) \cup S \cup \{x\}$ is an emulator of $E \subseteq M^k\}$. ii → iii → iv is immediate. For iv → i, we use the axiom of choice in the form that for every equivalence relation (D, R), there is a set containing exactly one element from each equivalence class. Let (D, R) be any equivalence relation. We can assume that D has at least two elements with are not R related. Let E = {x, y}, where x, y are not R related. The emulators of $E \subseteq (D, R)^2$ are exactly the subsets of D that contain at most one element from each equivalence class under R. Clearly any maximal emulator of $E \subseteq (D, R)^2$ contains exactly one element from each equivalence class of R. □

Finite Subset Emulation. Assuming M has finitely many components, every $E \subseteq M^k$ is an emulator of some finite $E' \subseteq M^k$ with $E' \subseteq E$. This is provable in RCA_0 for countable M with finitely many components.

Proof For the first claim, note that there are finitely many equivalence classes in any given dimension under M equivalence. Pick a representative from each M equivalence class of elements of E^2, and take E′ to be the set of all elements of E that are used. Working within RCA_0, we first construct the set V of elements of E^2 that are not M equivalent to any lessor element of E^2 (where lessor refers to an ad hoc enumeration of E^2 based on the enumeration of D). Then using finite Σ^0_1 separation, available in RCA_0, we form the set of all elements of E that are used. □

Emulation Transitivity. If S is an emulator of $E \subseteq M^k$ and E is an emulator of $E' \subseteq M^k$, then S is an emulator of $E' \subseteq M^k$. Let E be an emulator of $E' \subseteq M^k$ and E′ be an emulator of $E \subseteq M^k$. The emulators of $E \subseteq M^k$ are the same as the emulators of $E' \subseteq M^k$. The maximal emulators of $E \subseteq M^k$ are the same as the maximal emulators of $E' \subseteq M^k$. This is provable in RCA_0 for countable M.

Proof The first claim follows immediately from M equivalence being an equivalence relation. Now let E, E′ be mutual emulators in M^k. The second claim follows from the first claim.

For the third claim, first suppose S is a maximal emulator of $E \subseteq M^k$. Then S is an emulator of $E' \subseteq M^k$. Let $S' \supseteq S$ be an emulator of $E' \subseteq M^k$. Then S′ is an emulator of $E \subseteq M^k$, and so $S' = S$. Suppose S is a maximal emulator of $E' \subseteq M^k$. By the analogous argument, S is a maximal emulator of $E \subseteq M^k$. □

Maximal Emulation/3. ME/3. (RCA_0) Let M be countable with finitely many components. Every subset of M^k has a maximal emulator. The following are equivalent.

i. ACA_0.
ii. If M is countable then every subset of M^k has a maximal emulator containing any given emulator.
iii. In every equivalence relation M on N, every finite subset of N has a maximal emulator containing any given emulator.

Proof For the first claim, let $E \subseteq M^k$. By Finite Subset Emulation and Emulation Transitivity (second claim), we can assume that E is finite. Then use the usual greedy algorithm. For i → ii, use the same argument, starting with the given emulator, which creates the need for ACA_0 instead of just RCA_0 (because the given emulator may be infinite). ii → iii is obvious. Assume iii, and we derive ACA_0. It suffices to show that the range of every one-one $f{:}2N \to 2N+1$ exists. Let R be the equivalence relation on N given by R(n, m) if and only if $f(n) = m \vee f(m) = n \vee n = m$.

For iii → i, we use the structure $M = (N, R)$. The emulators of {0, 2} are the subsets of N that contain at most one element from each equivalence class of R. We use the emulator 2N of $\{0, 2\} \subseteq N$. Thus iii implies that there is a set S containing exactly one element from each equivalence class under R and also contains 2N. Thus S consists of 2N together with the odd numbers that are outside the range of f. From S, we obtain the range of f. □

We now present the most general formulation of Basic Emulation Theory as a Template.

Definition 2.3 Let $R \subseteq M^k \times M^k$. The R image of S, or (forward) image of S under R, is $R[S] = \{y: (\exists x \in S)(R(x, y))\}$.

Maximal Emulation Use Definition. MEU/DEF. $R \subseteq M^k \times M^k$ is ME usable if and only if for all subsets of M^k, some maximal emulator contains its R image.

In the countable case, actually forming images (as sets) requires ACA_0, and is not available in RCA_0. However, the notion "S contains its R image" is viewed as not presupposing that we actually form the image (as a set). So RCA_0 can be used to investigate ME usability in the countable context.

Theorem 2.2 *Let M have finitely many components.* $R \subseteq M^k \times M^k$ *is ME usable if and only if for all finite subsets of* M^k*, some maximal emulator contains its R image.* RCA_0 *proves this for countable M with finitely many components.*

Proof Let M be as given, and suppose that for all finite $E \subseteq M^k$, some maximal emulator of $E \subseteq M^k$ contains its R image. Let $E \subseteq M^k$. By Finite Subset Emulation, let E be an emulator of finite $E' \subseteq M^k$, where $E' \subseteq E$. By Emulation Transitivity, the maximal emulators of $E \subseteq M^k$ are the same as the maximal emulators of $E' \subseteq M^k$. Hence some maximal emulator of $E \subseteq M^k$ contains its R image. For the second claim, RCA_0 is enough because it was enough for Finite Subset Emulation and Emulation Transitivity. □

When verifying ME usability, we will generally use finite $E \subseteq M^k$ in accordance with Theorem 2.2, in order to emphasize the concrete aspects of Emulation Theory.

We now give a clarifying necessary condition for ME usability.

Maximal Emulation/4. ME/4. If $R \subseteq M^k \times M^k$ is ME usable then R is M preserving in the sense that $(\forall x, y)(R(x, y) \rightarrow x, y$ are M equivalent). RCA_0 proves this for countable M.

Proof Let $R \subseteq M^k \times M^k$ be ME usable. We claim that every M equivalence class $[x] \subseteq M^k$ is the unique maximal emulator of $[x] \subseteq M^k$. To see this, let S be an emulator of [x]. For $y \in S$, (y, y) is M equivalent to some $(x', x') \in [x]^2$. Hence for $y \in S$, x, y are M equivalent. So $S \subseteq [x]$. Since [x] is an emulator of [x], clearly the unique maximal emulator of $[x] \subseteq M^k$ is [x].

Now let R(x, y). [x] has a maximal emulator S containing its R image. Hence [x] contains its R image. Hence $y \in [x]$, and so x, y are M equivalent. □

Definition 2.4 Let $R \subseteq M^k \times M^k$. R is symmetric if and only if $(\forall x, y)(R(x, y) \leftrightarrow R(y, x))$. $R^{-1} = \{(y, x): R(x, y)\}$. The inverse image of S under R is $R^{-1}[S] = \{x: (\exists y \in S)(R(x, y))\}$. S is R invariant if and only if $(\forall x, y)(R(x, y) \rightarrow (x \in S \leftrightarrow y \in S))$. S is equivalent at x, y if and only if $x \in S \leftrightarrow y \in S$.

Theorem 2.3 *Let* $R \subseteq M^k \times M^k$*. S is R invariant if and only if S contains its image and inverse image under R. If R is symmetric, then S is R invariant if and only if S contains its R image.* RCA_0 *proves this for countable M.*

Proof Suppose $(\forall x, y)(R(x, y) \rightarrow (x \in S \leftrightarrow y \in S))$. If y is in the forward image of S under R then let R(x, y), $x \in S$. Then $y \in S$. If x is in the inverse image of S under R then let R(x, y), $y \in S$. Then $x \in S$. Conversely, suppose $S \supseteq R[S] \cup R^{-1}[S]$, and let R(x, y). If $x \in S$ then $y \in R[S]$ and so $y \in S$. If $y \in S$ then $x \in R^{-1}[S]$, and so $x \in S$. The rest is left to the reader. □

Maximal Emulation Invariant Use Definition. MEIU/DEF. $R \subseteq M^k \times M^k$ is ME invariantly usable if and only if for all subsets of M^k, some maximal emulator is R invariant.

Theorem 2.4 *$R \subseteq M^k \times M^k$ is ME invariantly usable if and only if $R \cup R^{-1}$ is ME usable. If $R \subseteq M^k \times M^k$ is symmetric then R is ME invariantly usable if and only if R is ME usable. RCA_0 proves this for countable M.*

Proof Left to the reader. □

Theorem 2.5 *(RCA_0) Let M be countable with finitely many components, and x, y $\in M^k$. The following are equivalent.*

i. *For finite subsets of M^k, some maximal emulator is equivalent at x, y $\in M^k$.*
ii. *{(x, y)} is ME invariantly usable.*
iii. *{(x, y), (y, x)} is ME usable.*

Proof Left to the reader. □

We shall see in Sect. 12.3.2 that the particularly simple form i, with its point equivalence, is already delicate.

The following provides basic tools for establishing results concerning ME usability and ME invariant usability.

Theorem 2.6 *The following hold.*

i. *Every subset of a (invariantly) ME usable $R \subseteq M^k \times M^k$ is (invariantly) ME usable.*
ii. *Let $f{:}M \rightarrow M'$ be an isomorphism from M onto M', E, $S \subseteq M^k$, and $R \subseteq M^k \times M^k$. S is an (maximal) emulator of $E \subseteq M^k$ if and only if f[S] is an (maximal) emulator of $f[E] \subseteq M'^k$. $R \subseteq M^k \times M^k$ is ME (invariantly) usable if and only if $f[R] \subseteq M'^k \times M'^k$ is (invariantly) usable.*
iii. *Let $g{:}M \rightarrow M$ be an automorphism of $EQR(M, k) \subseteq M^{2k}$. S is an (maximal) emulator of $E \subseteq M^k$ if and only if g[S] is an (maximal) emulator of $g[E] \subseteq M^k$. $R \subseteq M^k \times M^k$ is ME (invariantly) usable if and only if $g[R] \subseteq M^k \times M^k$ is ME (invariantly) usable.*

Here f, g always act coordinatewise.

Proof i, ii are left to the reader. For iii, let M, g, E, S, R be as given. Suppose S is an emulator of E. Then every (x, y) $\in S^2$ is M equivalent to some (z, w) $\in E^2$. Let (x, y) $\in f[S]^2$. Write (x, y) = (g(x′), g(y′)), x′, y′ $\in$ S. Let (z, w) $\in E^2$ be such that (x′, y′), (z, w) are M equivalent. Then (g(x′), g(y′)), (g(z), g(w)) are M equivalent. Hence (x, y), (g(z), g(w)) are M equivalent, and (g(z), g(w)) $\in g[E]^2$. Hence g[S] is an emulator of $g[E] \subseteq M^k$. The previous argument holds with the automorphism g^{-1}, and so g^{-1}[g[S)] is an emulator of $g^{-1}[g[E]] \subseteq M^k$. Hence S is an emulator of $E \subseteq M^k$.

Now suppose S is a maximal emulator of $E \subseteq M^k$. Let S′ be an emulator of $g[E] \subseteq M^k$ containing g[S]. Then $g^{-1}[S']$ is an emulator of $g^{-1}[g[E]] \subseteq M^k$ containing S, and so $g^{-1}[S'] \supseteq S$ is an emulator of $E \subseteq M^k$. Hence $g^{-1}[S'] = S$, and so S′ = g[S]. Hence S′ is a maximal emulator of $g[E] \subseteq M^k$. Finally suppose g[S] is a maximal emulator of $g[E] \subseteq M^k$. The previous argument holds for the automorphism g^{-1} of

EQR(M, k), and so $g^{-1}[g[S]]$ is a maximal emulator of $g^{-1}[g[E]] \subseteq M^k$. Hence S is a maximal emulator of $E \subseteq M^k$.

Next suppose $R \subseteq M^k \times M^k$ is ME usable. We now show that $g[R] \subseteq M^k \times M^k$ is ME usable. Let S be a maximal emulator of $g^{-1}[E] \subseteq M^k$, where S contains its R image. By the previous claim, g[S] is a maximal emulator of $E \subseteq M^k$. We claim that g[S] contains its g[R] image. To see this, let g[R](x, y), where x ∈ g[S]. Then $R(g^{-1}(x), g^{-1}(y))$, where $g^{-1}(x) \in S$, and since S contains its R image, we have $g^{-1}(y) \in S$, y ∈ g[S]. Now suppose $g[R] \subseteq M^k \times M^k$ is ME usable. The previous argument holds for the automorphism g^{-1} of EQR(M, k), and so $g^{-1}[g[R]] = R$ is ME usable. The claim for ME invariantly usable is verified by applying the previous claims to symmetric R. □

This completes our brief development of general Basic Emulation Theory.

12.3 Maximal Emulation on Q[0, 1]

We now develop Basic Emulation Theory in our special context Q[0, 1] = (Q[0, 1],<), where $Q[0, 1] = Q \cap [0, 1]$. We view the MESU/1 and MED/1 of Sect. 12.3.4 as the most immediately transparent statements independent of ZFC in this paper.

In Sect. 12.3.1, we present some background material concerning Maximal Emulation on Q[0, 1]. We present eight illustrative examples of $E \subseteq Q[0, 1]^2$ and their emulators and maximal emulators in order to help orient the reader.

In Sect. 12.3.2, we work with finite relations $R \subseteq Q[0, 1]^k \times Q[0, 1]^k$, with Theorems MEFU/1, 2, 3, MEOU/1, 2. MEFU/1 and MEOU/1, 2 are proved in RCA_0, whereas MEFU/2, 3 are proved in ACA′. We conjecture that MEFU/2, 3 cannot be proved in ACA_0.

In Sects. 12.3.3 and 12.3.4, we work with infinite relations $R \subseteq Q[0, 1]^k \times Q[0, 1]^k$, with a focus on the order theoretic $R \subseteq Q[0, 1]^k \times Q[0, 1]^k$. Here we view order theoretic R as "large" or "small" according to whether infinitely many or finitely many numbers are altered. In Sect. 12.3.3, we prove MELU/1, 2, 3, 4 in RCA_0.

In Sect. 12.3.4, we present MESU/1, 2, 3, and MED/1, 2, 3. MESU/1 and MED/1 are the most immediately transparent statements. All six are derived from MESU/2 and shown to imply MED/1, all within RCA_0. MESU/2 for dimension k = 2 is proved using a transfinite construction of uncountable length. Full MESU/2 is then proved in SRP^+. The proof is then modified to take place in WKL_0 + Con(SRP). In a reversal that will appear elsewhere in Concrete Mathematical Incompleteness (in preparation), we prove Con(SRP) in RCA_0 + MED/1. This establishes that all six statements MESU/1, 2, 3, MED/1, 2, 3 are provably equivalent to Con(SRP) over WKL_0. This also establishes that all six statements are independent of ZFC, assuming SRP is consistent.

We conjecture that MESU/2 for dimension 2 can be proved in RCA_0. We also conjecture that such a proof will be much more complicated than the proof given here using ω_1. However, see below.

In Sect. 12.3.6, we present a very natural extension of the notion of emulator to r-emulator. An emulator is just a 2-emulator. S is an emulator of E if and only if every element of S^2 is order equivalent to an element of E^2. S is an r-emulator of E if and only if every element of S^r is order equivalent to an element of E^r. Thus an emulator is just a 2-emulator. We revisit all statements quantifying over all r, and also for fixed r. Most of the previous results hold as long as r is not fixed to 1. In particular, MESU/1, 2, 3, MED/1, 2, 3 remain provably equivalent to Con(SRP) over WKL_0 when extended to r-emulators, r universally quantified.

We conjecture that MESU/1, 2, 3, MED/1, 2, 3 in dimension $k=2$, using r-emulators, quantifying over r, is not provable in ZFC\P or, equivalently in Z_2. We also conjecture that MESU/1, 2, 3, MED/1, 2, 3 in any fixed dimension $k \geq 3$, using r-emulators, quantifying over r, are all equivalent to Con(ZFC + "there exists a k-subtle cardinal") over WKL_0. In particular, we conjecture that MESU/1, 2, 3, MED/1, 2, 3 in dimension $k=3$, using r-emulators, quantifying over r, are all independent of ZFC, assuming ZFC + "there exists a subtle cardinal" is consistent.

From Theorem 3.1.6 and Corollary 3.1.7, it is clear that MEX/2, MEFU/2, 3, MELU/3, MELU/5, MESU/1, 2, 3, MED/1, 2, 3 are all implicitly Π^0_1 over WKL_0, and WKL_0 falsifiable via Gödel's Completeness Theorem (MESU/3 stated for order theoretic R). This is also clear if we use r-emulators, either quantifying over all $r \geq 1$, or fixing $r \geq 1$.

12.3.1 ME Usability

Definition 3.1.1 Q, Z, Z^+, N, $\Re$ are the sets of all rational numbers, integers, positive integers, nonnegative integers, and real numbers, respectively. We use variables p, q, with or without subscripts, over rationals, unless indicated otherwise. We use variables i, j, k, n, m, r, s, t, with or without subscripts, over positive integers, unless indicated otherwise. We use inequality chaining in the sense that, e.g., $p\ \alpha\ q\ \beta\ b \leftrightarrow p\ \alpha\ q \wedge q\ \beta\ b$, where $\alpha, \beta \in \{=, \neq, <, >, \leq, \geq\}$. $Q[(p, q)] = Q \cap [(p, q)]$, covering all four endpoint possibilities. Let $x \in Q^k$ and $y \in Q^n$. $(x, y) \in Q^{k+n}$ is the concatenation of x and y. min(x), max(x) are the least and greatest coordinates of x, respectively. For $1 \leq i \leq k$, x_i is the i-th coordinate of x. Thus x_i exists if and only if x is a tuple (finite sequence) of length $\geq i$. Let $S \subseteq Q^k$. $S|<p$, $S|\leq p$, $S|>p$, $S|\geq p$ is $S \cap (-\infty, p)^k$, $S \cap (-\infty, p]^k$, $S \cap (p, \infty)^k$, $S \subseteq [p,\infty)$, respectively. $Q^{k<} = \{x \in Q^k: x_1 < \cdots < x_k\}$, $Q^{k>} = \{x \in Q^k: x_1 > \cdots > x_k\}$.

We use the crucial order equivalence relation on $Q[0, 1]^k$. For example, (0.5, 0.7, 0.6) and (0.2, 1, 0.9) are order equivalent. This is the same as (Q,<) equivalence in the sense of M equivalence in Definition 2.2. By default, we have the order equivalence relation on $Q[0, 1]^k$, which is also the same as (Q[0, 1],<) equivalence. However, order equivalence plays such a fundamental role that we give the following equivalent definition.

Order Equivalence Definition $x, y \in Q[0, 1]^k$ are order equivalent if and only if for all $1 \leq i, j \leq k$, $x_i < x_j \leftrightarrow y_i < y_j$.

Definition 3.1.2 S is a maximal emulator of $E \subseteq Q[0, 1]^k$ if and only if $S \subseteq Q[0, 1]^k$ $\wedge$ every element of S^2 is order equivalent to an element of E^2 $\wedge$ this conjunction is false with S replaced by any proper superset of S.

Note that here E^2, S^2 are sets of 2k-tuples. We can equivalently break Definition 3.1.2 into two parts.

Definition 3.1.3 S is an emulator of $E \subseteq Q[0, 1]^k$ if and only if $S \subseteq Q[0, 1]^k$ and every element of S^2 is order equivalent to an element of E^2. S is a maximal emulator of $E \subseteq Q[0, 1]^k$ if and only if S is an emulator of $E \subseteq Q[0, 1]^k$ which is not a proper subset of an emulator of $E \subseteq Q[0, 1]^k$.

Note that Definitions 3.1.2 and 3.1.3 are special cases of ME/DEF/1, 2 of Sect. 12.2, as here we are using $M = (Q[0, 1], <)$.

We now give some illustrative examples of emulators and maximal emulators. Obviously, $\emptyset$ is vacuously an emulator of any $E \subseteq Q[0, 1]^k$, and is a maximal emulator of $\emptyset \subseteq Q[0, 1]^k$. EX1 is in dimension $k = 1$ and EX2-8 are in dimension $k = 2$.

EX1. $E \subseteq Q[0, 1]$. If $E = \emptyset$ then $\emptyset$ is the only emulator and it is the maximal emulator. If $|E| = 1$ then the emulators are subsets of Q[0, 1] of cardinality at most 1, and the ones of cardinality 1 are the maximal emulators. If $|E| \geq 2$ then the emulators are all of the subsets of Q[0, 1], and Q[0, 1] is the unique maximal emulator.

EX2. $E = \{(0, 0)\} \subseteq Q[0, 1]^2$. The emulators are $\emptyset$ and singletons $\{(p, p)\}$, $0 \leq p \leq 1$. The maximal emulators are these singletons.

EX3. $E = \{(0, 1)\} \subseteq Q[0, 1]^2$. The emulators are $\emptyset$ and singletons $\{(p, q)\}$, $0 \leq p < q \leq 1$. The maximal emulators are these singletons.

EX4. $E = \{(0, 0), (1, 1)\} \subseteq Q[0, 1]^2$. The emulators are the subsets of $\{(p, p): 0 \leq p \leq 1\}$. Exactly one is maximal, $\{(p, p): 0 \leq p \leq 1\}$.

EX5. $E = \{(0, 0), (0, 1)\} \subseteq Q[0, 1]^2$. The emulators are the sets that are contained in some $\{p\} \times Q[p, 1]$, $0 \leq p < 1$. The maximal emulators are the sets $\{p\} \times Q[p, 1]$, $0 \leq p \leq 1$.

EX6. $E = \{(0, 2/5), (1/5, 3/5), (2/5, 4/5), (3/5, 1)\}$. The emulators are the graphs of strictly increasing partial $f:Q[0, 1) \rightarrow Q(0, 1]$, where each defined $f(x) > x$. There are continuumly many maximal emulators of E.

EX7. $E = \{(p, q) \in Q[0, 1]^2: p < 1/2 < q\} \subseteq Q[0, 1]^2$. The emulators and maximal emulators are calculated in a self contained way in the proof of Lemma 3.2.2.

EX8. $E = \{(1/6, 1/4), (1/7, 1/3), (0, 1/5), (1/2, 1)\} \subseteq Q[0, 1]^2$. The emulators and maximal emulators are calculated in a self contained way in the proof of MELU/2 in Sect. 12.3.3.

Definition 3.1.4 Since we are using Q[0, 1] throughout Sect. 12.3, we make the following convention. If the dimension k of E or S has been given, then we can write E or S instead of $E \subseteq Q[0, 1]^k$ or $S \subseteq Q[0, 1]^k$. If the dimension k of R has been given, then we can write R instead of $R \subseteq Q[0, 1]^k \times Q[0, 1]^k$. These conventions also apply where E, S, R appear with superscripts or subscripts. Exceptions to this convention occur inside proofs where we use linear orderings other than Q[0, 1].

Here is some background information on the crucial order equivalence relation on the $Q[0, 1]^k$.

Theorem 3.1.1 *(RCA_0) There are finitely many equivalence classes under order equivalence on $Q[0, 1]^k$. The number, ot(k) is the same as the number of preferential arrangements, up to isomorphism, in the sense of (Gross 1962). ot(1) = 1, ot(2) = 3, ot(3) = 13, ot(4) = 75, ot(5) = 541, ot(6) = 4,683, ot(7) = 47,293, ot(8) = 545,835, ot(9) = 7,087,261, ot(10) = 102,247,563, ot(11) = 1,622,632,573, ot(12) = 28,091,567,595, ot(13) = 526,858,348,381, ot(14) = 10,641,342,970,443. (Q[0, 1], <) can be replaced by any dense linear ordering.*

Proof Here ot is read "order type". Note that $x, y \in Q[0, 1]^k$ are order equivalent if and only if $\{(i, j): x_i < x_j\} = \{(i, j): y_i < y_j\}$, and the number of sets of this form is obviously finite, trivially bounded by $2^{k\hat{}2}$. A preferential arrangement is commonly defined to be a connected and transitive relation R on a set V. By connectivity, R must be reflexive on V. The derived relation $x \sim y \leftrightarrow x\ R\ y \wedge y\ R\ x$ is an equivalence relation on V. Given $[x] \neq [y]$, we have $x\ R\ y \vee y\ R\ x$. Hence we have trichotomy for all [x], [y], namely exactly one of x R y, y R x, x = y holds. Thus R can be viewed as a reflexive linear ordering on the equivalence classes of ~.

With an underlying domain of k elements, the isomorphism types of these R's are in one-one correspondence with k-tuples from Q[0, 1] (or any given dense linear ordering) under order equivalence. To see this, let R be connected and transitive on {1, …, k}. By the previous paragraph, list the equivalence classes under ~ by $A_1, \ldots, A_n$, in increasing R order. The corresponding k-tuple is $t_1, \ldots, t_k$, where each t_i is the index of [i] in the list $A_1, \ldots, A_n$.

The displayed 14 values of ot are from Gross (1962). □

For additional work on ot(k), see Sloane (1964).

Theorem 3.1.2 *(RCA_0) Every $E \subseteq Q[0, 1]^k$ is an emulator of a finite subset. E has a recursive maximal emulator.*

Proof See Finite Subset Emulation and Theorem 2.2 and the first claim of ME/3, all in Sect. 12.2. □

It is clear that we obtain a maximal emulator in Theorem 3.1.2 of low computational complexity. It would be interesting to carefully investigate these from a computational complexity perspective.

Definition 3.1.5 $R \subseteq Q[0, 1]^k \times Q[0, 1]^k$ is ME usable if and only if for finite subsets of $Q[0, 1]^k$, some maximal emulator contains its R image. $R \subseteq Q[0, 1]^k \times Q[0, 1]^k$ is ME invariantly usable if and only if for finite subsets of $Q[0, 1]^k$, some maximal emulator is R invariant.

Note that Definition 3.1.5 is the special case of MEU/DEF and MEIU/DEF of Sect. 12.2, for M = (Q[0, 1], <). Here we use finite subsets as is justified by Theorem 2.5.

Throughout the rest of this paper, ME usable and ME invariantly usable will refer only to our Q[0, 1] context (i.e., M = (Q[0, 1], <)). Thus for $X \subseteq Q[0, 1]^k \times Q[0, 1]^k$, we write "$R \subseteq X$ is ME (invariantly) usable" to mean "$R \subseteq Q[0, 1]^k \times Q[0, 1]^k$ is ME (invariantly) usable and $R \subseteq X$". Thus this use of X in no way invalidates the intention that we are always working in the space Q[0, 1] = (Q[0, 1], <).

We will show that for the important class of order theoretic $R \subseteq Q[0, 1]^k \times Q[0, 1]^k$, "R is ME usable" is implicitly Π^0_1 over WKL_0 and WKL_0 falsifiable as defined below. But first we discuss the very robust notion of order theoretic, which is a special case of the notion of elementary subset that we introduced right after the discussion of ME/4 in Sect. 12.1.1.

Order Theoretic Definition $S \subseteq Q[0, 1]^k$ is order theoretic if and only if S is elementary in (Q[0, 1], <). I.e., S is of the form $\{x \in Q[0, 1]^k: \varphi\}$, where φ is a finite propositional combination of formulas $x_i < x_j$, $x_i < p$, $p < x_i$, with $1 \leq i, j \leq k$ and $p \in Q[0, 1]$.

Theorem 3.1.3 *$S \subseteq Q[0, 1]^k$ is order theoretic if and only if S is first order definable in (Q[0, 1], <) with parameters.*

Proof By the usual quantifier elimination for dense linear orderings. □

Definition 3.1.6 Let T be a first order theory whose language includes 0, S, +, x, < (on a sort for the nonnegative integers), which is recursively axiomatized and proves PFA. φ is implicitly Π^0_1 over T if and only if there is a Π^0_1 sentence ψ such that T proves $\varphi \leftrightarrow \psi$. φ is T falsifiable if and only if T proves "$\neg\varphi \rightarrow$ T proves $\neg\varphi$".

Note that the two outermost T proves, involves only the ordinary mention of prove. The single innermost T proves φ refers to a formalization of proofs within T.

Theorem 3.1.4 *(RCA_0) Let T be a recursively axiomatized theory that proves PFA. Every sentence implicitly Π^0_1 over T is T falsifiable. Furthermore assume T is finitely axiomatized. Every T falsifiable sentence is implicitly Π^0_1 over T augmented with induction for all formulas. In fact, induction only for formulas of quantifier complexity at most the maximum of T, φ is needed.*

Proof Let T be as given. Let φ be implicitly Π^0_1 over T. Let ψ be Π^0_1 where T proves $\varphi \leftrightarrow \psi$. Arguing in T, assume ¬φ. Then ¬ψ, and so T proves ¬ψ. Now T sees that T proves $\varphi \leftrightarrow \psi$. Therefore T sees that T proves ¬φ.

Now suppose φ is T falsifiable. We claim $\varphi \leftrightarrow \text{Con}(T + \varphi)$ is provable in T with induction. To see this, argue in T with induction. We see if φ is false then φ is refutable in T. Hence $\neg\varphi \rightarrow \neg\text{Con}(T + \varphi)$. Now suppose $\neg\text{Con}(T + \varphi)$. By cut elimination, we get a refutation of φ in T with a proof of quantifier complexity at most that of the maximum of the quantifier complexities of T, φ. We then perform an induction in T to derive ¬φ. □

According to Theorem 3.1.4, implicitly Π^0_1 and falsifiable are essentially the same notions.

The featured statements in Sect. 12.3 are implicitly Π^0_1 over WKL_0 via Gödel's Completeness Theorem, and therefore WKL_0 falsifiable. This can be easily seen through the following general result.

Lemma 3.1.5 *(EFA) Fix k ≥ 1, finite E ⊆ Q[0, 1]^k, and (an order theoretic presentation of) order theoretic R ⊆ Q[0, 1]^k × Q[0, 1]^k. The statement φ = "some maximal emulator S of E has R[S] ⊆ S" is implicitly Π^0_1 over WKL$_0$. Furthermore, the associated Π^0_1 forms and the equivalence proofs in WKL$_0$ can be constructed effectively from k, E and the order theoretic presentation of R, in a way that RCA$_0$ can verify.*

Proof Fix k, E, R as given. The associated Π^0_1 sentence will be Con(T[k, E, R]), where T(k, E, R) is a finitely axiomatized system associated with k, E and the given order theoretic presentation of R.

The parameters used to present R can be taken to be exactly the t + 2 rationals $0 < p_1 < \cdots < p_t < 1$. For any finite E ⊆ Q[0, 1]^k, we form the theory T[k, E, R] in first order predicate calculus with equality, the binary relation symbol <, constants 0, 1, p_1, …, p_t, and k-ary relation symbol S. The finitely many axioms of T[k, E, R] are

i. < is a strict dense linear ordering with left endpoint 0 and right endpoint 1.
ii. $0 < p_1 < \cdots < p_t < 1$.
iii. If x ∈ S and R(x, y), then y ∈ S.
iv. Any k-tuple x lies in S if and only if every element of $(S \cup \{x\})^2$ is order equivalent to an element of E^2.

For iii, we use the definition of R in <, =, 0, 1, p_1, …, p_t. For iv, we don't use actual elements of E, but simply use the enumerated order types of elements of E^2.

Arguing in RCA$_0$, suppose there is a model of T[k, E, R] with domain a subset of N. There is no problem formulating "there is a model of T[k, E, R]" as T has only finitely many axioms. Then there is a model of T[k, E, R] of the form (Q[0, 1], <, …), by isomorphism. Also by isomorphism, we can arrange that the constants p_1, …, p_t are actually p_1, …, p_t. So we have the model of T[k, E, R], (Q[0, 1], <, 0, 1, p_1, …, p_t, S). It is now clear that by iv, S is an emulator of E ⊆ Q[0, 1]^k. By iv, S is a maximal emulator of E ⊆ Q[0, 1]^k. By iii we see that R[S] ⊆ S, where this inclusion is formulated without actually forming R[S] as a set. Hence φ holds.

Arguing in RCA$_0$, suppose φ holds, and let S be a maximal emulator of E ⊆ Q[0, 1]^k, where R[S] ⊆ S. Then clearly (Q[0, 1], <, 0, 1, p_1, …, p_t, S) is a model of T[k, E, R], which we can make official via an isomorphism onto N. We want to derive Con(T[k, E, R]). If we do this in the most obvious way, we are going to be using induction with respect to a formula that involves the satisfaction predicate for the model of T, and that isn't even arithmetic. However, we can apply cut elimination for predicate calculus, which is available in RCA$_0$. Noting that the axioms of T are (universally quantified) Σ^0_1 formulas, we will only need the satisfaction relation for Σ^0_1 formulas and hence Σ^0_1 induction, which is available in RCA$_0$.

Now simply cite the formalized completeness theorem in WKL$_0$, which tells us that WKL$_0$ proves "if Con(T[k, E, R]) then T = T[k, E, R] has a model with domain a subset of ω". Thus we have a proof in WKL$_0$ of φ ↔ Con(T[k, E, R]). Thus we are using Con(T[k, E, R]) as our Π^0_1 sentence. The construction of Con(T[k, E, R]) and the equivalence proof is obviously effective from k, E and (the given order theoretic presentation of) R. □

Theorem 3.1.6 *(EFA) Consider the statement* $\varphi(k, E, R)$ = *"For finite* $E \subseteq Q[0, 1]^k$*, some maximal emulator S of* $E \subseteq Q[0, 1]^k$ *has* $R[S] \subseteq S$*".*

i. *If k, E, R are fixed in advance, where* $R \subseteq Q[0, 1]^k \times Q[0, 1]^k$ *is order theoretic, then* $\varphi(k, E, R)$ *is implicitly* Π^0_1 *over* WKL_0*.*
ii. *If k, R are fixed in advance, where* $R \subseteq Q[0, 1]^k \times Q[0, 1]^k$ *is order theoretic, then* $(\forall E \subseteq Q[0, 1]^k)(\varphi(k, E, R))$ *is implicitly* Π^0_1 *over* WKL_0*.*
iii. *In the equivalence proofs for implicitly* Π^0_1 *in i, ii, the forward direction can be taken to be in* RCA_0 *and the backward direction can be taken to be in* WKL_0 *(where the tree axiom has no set parameters).*
iv. *In the falsifiability proofs, the first T can be taken to be* WKL_0 *(where the tree axiom has no set parameters) and the second T can be taken to be* RCA_0*.*

Furthermore, the associated Π^0_1 forms and equivalence proofs can be constructed effectively from the fixed parameters, in such in a way that EFA can verify.

Proof Apply Lemma 3.1.5 and its proof. This establishes i–iii. For iv, using WKL_0 for the first T, argue in WKL_0. Suppose $\varphi(k, E, R)$ is false. Then Con(T[k, E, R]) is false. So Con(T[k, E, R]) is refutable in RCA_0 (even in EFA). Hence $\varphi(k, E, R)$ is refutable in RCA_0, since RCA_0 proves $\varphi(k, E, R) \rightarrow$ Con(T[k, E, R]). □

Corollary 3.1.7 *(EFA) Consider the statement* $\varphi(k, R)$ = *"*$R \subseteq Q[0, 1]^k \times Q[0, 1]^k$ *is ME usable".*
i. If k, R are fixed in advance, where $R \subseteq Q[0, 1]^k \times Q[0, 1]^k$ *is order theoretic, then* $\varphi(k, R)$ *is implicitly* Π^0_1 *over* WKL_0*.*
iii, iv and the last sentence of Theorem 3.1.6 also apply here.

Proof The statement in quotes is the same as "For all finite $E \subseteq Q[0, 1]^k$, some maximal emulator S of $E \subseteq Q[0, 1]^k$ has $R[S] \subseteq S$" and so we can apply Theorem 3.1.6. □

12.3.2 Finite Relations

We now address the ME usability of finite $R \subseteq Q[0, 1]^k \times Q[0, 1]^k$. The trivial case k = 1 is dispensed with later by MELU/1 in Sect. 12.3.3, even for arbitrary R.

Order Preserving Definition. $R \subseteq Q[0, 1]^k \times Q[0, 1]^k$ is order preserving if and only if $(\forall x, y)(R(x, y) \rightarrow$ x, y are order equivalent).

Maximal Emulation Necessary Use. MENU. If $R \subseteq Q[0, 1]^k \times Q[0, 1]^k$ is ME usable then R is order preserving.

Proof of MENU In RCA_0. By ME/4 in Sect. 12.2. □

In light of MENU, we need be concerned only with order preserving $R \subseteq Q[0, 1]^k \times Q[0, 1]^k$ throughout the rest of Sect. 12.3.

Are we already done in the sense that every finite order preserving $R \subseteq Q[0, 1]^k \times Q[0, 1]^k$ is ME usable? We shall see that this is close to being true, but the endpoints 0, 1 create issues. See MEFU/3 below.

Let's first look at two very simple examples of ME usability. These examples are so simple that we conveniently state them directly without using the ME usability terminology.

Maximal Emulation Example/1. MEX/1. For finite subsets of $Q[0, 1]^2$, some maximal emulator is equivalent at (1/2, 1/3), (1/3, 1/4).

Maximal Emulation Example/2. MEX/2. For finite subsets of $Q[0, 1]^2$, some maximal emulator is equivalent at (1, 1/2), (1/2, 1/3).

By Theorem 2.5, MEX/1, 2 assert, respectively, that the two element relations {((1/2, 1/3), (1/3, 1/4)), ((1/3, 1/4), (1/4, 1/3))} and {((1, 1/2), (1/2, 1/3)), ((1/2, 1/3), (1, 1/2))} are ME usable, and equivalently, that the one element relations {((1/2, 1/3), (1/3, 1/4))} and {((1/3, 1/2), (1/2, 1))} are ME invariantly usable. The first pair of relations are of cardinality 2, and the second pair of relations are of cardinality 1.

Proof of MEX/1 In RCA_0. Let $E \subseteq Q[0, 1]^2$ be finite, and let S be a maximal emulator of E. We now find $1 > p > q > r > 0$ such that $(p, q) \in S \leftrightarrow (q, r) \in S$.

case 1. $(1/2, 1/3) \in S$. We can assume $(1/3, 1/4)(1/3, 1/5) \notin S$. We can assume $(1/4, 1/5), (1/5, 1/6) \in S$. Then we are done.

case 2. $(1/2, 1/3) \notin S$. We can assume $(1/3, 1/4), (1/3, 1/5) \in S$. We can assume $(1/4, 1/5), (1/5, 1/6) \notin S$. Then we are done.

Now we map S onto S′ via any increasing bijection of Q[0, 1] onto Q[0, 1] mapping p, q, r to 1/2, 1/3, 1/4. This preserves being a maximal emulator of E. □

Note that this argument will not work for MEX/2 because of the endpoint 1. Before addressing MEX/2, we first give the appropriate general form of MEX/1.

Maximal Emulation Finite Use/1. MEFU/1. Any finite order preserving $R \subseteq Q(0, 1)^k \times Q(0, 1)^k$ is ME usable.

Proof of MEFU/1 In RCA_0. Let R be as given and let $E \subseteq Q[0, 1]^k$ be finite. Let $p_1 < \cdots < p_n \in Q(0, 1)$ be the rationals appearing in R. Let S be a maximal emulator of E. We now apply the usual finite Ramsey theorem from Ramsey (1930) as follows. Let $V \subseteq Q(0, 1)$ be finite and sufficiently large. Let $V' \subseteq V$, $|V'| = n$, where membership in S of k-tuples from V′ depend only on their order type. Let $f: Q[0, 1] \rightarrow Q[0, 1]$ be a strictly increasing bijection mapping V′ onto $\{p_1, \ldots, p_n\}$. By Theorem 2.6, f[S] is a maximal emulator of E. Note that membership in f[S] of k-tuples from $\{p_1, \ldots, p_n\}$ depends only on their order type. We now claim that f[S] contains its R image. To see this, let $R(x, y)$, $x \in S$. Then $x, y \in \{p_1, \ldots, p_n\}^k$, and x, y are order equivalent. Hence $y \in S$. This argument will also show that R is ME invariantly usable, although we can also get this by using the relation $R'(x, y) \leftrightarrow R(x, y) \vee R(y, x)$, which must also be finite and order preserving. □

We now give a proof of MEX/2 before taking up much more general results. Additional ideas are required, and the proof is not given in RCA_0.

Proof of MEX/2 In ACA_0. Let $E \subseteq Q[0, 1]^2$ be finite. Construct sets $S_1 \subseteq \cdots \subseteq S_5$ in five steps, where each $S_i \subseteq Q[0, i]^2$ is a maximal emulator of $E \subseteq Q[0, i]^2$. We claim that each S_i, S_{i+1} agree on $Q[0, i]^2$. Let $x \in Q[0, i]^2$. If $x \notin S_i$ then $S_i \cup \{x\}$ is not an emulator of $E \subseteq Q[0, i]^2$, and hence $S_{i+1} \cup \{x\}$ is not an emulator of $E \subseteq Q[0, i+1]^2$, and therefore $x \notin S_{i+1}$. (Here the same E is viewed as a subset of the various $Q[0, i]^2$).

By the argument by cases in the proof of MEX/1, let $a>b>c$ be from $\{1, 2, 3, 4, 5\}$, where $(a, b) \in S_5 \leftrightarrow (b, c) \in S_5$. By the above claim, $(a, b) \in S_a \leftrightarrow (b, c) \in S_a$. Now map $S_a \cap Q[0, a]^2$ onto $S' \cap Q[0, 1]^2$ by an increasing bijection mapping a, b, c to 1, 1/2, 1/3. Then S' is a maximal emulator of $E \subseteq Q[0, 1]^2$ and $(1, 1/2) \in S \leftrightarrow (1/2, 1/3) \in S$. □

Here is our first generalization of MEX/2.

Definition 3.2.1 Let $(x, y) \in Q[0, 1]^k \times Q[0, 1]^k$ and $R \subseteq Q[0, 1]^k \times Q[0, 1]^k$. p is present in (x, y) if and only if p is a coordinate of x or y. p is altered in (x, y) if and only if there exists i such that $p = x_i \neq y_i$ or $p = y_i \neq x_i$. p is present in R if and only if there exists i such that $p = x_i$ or $p = y_i$. p is altered by R if and only if p is altered in some element of R. We also write "p appears in (x, y)", "(x, y) alters p", "p appears in R", and "R alters p".

Maximal Emulation Finite Use/2. MEFU/2. Any finite order preserving $R \subseteq Q(0, 1]^k \times Q(0, 1]^k$ is ME usable.

Here is a second, stronger, generalization of MEX/2, which we prove.

Maximal Emulation Finite Use/3. MEFU/3. Any finite order preserving $R \subseteq Q[0, 1]^k \times Q[0, 1]^k$ not altering both of 0, 1 is ME usable.

Obviously MEFU/3 → MEFU/2 → MEFU/1 in RCA_0.

Lemma 3.2.1 *(RCA_0) Suppose MEFU/3 holds with "not alter 0". Then MEFU/3 holds with "not alter 1".*

Proof Assume MEFU/3 holds with "not alter 0". Let finite order preserving $R \subseteq Q[0, 1]^k \times Q[0, 1]^k$ not alter 1, and finite $E \subseteq Q[0, 1]^k$ be given. We use the bijection $f:Q[0, 1] \to Q[0, 1]$ given by $f(p) = 1 - p$. We claim that f is an automorphism of the order equivalence relation EQR(Q[0, 1], k) on $Q[0, 1]^k$ (as a subset of $Q[0, 1]^{2k}$). To see this, we need to verify that $(p_1, \ldots, p_k)$, $(q_1, \ldots, q_k)$ are order equivalent if and only if $(1 - p_1, \ldots, 1 - p_k)$, $(1 - q_1, \ldots, 1 - q_k)$ are order equivalent. Suppose $(p_1, \ldots, p_k)$, $(q_1, \ldots, q_k)$ are order equivalent. Let $1 \le i, j \le k$. We want $(1 - p_i < 1 - p_j \leftrightarrow 1 - q_i < 1 - q_j)$. I.e., $(p_j < p_i \leftrightarrow q_j < q_i)$, which follows from $(p_1, \ldots, p_k)$, $(q_1, \ldots, q_k)$ being order equivalent. The converse is proved analogously.

By Theorem 2.6, we see that R is ME usable if and only if $1 - R$ is ME usable. We claim that $1 - R$ is order preserving and does not alter 0. Suppose $1 - R(x, y)$. Then $R(1 - x, 1 - y)$, and so $1 - x$, $1 - y$ are order equivalent and x, y are order equivalent. Suppose $1 - R$ alters 0, and let $1 - R(x, y)$, where 0 lies in the two element set $\{x_i,$

y_i}. Then 1 lies in the two element set {$1 - x_i$, $1 - y_i$} and R($1 - x_i$, $1 - y_i$). This contradicts that R does not alter 1.

Now since MEFU/3 holds with "not alter 0", we see that 1 − R is ME usable. Hence R is ME usable. □

Proof of MEFU/3 In ACA′. By Lemma 3.2.1, it suffices to prove MEFU/3 with "does not alter 0". Let finite $R \subseteq Q[0, 1]^k \times Q[0, 1]^k$ not alter 0, and finite $E \subseteq Q[0, 1]^k$ be given. Let the rationals appearing in R be among $0 < p_1 < \cdots < p_n < 1$. Construct $S_1 \subseteq S_2 \subseteq \cdots \subseteq S_m$ in m steps where each S_i is a maximal emulator of $E \subseteq Q[0, i]^k$. Here m is so large relative to k, n that we can use it here with the usual finite Ramsey theorem, (Ramsey 1930). As in the proof of MEX/2, each S_i, S_{i+1} agree on $Q[0, i]^k$. By the finite Ramsey theorem, let V be an $n+1$ element subset of {1, …, m} such that membership of k-tuples from V ∪ {0} in S_m depends only on their order type with constant 0. Hence membership of k-tuples from V ∪ {0} in $S_{max(V)}$ depends only on their order type with constant 0.

Now let h:V ∪ {0} → {0, p_1, …, p_n, 1} be the unique increasing bijection. Extend h to an increasing bijection h′:Q[0, max(V)] → Q[0, 1]. It is clear that membership of k-tuples from {0, p_1, …, p_n, 1} in h[$S_{max(V)}$] depends only on their order type with constant 0. Also, by Theorem 2.6, h′[$S_{max(V)}$] is a maximal emulation of h′[E] ⊆ Q[0, 1].

We claim that $S_{max(V)}$ contains its h'^{-1}[R] image. To see this, let h'^{-1}[R](x, y), x ∈ $S_{max(V)}$. Then R(h′(x), h′(y)), and so h′(x), h′(y) are order equivalent with constant 0. Hence x, y are order equivalent with constant 0. Therefore x ∈ $S_{max(V)}$ ↔ y ∈ $S_{max(V)}$, and since x ∈ $S_{max(V)}$, we have y ∈ $S_{max(V)}$.

It now follows that h′[$S_{max(V)}$] contains its R image. So h′[$S_{max(V)}$] is a maximal emulation of $h'[E] \subseteq Q[0, 1]^k$ containing its R image. Now h′[E] is an emulation of $E \subseteq Q[0, 1]^k$ and E is an emulation of $h'[E] \subseteq Q[0, 1]^k$. So by Emulation Transitivity in Sect. 12.2, h′[$S_{max(V)}$] is also a maximal emulation of $E \subseteq Q[0, 1]^k$. □

We conjecture that MEFU/2, 3 are not provable in ACA_0. Note that we have proved MEFU/1 in RCA_0.

If finite R alters both 0, 1, then open issues arise. However, we have the following.

Maximal Emulation Singleton Use/1. MEOU/1. Any order preserving $R \subseteq Q[0, 1]^k \times Q[0, 1]^k$ of cardinality 1 is ME usable.

Proof of MEOU/1 In RCA_0. Let R = {(x, y)} be order preserving. I.e., x, y are order equivalent. Let $E \subseteq Q[0, 1]^k$ be finite.

case 1. E has an element order equivalent to y. Then {y} is an emulator of $E \subseteq Q[0, 1]^k$, and so extends to a maximal emulator S of $E \subseteq Q[0, 1]^k$. Since y ∈ S, S contains its R image, which is {y} if x ∈ S and ∅ if x ∉ S.

case 2. E has no element order equivalent to y. Let S be a maximal emulator of $E \subseteq Q[0, 1]^k$. Then x, y ∉ S since x, y are order equivalent. Hence S contains its R image, which is ∅. □

MEOU/1 is false for cardinality 1 even in dimension k = 2 and R of cardinality 2. In fact, we have the following.

Lemma 3.2.2. *(RCA_0) {((0, 1/2), (1/2, 1))} is not ME invariantly usable.*

Proof Let $E = \{(p, q) \in Q[0, 1]^2: p < 1/2 < q\}$. We claim that S is an emulator of $E \subseteq Q[0, 1]^2$ if and only if $S \subseteq Q[0, 1]^{2<}$ and the first coordinate of every element of S is less than the second coordinate of every element of S. To see this, suppose S is an emulator of $E \subseteq Q[0, 1]^2$. If $x \in S$ then (x, x) is order equivalent to some (y, y) in E^2, and so $x \in Q[0, 1]^{2<}$. Hence $S \subseteq Q[0, 1]^{2<}$. If $x, y \in S$ then (x, y) is order equivalent to some (z, w) in E^2, and so $x_1 < y_2$. Conversely, suppose $S \subseteq Q[0, 1]^{2<}$, where the first coordinate of every element of S is less than the second coordinate of every element of S. Let $x, y \in S$. Note that $x_1 < y_1 = x_2 < y_2$.

Let $0 \leq p_1, q_1 < 1/2 < p_2, q_2 \leq 1$, where (x_1, y_1), (p_1, q_1) are order equivalent, and (x_2, y_2), (p_2, q_2) are order equivalent. Then (x, y) and $((p_1, q_1), (p_2, q_2))$ are order equivalent.

Let S be a maximal emulator of E. Obviously S is nonempty. Let α be the sup of the first terms of the pairs in S and β be the inf of the second terms of the pairs in S. Clearly $\alpha \leq \beta$.

If $\alpha < \beta$ then $S \cup \{(p, q)\}$ is an emulator of E where $\alpha < p < q < \beta$, violating the maximality of S. Hence $\alpha = \beta$.

case 1. α is irrational. Then all first terms are < and all second terms are > α. Hence $S = \{(p, q): p < \alpha < q\}$.

case 2. $\alpha = \beta$ is rational. If $\alpha = 0$ then $S = \{(0, q): 0 < q \leq 1\}$, and so $(0, 1/2) \in S \wedge (1/2, 1) \notin S$, violating R invariance. If $\alpha = 1$ then $S = \{(p, 1): 0 \leq p < 1\}$, and so $(0, 1/2) \notin S \wedge (1/2, 1) \in S$, violating R invariance. Now suppose $0 < \alpha < 1$. If $0 \leq p < \alpha < q \leq 1$ then $S \cup \{(p, q)\}$ is an emulator of E, and so $(p, q) \in S$. What is not clear is which (α, q) and (p, α) lie in S. If none of these lie in S then $S \cup \{(\alpha, p)\}$ is an emulator of E. Hence at least one of these lie in S. Suppose some (α, q) lies in S. Then no (p, α) lies in S. Hence all (α, q), $\alpha < q \leq 1$, lie in S. Alternatively, suppose some (p, α) lies in S. Then no (α, q) lies in S, and hence all (p, α), $0 \leq p < \alpha$, lie in S.

This determines exactly what the maximal emulators of E are. They are the sets S =

i. $\{(p, q) \in Q[0, 1]^2: 0 \leq p \leq \alpha < q \leq 1\}$, where α is a real number in [0, 1).
ii. $\{(p, q) \in Q[0, 1]^2: 0 \leq p < \alpha \leq q \leq 1\}$, where α is a real number in (0, 1].

case i. $(0, 1/2) \in S$ if and only if $\alpha < 1/2$, and $(1/2, 1) \in S$ if and only if $1/2 \leq \alpha < 1$ if and only if $1/2 \leq \alpha$. Hence S is not equivalent at (0, 1/2), (1/2, 1).

case ii. $(0, 1/2) \in S$ if and only if $0 < \alpha \leq 1/2$ if and only if $\alpha \leq 1/2$, and $(1/2, 1) \in S$ if and only if $1/2 < \alpha$. Hence S is not equivalent at (0, 1/2), (1/2, 1). □

We now present a method for establishing that {(x, y)} is ME invariantly usable.

Lemma 3.2.3 *(RCA_0) Suppose $x, y \in Q[0, 1]^k$ are order equivalent and there exists $z \in Q[0, 1]^k$ such that (x, y), (x, z), (z, y) are order equivalent (as 2k-tuples). Then $\{(x, y)\} \in Q[0, 1]^2 \times Q[0, 1]^2$ is ME invariantly usable.*

Proof Let x, y, z be as given. Let $E \subseteq Q[0, 1]^2$ be finite. We find a maximal emulator of E that is equivalent at x, y.

case 1. There exist x′, y′ ∈ E such that (x′, y′), (x, y) are order equivalent. Then {x, y} is an emulator of E which extends to a maximal emulator of E, which is obviously equivalent at x, y.

case 2. Otherwise. If E does not have an element order equivalent to x, then any maximal emulator of E is equivalent at x, y. Suppose otherwise. Clearly x, y, z are order equivalent and so {z} is an emulator of E. Let S be a maximal emulator of E with z ∈ S. Since (x, z), (z, y) are not order equivalent to any element of E^2, clearly x, y ∉ S. Hence S is equivalent at x, y. □

We now have the following complete determination of the ME invariant usability of singletons in dimension k = 2.

Maximal Emulation Singleton Use/2. MEOU/2. Let x, y ∈ $Q[0, 1]^2$ be order equivalent. The following are equivalent.

i. {(x, y)} is ME invariantly usable.
ii. {(x, y), (y, x)} is ME usable.
iii. For finite subsets of $Q[0, 1]^2$, some maximal emulator is equivalent at x, y.
iv. {x, y} is not any {(0, p), (p, 1)} and not any {(1, p), (p, 0)}, 0 < p < 1.

Proof In RCA_0. The equivalence of i, ii, iii is immediate (or use Theorems 2.3, 2.4). Assume i. By Lemma 3.2.2, {((0, 1/2), (1/2, 1))} is not ME invariantly usable. By Theorem 2.6, ME invariant usability (and ME usability) remains unchanged under increasing automorphisms of Q[0, 1], and so {((0, p), (p, 1))}, 0 < p < 1, is not ME invariantly usable. Also note that coordinate reversal is an automorphism of EQR(Q[0, 1]). Hence by Theorem 2.6, ME invariant usability (and ME usability) remains unchanged under coordinate reversal, and so {((1, p), (p, 0))}, 0 < p < 1, is not ME invariantly usable. This establishes iv.

Assume iv. If not both 0, 1 are altered in {(x, y)} then by MEFU/3, {(x, y)} is ME invariantly usable. So we assume that both 0, 1 are altered in (x, y). We derive i. We now split according to the fact that 0 is among x_1, x_2, y_1, y_2.

case 1. (x, y) = ((0, a), (b, c)). Suppose b = 0. Since 0, 1 are both altered, {a, c} = {0, 1}. This, however, violates that x, y are order equivalent. If b = 1 then (x, y) = ((0, a), (1, c)), and byorder equivalence, (x, y) = ((0, 0), (1, 1)). Since ((0, 0), (1, 1)), ((0, 0), (1/2, 1/2)), ((1/2, 1/2))(1, 1)) are order equivalent, {(x, y)} is ME invariantly usable by Lemma 3.2.3. So we can assume 0 < b < 1. 1 is altered in (x, y), we have a = 1 ∨ c = 1.

case 1.1. a = 1. Then (x, y) = ((0, 1), (b, c)), where 0 < b < c ≤ 1 by order equivalence. Since 1 is altered, we have 0 < b < c < 1. Since ((0, 1), (b, c)), ((0, 1), (b′, c′)), ((b, c), (b′, c′)) are order equivalent, assuming b < b′ < c′ < c, we see that {(x, y)} is ME invariantly usable by Lemma 3.2.3.

case 1.2. c = 1. Then (x, y) = ((0, a), (b, 1)), where 0 < a, b < 1 by order equivalence. By iv, a ≠ b. If a < b then ((0, a), (b, 1)), ((0, a), (c, d)), ((c, d), (b, 1)) are order equivalent for a < c < d < b. If a > b then ((0, a), (b, 1)), ((0, a), (c, d)), ((c, d), b, 1)) are also order equivalent for 0 < c < b < a < d < 1. Hence {(x, y)} is ME invariantly usable by Lemma 3.2.3.

case 2. (x, y) = ((a, 0), (b, c)). Let (x′, y′) be the result of flipping the two coordinates in both x and y. Then both 0, 1 are altered in (x′, y′), and (x′, y′) = ((0, a), (c, b)). Apply case 1 to obtain that {(x′, y′)} is ME invariantly usable. As explained earlier, flipping the two coordinates in both x′, y′ preserves ME invariant usability (using Theorem 2.6), so we see that {(x, y)} is ME invariantly usable.

case 3. (x, y) = ((b, c), (0, a)). Obviously {(x, y)} is ME invariantly usable if and only if {(y, x)} is ME invariantly usable. Apply case 1 to (y, x).

case 4. (x, y) = ((b, c), (a, 0)). Obviously {(x, y)} is ME invariantly usable if and only if {(y, x)} is ME invariantly usable. Apply case 2 to (y, x). □

We conjecture that a complete determination of the ME invariant usability of singletons in higher dimensions can be accomplished in RCA_0 (in ourQ[0, 1] context). We further conjecture that a complete determination of the ME usability (and hence the ME invariant usability) of the finite $R \subseteq Q[0, 1]^k \times Q[0, 1]^k$ can be carried out in ACA′.

12.3.3 Large Relations

We begin with the easy complete determination of the ME usable R ⊆ Q[0, 1] × Q[0, 1]. Note that every R ⊂ Q[0, 1] × Q[0, 1] is triviallyorder preserving.

Lemma 3.3.1 *(RCA_0) The maximal emulators of ∅ ⊆ Q[0, 1] is just ∅. The maximal emulators of any {p} ⊆ Q[0, 1] are all of the singletons. The maximal emulator of E ⊆ Q[0, 1], |E| ≥ 2, is just Q[0, 1].*

Proof Left to the reader. □

Maximal Emulation Large Use/1. MELU/1. R ⊆ Q[0, 1] × Q[0, 1] is ME usable if and only if there is some p ∈ Q[0, 1] not altered by R.

Proof In RCA_0. Suppose R does not alter p ∈ Q[0, 1]. Let E ⊆ Q[0, 1] be finite. By Lemma 3.3.1, there is a maximal emulator of E ⊆ Q[0, 1] among the three sets ∅, {p}, Q[0, 1], each of which contain their R image. Hence R is ME usable.

Suppose R is ME usable. Let S be an emulator of {0} ⊆ Q[0, 1] which contains its R image. By Lemma 3.3.1, let S = {p}. So any q with R(p, q) must be p. I.e., p is not altered by R. □

In particular, MELU/1 tells us that the "large" relations Q[0, 1) × Q[0, 1) and Q(0, 1] × Q(0, 1] are ME usable. That's why we use "large" in the header of MELU/1.

More generally, we consider order theoretic $R \subseteq Q[0, 1]^k \times Q[0, 1]^k$ to be "large" if infinitely many numbers are altered, and "small" if finitely many numbers are altered. We are less clear about what we want to mean by "large" and "small" for general R, but the development in this paper is driven by the order theoretic R.

Here we consider only the particularly simple "large" relations of the form $V \times V \subseteq Q[0, 1]^k \times Q[0, 1]^k$.

Theorem 3.3.2 *(RCA_0) Let $V \subseteq Q[0, 1]^k$. The following are equivalent.*

i. *$V \times V \subseteq Q[0, 1]^k \times Q[0, 1]^k$ is ME invariantly usable.*
ii. *$V \times V \subseteq Q[0, 1]^k \times Q[0, 1]^k$ is ME usable.*
iii. *For subsets of $Q[0, 1]^k$, some maximal emulator contains or is disjoint from V.*

If any of these hold then all x, y $\in$ V are order equivalent.

Proof i $\rightarrow$ ii is immediate. Assume ii and let $E \subseteq Q[0, 1]^k$ be finite. Let S be a maximal emulator of $E \subseteq Q[0, 1]^k$ that contains its $V \times V$ image. If S meets V then E contains V. Hence S contains or is disjoint from V. Now suppose iii and let $E \subseteq Q[0, 1]^k$ be finite. Let S be a maximal set that either contains or is disjoint from V. In the first case, S is $V \times V$ invariant. In the second case, S is $V \times V$ invariant, vacuously.

For the last claim, if any of i–iii hold then $V \times V$ is order preserving (MENU of Sect. 12.3.1), and hence any x, y $\in$ V are order equivalent. □

MELU/1 fails badly in the highly nontrivial environment of dimension $k = 2$.

Maximal Emulation Large Use/2. MELU/2. $Q(0, 1)^{2<} \times Q(0, 1)^{2<}$ is not ME usable. It is order preserving, order theoretic, and 0, 1 are not present.

Proof In RCA_0. Let $R = Q(0, 1)^{2<} \times Q(0, 1)^{2<}$. Let $E = \{(1/6, 1/4), (1/7, 1/3), (0, 1/5), (1/2, 1)\}$. The idea behind E is that no coordinate of an element of E is a coordinate of any other element of E, and the pairs of elements are in general position relative to that restriction. Then S is an emulator of $E \subseteq Q[0, 1]^2$ if and only if $S \subseteq Q[0, 1]^{2<}$ and no coordinate of an element of S is a coordinate of any other element of S.

Let S be a maximal emulator of $E \subseteq Q[0, 1]^2$. Suppose $p < q$ are both not present in S. Then $S \cup \{(p, q)\}$ is an emulator of $E \subseteq Q[0, 1]^2$, contradicting maximality. Hence all numbers in Q[0, 1] are present in S except for possibly one number. In particular, S has at least three elements, and so S has an element (p, q), where $0 < p < q < 1$. If S contains its R image, then $(p, q') \in S$ for any $p < q < q' < 1$, contradicting that E has no has no repeated numbers. Hence S does not contain its R image. □

Maximal Emulation Large Use/3. MELU/3. $Q[1/2, 1)^{2<} \times Q[1/2, 1)^{2<}$ is ME usable.

Proof In RCA_0. Let $E \subseteq Q[0, 1]^2$ be finite. We find a maximal emulator S of E which contains or is disjoint from $Q[1/2, 1)^{2<}$, which suffices according to Theorem 3.3.2. The first case that applies is operational.

case 1. E has no (p, q), $p < q$. Let S be a maximal emulator of $E \subseteq Q[0, 1]^2$. Then S is disjoint from $Q[1/2, 1)^{2<}$.
case 2. E has no (p, q), (p′, q′), $p < q < p' < q'$. Let S be a maximal emulator of $E \subseteq Q[0, 1]^2$ containing (0, 1/3). Then S is disjoint from $Q[1/2, 1)^{2<}$.
case 3. E has no (p, q), (p′, q′), $p' < p < q < q'$. Let S be a maximal emulator of $E \subseteq Q[0, 1]^2$ containing (1/3, 1). Then S is disjoint from $Q[1/2, 1)^{2<}$.

case 4. E has no (p, q), (p′, q), $p<p'<q$. Let $C=\{(p, 1-2p): 0\leq p\leq 1/4\}$. Since case 3 does not apply, C is an emulator of $E\subseteq Q[0, 1]^2$. Let S be a maximal emulator of $E\subseteq Q[0, 1]^2$ containing C. Then every number in [1/2, 1] is the right endpoint of a unique element of C, and that unique element of C has left endpoint $\leq 1/4$. Therefore S is disjoint from $Q[1/2, 1)^{2<}$.
case 5. E has no (p, q), (p, q′), $p<q<q'$. Clearly $Q[0, 1)\times\{1\}$ is an emulator of $E\subseteq Q[0, 1]^2$ since case 4 does not apply. Let S be a maximal emulator of $E\subseteq Q[0, 1]^2$ containing $Q[0, 1]\times\{1\}$. Then S is disjoint from $Q[1/2, 1)^{2<}$.
case 6. E has no (p, q), (p′, q′), $p<p'<q<q'$. Let $B=\{(1/3, q): 1/3<q\leq 1$. Since case 5 does not apply, B is an emulator of $E\subseteq Q[0, 1]^2$. Let S be a maximal emulator of $E\subseteq Q[0, 1]^2$ containing B. Because $(1/3, 1)\in B$, clearly S is disjoint from $Q[1/2, 1)^{2<}$.
case 7. Otherwise. Then every element of $Q[0, 1]^{2<}\times Q[0, 1]^{2<}$ is order equivalent to an element of E^2. Hence $Q[0, 1]^{2<}$ is an emulator of $E\subseteq Q[0, 1]^2$, and so some maximal emulator S of E contains $Q[0, 1]^{2<}$. □

The situation changes considerably with dimension $k\geq 3$.

Maximal Emulation Large Use/4. MELU/4. For $k\geq 3$, $Q[1/3, 1/2]^{k<}\times Q[1/3, 1/2]^{k<}$ is not ME usable.

Proof In RCA_0. Fix $k\geq 3$. Let $x_1, y_1, \ldots, x_n, y_n\in Q[0, 1]^{k<}$ be such that

i. For each i, it is not the case that x_i, y_i have the same first and last coordinates.
ii. Each x_{i+1}, y_{i+1} has all 2 k combined coordinates greater than all 2 k combined coordinates in x_i, y_i.
iii. Every $(x, y)\in Q[0, 1]^{k<}\times Q[0, 1]^{k<}$, where it is not the case that x, y have the same first and last coordinates, is order equivalent (as a 2k-tuple) to some (x_i, y_i).

Let $E=\{x_1, \ldots, x_n, y_1, \ldots, y_n\}$. We claim that S is an emulator of E if and only if $S\subseteq Q[0, 1]^{k<}$, and for all distinct $x, y\in S$, it is not the case that x, y have the same first and last coordinates. Suppose S is an emulator of E, and let $x, y\in S$ be distinct, where x, y have the same first and last coordinates. Then (x, y) cannot be order equivalent to any (x_i, y_i), and also (x, y) cannot be order equivalent to any (x_i, y_j), $i\neq j$. So S is not an emulator of E. Conversely, suppose $S\subseteq Q[0, 1]^{k<}$, where for all distinct x, y $\in$ S, it is not the case that x, y have same first and last coordinate. Then every (x, y) $\in S^2$ is order equivalent to some $(x_i, y_i)\in E$ by iii.

Let S be a maximal emulator of E. We claim that for all $p<q$ there exists $p<b_1<\cdots<b_{k-2}<q$ such that $(p, b_1, \cdots, b_{k-2}, q)\in S$. Suppose this is false for $p<q$. Choose $p<c_1<\cdots<c_{k-2}<q$. Then $S\cup\{(p, c_1, \ldots, c_{k-2}, q)\}$ is an emulator of E, and so $(p, c_1, \ldots, c_{k-2}, q)\in S$, which is a contradiction. Now let $(1/3, b_1, \ldots, b_{k-2}, 1/2)\in S$. S cannot contain its R image since for $1/3<b_1', b_2, \ldots, b_{k-2}<1/2$, $1/3<b_1'<b_2$, we have $(1/3, b_1', b_2, \ldots, b_{k-2}, 1/2)\in S$. □

We are still very far from obtaining any kind of complete determination of the order theoretic usable $R\subseteq Q[0, 1]^k\times Q[0, 1]^k$, even in dimension $k=2$. MELU/2, 3,

4 indicate that, roughly speaking, large order theoreticR are not usable in dimension $k \geq 3$ but somewhat usable in dimension $k = 2$. Just as one example of the many issues remaining here with large order theoretic R, consider the following possibility.

Maximal Emulation Large Use/5. MELU/5. All order preserving order theoretic $R \subseteq Q[1/3, 1/2]^2 \times Q[1/3, 1/2]^2$ are usable.

We do not know the status of MELU/5, and its stronger forms with $Q[1/2, 1)^2 \times Q[1/2, 1)^2$ and even $Q[1/2, 1]^2 \times Q[1/2, 1]^2$. But note that MELU/5 with dimension $k \geq 3$ is strongly refuted by MELU/4.

12.3.4 Small Relations

Here we first confront independence from ZFC. The most immediately transparent examples presented here are MESU/1 and MED/1.

We start with the most obvious parameterization of finite R, obtained by simply adding a new dummy variable.

Parameterization Definition Let $R \subseteq Q[0, 1]^k \times Q[0, 1]^k$. The parameterization of R is the relation $R' \subseteq Q[0, 1]^{k+1} \times Q[0, 1]^{k+1}$ given by $R'(x, y) \leftrightarrow R((x_1, \ldots, x_k), (y_1, \ldots, y_k)) \wedge x_{k+1} = y_{k+1}$.

Parameterizations are too strong to be used for ME usability. This is because parameterizations are generally not even order preserving - a necessary condition for ME usability – see MENU in Sect. 12.3.2.

Theorem 3.4.1 *(RCA_0) Let $R \subseteq Q[0, 1]^k \times Q[0, 1]^k$. The following are equivalent.*

i. The parameterization of R is ME usable.
ii. The parameterization of R is order preserving.
iii. R alters no numbers.

Proof Let R be as given. $i \rightarrow ii$, and $iii \rightarrow i$ are immediate. Assume ii and let R' be the parameterization of R. Suppose R alters some number, and let R(x, y) where $x_i \neq y_i$. Clearly $R'((x, x_i), (y, x_i))$, and so $(x, x_i), (y, x_i)$ are order equivalent. Therefore $x_i = y_i$, which is a contradiction. □

The obvious way to fix parameterizations is to use the weaker notion of lower parameterization, which does behave well with order preservation.

Lower Parameterization Definition Let $R \subseteq Q[0, 1]^k \times Q[0, 1]^k$. The lower parameterization of R is the relation $R' \subseteq Q[0, 1]^{k+1} \times Q[0, 1]^{k+1}$ given by $R'(x, y) \leftrightarrow R((x_1, \ldots, x_k), (y_1, \ldots, y_k)) \wedge x_{k+1} = y_{k+1} < x_1, \ldots, x_k, y_1, \ldots, y_k$.

Theorem 3.4.2 *(RCA_0) The lower parameterization of any $R \subseteq Q[0, 1]^k \times Q[0, 1]^k$ is order preserving if and only if $R\,|{>}0$ is order preserving.*

Proof Let $R' \subseteq Q[0, 1]^{k+1} \times Q[0, 1]^{k+1}$ be the lower parameterization of $R \subseteq Q[0, 1]^k \times Q[0, 1]^k$. Suppose R' is order preserving. Let $R|{>}0(x, y)$. Then $R'((x, 0), (y, 0))$, and so $(x, 0)$, $(y, 0)$ are order equivalent. Hence x, y are order equivalent. Conversely, suppose $R\,|{>}0$ is order preserving, and let $R'(x, y)$. Then $R((x_1, \ldots, x_k), (y_1, \ldots, y_k)) \wedge x_{k+1} = y_{k+1} < x_1, \ldots, x_k, y_1, \ldots, y_k$. It is clear that x, y are order equivalent. □

Maximal Emulation Small Use/1. MESU/1. The lower parameterization of any order preserving finite $R \subseteq Q[0, 1]^k \times Q[0, 1]^k$ is ME usable.

We shall see that MESU/1 is neither provable nor refutable from ZFC, assuming SRP is consistent. We first discuss some weaker and stronger forms of MESU/1. On the weaker side, we use what we call drop equivalence. This allows us to conveniently state a very specific special case of MESU/1 without using the ME use terminology, while still having independence from ZFC.

Drop Equivalence Definition Let $S \subseteq Q[0, 1]^k$. S is drop equivalent at x, y if and only if $x, y \in Q[0, 1]^k \wedge x_k = y_k \wedge (\forall p \in Q[0, x_k))(S(x_1, \ldots, x_{k-1}, p) \leftrightarrow S(y_1, \ldots, y_{k-1}, p))$.

We can think of x, y as raindrops in the space $Q[0, 1]^k$, at the same height $x_k = y_k$ over the ground. As they fall to the ground in tandem, they generally go in and out of a given set $S \subseteq Q[0, 1]^k$. Drop equivalence asserts that as they fall in tandem, one is in S if and only if the other is in S. I.e., x, y have the same pattern of membership in S as they fall in tandem.

Maximal Emulation Drop/1. MED/1. For finite subsets of $Q[0, 1]^k$, some maximal emulator is drop equivalent at $(1, 1/2, \ldots, 1/k)$, $(1/2, \ldots, 1/k, 1/k)$.

Thus we have presented the most immediately transparent statements independent of ZFC in this paper - MESU/1 and MED/1.

Theorem 3.4.3 *(RCA_0) MESU/1 → MED/1. For $k \geq 1$, MESU/1 for dimension k implies MED/1 for dimension $k+1$.*

Proof It is clear that MED/1 for k + 1 asserts the ME usability of $R \subseteq Q[0, 1]^{k+1} \times Q[0, 1]^{k+1}$, where $R(x, y) \leftrightarrow R'(x, y) \vee R'(y, x)$, and $R'(x, y) \leftrightarrow (x_1, \ldots, x_k) = (1/1/2, \ldots, 1/k) \wedge (y_1, \ldots, y_k) = (1/2, \ldots, 1/k+1) \wedge x_{k+1} = y_{k+1} < 1/(k+1)$. Note that R is the lower parameterization of $\{((1, \ldots, 1/k), (1/2, \ldots, 1/(k+1))), ((1/2, \ldots, 1/(k+1)), (1, \ldots, 1/k))\}$, which is clearly order preserving and finite, ready for use in MESU/1 for k. □

We now state a far reaching extension of MESU/1 using certain critical equivalence relations.

Definition 3.4.1 Let $A \subseteq Q[0, 1]$. The relation $R_k(A) \subseteq Q[0, 1]^k \times Q[0, 1]^k$ is given by $R_k(A)(x, y)$ if and only if

i. x, y are order equivalent.
ii. If $x_i \neq y_i$ then all $x_j \geq x_i$ and $y_j \geq y_i$ lie in A.

Lemma 3.4.4 *(RCA_0) $R_k(A)$ is order preserving. Let $R_k(A)(x, y)$. $(\forall i, j)(x_i \neq y_i \rightarrow x_i, y_i \in A)$. $(\forall i)(x_i \in A \leftrightarrow y_i \in A)$.*

Proof Fix $A \subseteq Q[0, 1]$. Obviously $R_k(A)$ is order preserving. Let $R_k(A)(x, y)$. If $x_i \neq y_i$ then all $x_j \geq x_i$ and $y_j \geq y_i$ lie in A, and in particular, $x_i, y_i \in A$. If $x_i = y_i$ then we are done. If $x_i \neq y_i$ then $x_i, y_i \in A$ and we are done. □

Theorem 3.4.5 *(RCA_0) Each $R_k(A)$ is an order preserving equivalence relation on $Q[0, 1]^k$. If $|A| \leq 1$ then $R_k(A)$ alters no numbers. If $|A| \geq 2$ then $R_k(A)$ alters exactly the elements of A. If A is finite then $R_k(A)$ alters finitely many numbers and $R_k(A)$ is order theoretic.*

Proof Obviously $R_k(A)$ is reflexive and symmetric. For transitivity, let $R_k(A)(x, y)$, $R_k(A)(y, z)$. We want $R_k(A)(x, z)$. Clearly x, y, z are order equivalent. Let $x_i \neq z_i$. Clearly $x_i \neq y_i \vee y_i \neq z_i$.

case 1. $x_i \neq y_i$. Then every $x_j \geq x_i$ lies in A. Let $z_j \geq z_i$. By order equivalence, $y_j \geq y_i \wedge y_j \in A$. By Lemma 3.4.4, $z_j \in A$.

case 2. $y_i \neq z_i$. Then every $z_j \geq z_i$ lies in A. Let $x_j \geq x_i$. By order equivalence, $y_j \geq y_i \wedge y_j \in A$. By Lemma 3.4.4, $x_j \in A$.

This establishes the first claim.

Suppose $|A| \leq 1$. If $R_k(A)$ alters p, q then by Lemma 3.4.4, $p, q \in A$ and $p = q$. Hence $R_k(A)$ alters no numbers. Suppose $|A| \geq 2$. Let $p \in A$. Then $R_k(A)((p, \ldots, p)(q, \ldots, q))$ where $p \neq q \in A$. Conversely, if $R_k(A)$ alters p then by Lemma 3.4.4, $p \in A$.

Let $A \subseteq Q[0, 1]$ be finite. Then obviously $R_k(A)$ is order theoretic with parameters from A. By the previous claims, $R_k(A)$ alters at most the elements of A, which is finite. □

Maximal Emulation Small Use/2. MESU/2. For finite $A \subseteq Q(0, 1]$, $R_k(A) \subseteq Q[0, 1]^k \times Q[0, 1]^k$ is ME usable.

We now give a finiteness condition on $R \subseteq Q[0, 1]^k \times Q[0, 1]^k$.

Finiteness Condition $R \subseteq Q[0, 1]^k \times Q[0, 1]^k$ has the Finiteness Condition if and only if there are finitely many p appearing in some element of R that alters some $q \leq p$.

Theorem 3.4.6 *(RCA_0) An order preserving $R \subseteq Q[0, 1]^k \times Q[0, 1]^k$ has the Finiteness Condition if and only if it is contained in some $R_k(A)$, A finite. An order preserving $R \subseteq Q[0, 1]^k \times Q[0, 1]^k$ has the Finiteness Condition and does not alter 0 if and only if it is contained in some $R_k(A)$, $0 \notin A$, A finite.*

Proof For the first claim, let $R \subseteq Q[0, 1]^k \times Q[0, 1]^k$ be as given. Let A be the set of all p such for some $(x, y) \in R$, p is both present in (x, y) and at least as large as some number altered by (x, y). Then A is finite. Let R(x, y). Then x, y are order equivalent. Suppose $x_i \neq y_i$. Then obviously all $x_j \geq x_i$ and $y_j \geq y_i$ lie in A. Hence $R_k(A)(x, y)$. Conversely, suppose $R \subseteq R_k(A)$, A finite. We claim that all p appearing in some element of $R_k(A)$ that alters some $q \leq p$ lie in A. To see this, let $R_k(A)(x, y) \wedge x_i \neq y_i \wedge q \in \{x_i, y_i\} \wedge q \leq p = x_j$. Then $p \in A$. Therefore $R_k(A)$ has the Finiteness Condition. Since $R \subseteq R_k(A)$, clearly R has the Finiteness Condition.

For the second claim, let order preserving $R \subseteq Q[0, 1]^k \times Q[0, 1]^k$ have the Finiteness Condition and not alter 0. Repeat the proof of the first claim, noting that the A constructed there must not contain 0. Conversely, note that $R_k(A)$, $0 \notin A$, A finite, has the Finiteness Condition and does not alter 0. Therefore R has the Finiteness Condition and does not alter 0. □

Maximal Emulation Small Use/3. MESU/3. Every order preserving $R \subseteq Q[0, 1]^k \times Q[0, 1]^k$ with the Finiteness Condition, not altering 0, is ME usable.

Theorem 3.4.7 *(RCA_0) MESU/2 ↔ MESU/3. For $k \geq 1$, MESU/2 for dimension k ↔ MESU/3 for dimension k.*

Proof The first claim follows from the second claim. Assume MESU/2 for k, and let $R \subseteq Q[0, 1]^k \times Q[0, 1]^k$ be order preserving with the Finiteness Condition, not altering 0. By Theorem 3.4.6, let $R \subseteq R_k(A)$, A finite, $0 \notin A$. By MESU/2 for k, $R_k(A)$ is ME usable. Hence by Theorem 2.6, R is ME usable. Assume MESU/3 for k, and let $A \subseteq Q(0, 1]$ be finite. By Theorem 3.4.6, $R_k(A)$ is order preserving and has the Finiteness Condition and does not alter 0. Hence $R_k(A)$ is ME usable. □

In MED/1 above, it is natural to ask which numbers can be used.

Droppable Tuples Definition $x, y \in Q[0, 1]^k$ are droppable if and only if

i. $x_k = y_k$.
ii. $(x_1, \ldots, x_{k-1}), (y_1, \ldots, y_{k-1})$ are order equivalent.
iii. For all $1 \leq i \leq k$, $(x_i < x_k \vee y_i < y_k) \rightarrow x_i = y_i$.

Maximal Emulation Drop/2. MED/2. Let $x, y \in Q[0, 1]^k$. The following are equivalent.

i. For finite subsets of $Q[0, 1]^k$, some maximal emulator is drop equivalent at x, y.
ii. x, y are droppable or $x_k = y_k = 0$.

In fact, we have the following multiple form.

Maximal Emulation Drop/3. MED/3. Let $x_1, y_1, \ldots, x_n, y_n \in Q[0, 1]^k$. The following are equivalent.

i. For finite subsets of $Q[0, 1]^k$, some maximal emulator is drop equivalent at every x_i, y_i.
ii. For finite subsets of $Q[0, 1]^k$ and $1 \leq i \leq k$, some maximal emulator is drop equivalent at (x_i, y_i)
iii. For all i, x_i, y_i is droppable or $(x_i)_k = (y_i)_k = 0$.

Theorem 3.4.8 *(RCA_0) Each of MESU/1, 2, 3, MED/1, 2, 3 follows from MESU/2 and implies MED/1. For $k \geq 1$, MESU/1 for dimension k, MESU/2, 3, MED/1, 2, 3 for dimension $k+1$ each follow from MESU/2 for dimension $k+1$ and each imply MED/1 for dimension $k+1$. Each of MESU/2, 3, MED/1, 2, 3 hold for dimension $k=1$.*

Proof Fix $k \geq 1$. The first claim follows from the second claim. Assume MESU/2 for $k+1$. We have MESU/3 for $k+1$ by Theorem 3.4.7. For MESU/1 for k, let $R \subseteq Q[0, 1]^k \times Q[0, 1]^k$ be finite and order preserving. Let A be the set of numbers appearing in R. We claim that the lower parameterization R' of R is contained in $R_{k+1}(A)$. To see this, let $R'(x, y)$. Then $R(x_1, \ldots, x_k) \wedge R(y_1, \ldots, y_k) \wedge x_{k+1} = y_{k+1} < x_1, \ldots, x_k, y_1, \ldots, y_k \wedge x_1, \ldots, x_k, y_1, \ldots, y_k \in A$. If $x_i \neq y_i$ then $1 \leq i \leq k$ and obviously all $x_j \geq x_i$ and $y_j \geq y_i$ lie in A. Hence $R_{k+1}(A)(x, y)$, and so by MESU/2 for $k+1$, R' is ME usable.

For MED/3 for $k+1$, let $x_1, y_1, \ldots, x_n, y_n \in Q[0, 1]^{k+1}$. We show $i \rightarrow ii \rightarrow iii \rightarrow i$. $i \rightarrow ii$ is immediate, and we now show $ii \rightarrow iii$ (for $k+1$), even without using MESU/2 for $k+1$. Assume ii for $k+1$. By the drop equivalencein ii, we have $(x_i)_{k+1} = (y_i)_{k+1}$. Assume $(x_i)_{k+1} = (y_i)_{k+1} = 0$ is false. Then $(x_i)_{k+1} = (y_i)_{k+1} > 0$. To show that x_i, y_i is droppable, we first restate a consequence of drop equivalence in terms of ME usability. Let $R<x_i, y_i> \subseteq Q[0, 1]^{k+1} \times Q[0, 1]^{k+1}$ given by $R(z, w) \leftrightarrow (z_1, \ldots, z_k) = ((x_i)_1, \ldots, (x_i)_k) \wedge (w_1, \ldots, w_k) = ((y_i)_1, \ldots, (y_i)_k) \wedge z_{k+1} = w_{k+1} < (x_i)_k = (y_i)_k$. By ii, $R<x_i, y_i>$ is ME usable, and so by MENU, $R<x_i, y_i>$ is order preserving. Hence $(x_1, \ldots, x_k, 0), (y_1, \ldots, y_k, 0)$ are order equivalent, and therefore $(x_1, \ldots, x_k), (y_1, \ldots, y_k)$ are order equivalent. Now suppose $(x_i)_j < (x_i)_{k+1} \vee (y_i)_j < (x_i)_{k+1}$. Then $((x_i)_1, \ldots, (x_i)_k, (x_i)_j), ((y_i)_1, \ldots, (y_i)_k, (x_i)_j)$ are $R<x_i, y_i>$ related and order equivalent, and therefore $(y_i)_j = (x_i)_j$. This verifies that x_i, y_i is droppable.

Finally, assume iii in MED/3 for $k+1$. We show i in MED/3 for $k+1$ using MESU/2. We first claim that each $R<x_i, y_i> \subseteq R_{k+1}(A)$, where $A = \{(x_i)_1, \ldots, (x_i)_k, (y_i)_1, \ldots, (y_i)_k\} \backslash Q[0, x_{k+1})$. If $(x_i)_{k+1} = (y_i)_{k+1} = 0$ then we have i vacuously. Assume x_i, y_i is droppable. The relation $R<x_i, y_i>$ defined above, associated with x_i, y_i, is order preserving, using that x_i, y_i is droppable. We claim that each $R<x_i, y_i> \subseteq R_k(A)$, where $A = \{(x_i)_1, \ldots, (x_i)_k, (y_i)_1, \ldots, (y_i)_k\} \backslash Q[0, x_{k+1})$. We need to examine $((x_i)_1, \ldots, (x_i)_k, p), ((y_i)_1, \ldots, (y_i)_k, p), 0 \leq p < (x_i)_{k+1} = (y_i)_{k+1}$. Let $(x_i)_j \neq (y_i)_j$. By droppability, $(x_i)_j$, $(y_i)_j \geq x_{k+1} = y_{k+1}$. Also the $x_n \geq (x_i)_j$ and $y_n \geq (y_i)_j$ lie in A. The claim is established.

It is now clear that $\cup_i R<x_i, y_i> \subseteq R_{k+1}(A)$, and so $\cup_i R<x_i, y_i>$ is ME invariantly usable. It is now clear that i of MED/3 holds.

To obtain MED/1, 2 for $k+1$, obviously MED/3 for $k+1 \rightarrow$ MED/2 for $k+1 \rightarrow$ MED/1 for $k+1$.

So see that MESU/1 for k, MESU/2, 3, MED/1, 2, 3 for $k+1$ each imply MED/1 for $k+1$, it suffices to show that MESU/1 for k and MESU/3 for $k+1$ imply MED/1 for $k+1$. Note that MED/1 for $k+1$ asserts that the following $R \subseteq Q[0, 1]^{k+1} \times Q[0, 1]^{k+1}$ is ME invariantly usable. $R(x, y) \leftrightarrow (x_1, \ldots, x_k) = (1, 1/2, \ldots, 1/k) \wedge (y_1, \ldots, y_k) = (1/2, \ldots, 1(k+1) \wedge x_{k+1} = y_{k+1} < 1/(k+1)$. Since R is contained in the lower parameterization of a finite order preserving $R' \subseteq Q[0, 1]^k \times Q[0, 1]^k$, and also is order preserving with the Finiteness Condition and not altering 0, we can simply apply MESU/1 for k and MESU/3 for $k+1$.

For the final claim, MED/1, 2, 3 are vacuous for $k = 1$. For MESU/2 for $k = 1$, note that for finite $A \subseteq Q[0, 1]$, $R_1(A)$ is ME usable by, e.g., MELU/1 of Sect. 12.3.3. For MESU/3, use Theorem 3.4.7. □

We now obtain dimension reduction as in Theorem 3.4.10 below.

Definition 3.4.2 Let $R \subseteq Q[0, 1]^k \times Q[0, 1]^k$ and $1 \leq i < k$. $\gamma(R, i) \subseteq Q[0, 1]^{k+1} \times Q[0, 1]^{k+1}$ is given by $\gamma(R, i)(x, y) \leftrightarrow R((x_1, \ldots, x_i, x_{i+2}, \ldots, x_k), (y_1, \ldots, y_i, y_{i+2}, \ldots, y_k)) \wedge x_i = x_{i+1} \wedge y_i = y_{i+1}$.

Lemma 3.4.9 *(RCA_0) Let $R \subseteq Q[0, 1]^k \times Q[0, 1]^k$ and $1 \leq i < k$. R is ME usable if and only if $\gamma(R, i)$ is ME usable.*

Proof Left to the reader. □

Theorem 3.4.10 *For all $k \geq 1$, any of MESU/1, 2, 3, MED/1, 2, 3 for dimension $k+1$ implies that same statement for dimension k.*

Proof Assume MESU/1 for dimension $k+1$. Let $R \subseteq Q[0, 1]^k \times Q[0, 1]^k$ be finite and order preserving. We want to show that the lower parameterization $LP(R) \subseteq Q[0, 1]^{k+1} \times Q[0, 1]^{k+1}$ is ME usable. By Lemma 3.4.9, it suffices to show that $\gamma(LP(R), k)$ is ME usable. But $\gamma(LP(R), k) = LP(\gamma(R, k-1))$. Since $\gamma(R, k-1) \subseteq Q[0, 1]^{k+1} \times Q[0, 1]^{k+1}$ is finite and order preserving, $LP((R, k-1))$ is ME usable by MESU/1 for dimension $k+1$.

Assume MESU/2 for dimension $k+1$. Let $A \subseteq Q(0, 1]$ be finite. We want to show that $R_k(A) \subseteq Q[0, 1]^k \times Q[0, 1]^k$ is ME usable. Now $\gamma(R_k(A), k) \subseteq R_{k+1}(A)$, and so $\gamma(R_k(A), k) \subseteq Q[0, 1]^{k+1} \times Q[0, 1]^{k+1}$ is ME usable. By Lemma 3.4.9, $R_k(A)$ is ME usable.

Assume MESU/3 for dimension $k+1$. Apply Theorem 3.4.7 to the previous claim.

Assume MED/2 for dimension $k+1$. Let $x, y \in Q[0, 1]^k$. We have seen that in MED/2 for dimension k, $i \rightarrow ii$ is provable in RCA_0 in the proof of Theorem 3.4.8 (without assuming MED/2 for dimension $k+1$). Assume ii in MED/2. If $x_k = y_k = 0$ then i in MED/2. Now assume x, y are droppable. Let $x' = (x_1, \ldots, x_{k-1}, x_{k-1}, x_k)$, $y' = (y_1, \ldots, y_{k-1}, y_{k-1}, y_k)$. Then $x', y' \in Q[0, 1]^{k+1}$ are droppable. Let $R \subseteq Q[0, 1]^k \times Q[0, 1]^k$ be the symmetric relation associated with drop equivalence at x, y. Then $\gamma(R, k-1) \subseteq Q[0, 1]^{k+1} \times Q[0, 1]^{k+1}$ is the symmetric relation associated with drop equivalence at x', y'. By MED/2 for dimension $k+1$, $\gamma(R, k-1)$ is ME usable. Hence by Lemma 3.4.9, R is ME usable. Therefore i in MED/2 holds.

Assume MED/3 for dimension $k+1$. Let $x_1, y_1, \ldots, x_n, y_n \in Q[0, 1]^k$. We have seen that in MED/3 for dimension k, $i \rightarrow ii \rightarrow iii$ is provable in RCA_0 in the proof of Theorem 3.4.8. (without assuming MED/3 for dimension $k+1$). Assume iii in MED/3. For $1 \leq i \leq k$, let x_i', y_i' be defined as in the previous paragraph. Let $R \subseteq Q[0, 1]^k \times Q[0, 1]^k$ be the union of the symmetric relations associated with drop equivalence at the x_i, y_i that do not have $(x_i)_k = (y_i)_k = 0$. Then $\gamma(R, k-1) \subseteq Q[0, 1]^{k+1} \times Q[0, 1]^{k+1}$ is the symmetric relation associated with drop equivalence in the multiple form for these x_i', y_i'. By MED/3 for dimension $k+1$, $\gamma(R, k-1)$ is ME usable. Hence by Lemma 3.4.9, R is ME usable. Therefore i in MED/3 holds.

The claim for MED/1 will be addressed elsewhere in Concrete Mathematical Incompleteness (in preparation). □

The following result shows that MESU/1, 2, 3, MED/1, 2, 3 are far beyond the reach of ZFC.

Theorem 3.4.11 *RCA_0 proves MED/1 → Con(SRP). The following is provable in EFA. $(\forall n)(\exists k)$(RCA_0 proves (MED/1 for dimension k) → Con(SRP[n])).*

Proof This reversal will appear in Concrete Mathematical Incompleteness (in preparation). □

12.3.5 Exotic Proof

This entire section is devoted to the proof of MESU/2 in dimension k. As a Corollary, we obtain all six of MESU/1, 2, 3, MED/1, 2, 3 through Theorems 3.4.8 and 3.4.10. We have already dispensed with the trivial case $k = 1$, and thus focus on $k \geq 2$. The key technique for proving MESU/2 is to use a certain transfinite extension of our space $Q[0, 1] = (Q[0, 1], <)$. The length of the transfinite constructions that we will use will be roughly some strategically chosen limit ordinal κ. At some point in the proof, we will need that κ is an uncountable regular cardinal, the smallest of which is ω_1. Then soon later, difficulties arise, and the proof starts to work only for dimension $k = 2$. At that point in the proof, we will continue to assume that κ is an uncountable regular cardinal (such as ω_1) for dimension $k = 2$. However, to carry on the proof for dimensions $k \geq 3$, we will assume that κ is a $(k - 2)$-subtle cardinal. The rest of the proof of MESU/2 proceeds normally and uniformly in k.

Thus the proof of MESU/2 for dimension $k = 2$ goes through naturally using $\kappa = \omega_1$, and therefore well within ZFC, and, with standard modifications, in Z or even in Z_3. For dimension $k \geq 3$ we use ZFC + $(\exists\kappa)(\kappa$ is $(k - 2)$-subtle), which is a weak fragment of SRP[$k - 1$]. See Appendix A. We conjecture that MESU/2 is provable in RCA_0 for dimension $k = 2$, and that such a proof will be far more complicated but of a totally different character than the proof given here using ω_1, involving a painstaking combinatorial analysis of maximal emulation in two dimensions. However, we conjecture that the natural extension of MESU/2 for dimension $k = 2$, discussed in Sect. 12.3.6, cannot be proved in ZFC\P or Z_2.

Towards the proof of MESU/2, fix $k \geq 2$, finite $E \subseteq Q[0, 1]^k$, and finite $A \subseteq Q(0, 1]$. We construct a maximal emulator S of $E \subseteq Q[0, 1]^k$ that is $R_k(A)$ invariant. We can assume without loss of generality that $A = \{q_1 < \cdots < q_n = 1\}$, where $n \geq 1$ and $q_1 > 0$. We also assume, without loss of generality, that $E \subseteq Q(0, 1)^k$. Let κ be a limit ordinal. At certain points in the proof we will be making further requirements on κ. Ordinals are treated in the usual set theoretic way as the set of their predecessors. This is the epsilon connected transitive set definition. < is used to compare ordinals, but also to compare rationals. For other comparisons, we adorn the < symbol.

Definition 3.5.1 $T[\kappa] = \kappa{+}1 \times Q[0, 1)$. For $x \in T[\kappa]$, $ord(x) = x_1$. $<_{T[\kappa]}$ is the linear ordering on $T[\kappa]$ given by $x <_{T[\kappa]} y$ if and only if $ord(x) < ord(y) \vee (ord(x) = ord(y) \wedge x_2 < y_2)$. $x \leq_T y \leftrightarrow (x <_T y \vee x = y)$. S is a $T[\kappa]$-emulator of E if and only if $S \subseteq T[\kappa]^k$, and every $(x, y) \in S^2$ is order equivalent to some $(z, w) \in E^2$ (using $<_{T[\kappa]}$ and numerical <). S is a maximal $T[\kappa]$-emulator of E if and only if S is a $T[\kappa]$-emulator of E which is not a proper subset of any T[κ]-emulator of E. For $x \in T[\kappa]^k$, ord(x) = max(ord(x_1), …, ord(x_k)). λ always denotes a limit ordinal.

Note that in the above definition, we use a notion of emulator (T[κ]-emulator) in which the emulators live in a different space than the sets being emulated. We have avoided this earlier in the paper, but here it is very convenient. And here, and later, since there are so many linear orderings being defined, we parenthetically mention which linear orderings are being presently used.

We view $\kappa+1 \times \{0\}$ as the preferred closed subset of T[κ] under $<_{T[\kappa]}$.

Definition 3.5.2 Fix an effective enumeration $0 = p_0, p_1, \ldots$ of Q[0, 1), without repetition. <‘ is the linear ordering ordering on Q[0, 1) of type ω given by the p's. <* is the linear ordering on T[κ] given by x <* y if and only if ord(x) < ord(y) ∨ (ord(x) = ord(y) ∧ x_2 <‘ y_2. <** is the linear ordering on $T[\kappa]^k$ given by x <** y if and only if max(x) <* max(y) ∨ (max(x) = max(y) ∧ x is lexicographically earlier than y using <* on the k coordinates), where the four max's here are with respect to <*. x ≤* y ↔ x <* y ∨ x = y. x ≤** y ↔ x <** y ∨ x = y. The greedy T[κ]-emulator of E, GE(E, T[κ]) ⊆ $T[\kappa]^k$, is defined by transfinite recursion on <** given by the equation GE(E, T[κ]) = {x ∈ $T[\kappa]^k$: GE(E, T[κ])|<**x ∪ {x} is a T[κ]-emulator of E}.

Although we write $<_{T[\kappa]}$ to distinguish it from < (numerical comparison of rationals), we decided to write just <* and <** for readability.

Note that Definition 3.5.2 relies on the easily established fact that <* and <** are well orderings. Note also that we are using two orderings on T[κ], the dense $<_{T[\kappa]}$ and the well ordering <*. Here the T[κ]-emulator is constructed along the particular extension <** of <* to $T[\kappa]^k$.

Lemma 3.5.1 *The following hold. Below, α < κ.*

i. *GE(E, T[κ]) is uniquely defined, by the equation in Definition 3.4.4.*
ii. *GE(E, T[κ]) is a maximal T[κ]-emulator of E.*
iii. *For x, y ∈ T[κ], x ≤* y → ord(x) ≤ ord(y).*
iv. *For x, y ∈ $T[\kappa]^k$, x ≤** y → ord(x) ≤ ord(y).*
v. *For x ∈ T[κ], x $<_{T[\kappa]}$ (α, 0) ↔ x <* (α, 0) ↔ ord(x) < α.*
vi. *For x ∈ $T[\kappa]^k$, (∀i)(x_i $<_{T[\kappa]}$ (α, 0)) ↔ ord(x) < α ↔ x <** ((α, 0), (0, 0), …, (0, 0)).*

Proof i is left to the reader. For ii, GE(E, T[κ]) = $\cup_x$GE(E, T[κ])|<**x, and so is the union of T[κ]-emulators of E. Hence GE(E, T[κ]) is a T[κ]-emulator of E. For maximality, suppose GE(E, T[κ]) ∪ {x} is a T[κ]-emulator of E. Then GE(E, T[κ])|<**x ∪ {x} is a T[κ] emulator of E, putting x in GE(E, T[κ]).

iii is immediate from the definition of <*. For iv, let x ≤** y. Then max(x) ≤* max(y), where the max's use ≤*. Now apply iii.

For v, let x $<_{T[\kappa]}$ (α, 0). ord(x) = α is impossible since x is lexicographically earlier than (α, 0). Hence ord(x) < α.

For vi, we show $(\forall i)(x_i <_{T[\kappa]} (\alpha, 0)) \rightarrow \mathrm{ord}(x) < \alpha \rightarrow x <^{**} ((\alpha, 0), (0, 0), \ldots, (0, 0)) \rightarrow (\forall i)(x_i <_{T[\kappa]} (\alpha, 0))$. The first implication is by vii. The second implication is clear by comparing the max's of both sides, max with respect to $<^*$. For the third implication, suppose $x <^{**} ((\alpha, 0), (0, 0), \ldots, (0, 0))$. If $\max(x) = (\alpha, 0)$ then x is lexicographically earlier than $((\alpha, 0), (0, 0), \ldots, (0, 0))$, which is impossible. If $\max(x) <^* (\alpha, 0)$, max with respect to $<^*$, then $\mathrm{ord}(x) < \alpha$ and so $(\forall i)(x_i <_{T[\kappa]} (\alpha, 0))$. □

Definition 3.5.3 Let $x \in T[\kappa]^k$. $x \backslash \kappa$ is the tuple of length $\leq k$ obtained by deleting all coordinates of x whose first term is κ, from x. For this purpose only, we allow the 0-tuple, whose ord is taken to be 0. $x(\kappa|\alpha)$ and $x(\alpha|\kappa)$ are the results of replacing κ by α and α by κ, respectively, throughout x, where $\alpha \leq \kappa$. These ordinal replacements are done at the first terms of the coordinates of x only.

Lemma 3.5.2 *Let κ be an uncountable regular cardinal. There is a closed unbounded $C \subseteq \kappa$ of limit ordinals such that the following holds. Let $x \in T[\kappa]^k$. If $\mathrm{ord}(x \backslash \kappa) < \lambda \in C$, then $x \in GE(E, T[\kappa]) \leftrightarrow x(\kappa|\lambda) \in GE(E, T[\kappa])$. If $\mathrm{ord}(x) \leq \lambda \in C$, then $x \in GE(E, T[\kappa]) \leftrightarrow x(\lambda|\kappa) \in GE(E, T[\kappa])$.*

Proof Let κ be as given. We form the structure $M = (\kappa+1, Q[0, 1), E, <_{\kappa+1}, <_{Q[0, 1)}, <', GE(E, T[\kappa]))$, where

i. The two domains are κ+1 and Q[0, 1).
ii. E is the current E as a k-ary predicate on Q[0, 1).
iii. $<_{\kappa+1}$ is the usual ordering on κ+1.
iv. $<_{Q[0, 1]}$ is the usual numerical ordering on Q[0, 1).
v. $<'$ is the ordering of Q[0, 1) given by the p's from Definition 3.5.2.
vi. GE(E, T[κ]) is used as a 2k-ary predicate whose 2 k arguments are of sorts κ+1, Q[0, 1), κ+1, Q[0, 1), …, κ+1, Q[0, 1).

There is a crucial sentence φ that holds in M. φ asserts that GE(E, T[κ]) is a greedy T-emulator of E (which we know is unique), and is formulated as follows. A given 2 k-tuple x lies in (this form of) GE(E, T[κ]) if and only if for all $y \leq^{**} x$ from GE(E, T[κ]), (x, y) is order equivalent to some $(z, w) \in E^2$ (using $<_{T[\kappa]}$ and $<_{Q[0, 1)}$). The $\leq^{**}$ here is our usual $\leq^{**}$, and is formulated in the obvious way using $<_{\kappa+1}$ and $<'$. The $<_{T[\kappa]}$ here is formulated in the obvious way using $<_{\kappa+1}$ and $<_{Q[0, 1)}$.

We will ignore these differences in presentation, and effectively regard the present 2k-ary predicate GE(E, T[κ]) as the same as GE(E, T[κ]).

By standard techniques from elementary model theory, we form a transfinite sequence of elementary substructures M_α, $\alpha < \kappa$, of M, whose first domains have the following property: the set of first coordinates of its elements is of the form $\gamma_\alpha \cup \{\kappa\}$. Furthermore, the set of these γ_α, $\alpha < \kappa$, forms a closed unbounded set of limit ordinals $< \kappa$. I.e., each $\gamma_\alpha < \gamma_{\alpha+1}$ and each γ_λ, $\lambda < \kappa$, is the sup of the γ_α, $\alpha < \lambda$. This construction crucially relies on κ being an uncountable regular cardinal. Clearly κ appears in all M_α, and the second domains are all Q[0, 1).

To clarify this picture for the reader, it is automatic that κ appears in the first domain of an elementary substructure of M, and the second domain is just Q[0, 1), but generally there could be lots of gaps in the ordinals appearing in the first domain, whereby many ordinals less than many ordinals appearing in the first domain are missing from that first domain. In the above construction, we have arranged that all such gaps have been filled in (of course, there must be one gigantic gap below κ itself). Also the use of elementary substructures is very convenient but clearly overkill, as we need only weak forms of elementarity.

Write $M_\alpha = (\{\kappa\} \cup \gamma_\alpha, Q[0, 1), E, <_{\kappa+1}|, <_{Q[0,1)}, <', GE(E, T[\kappa])|)$, where $|$ is used to restrict to the first domain $\{\kappa\} \cup \gamma_\alpha$. Let $j: \{\kappa\} \cup \gamma_\alpha \cup Q[0, 1) \to \gamma_\alpha+1 \cup Q[0, 1)$ be the identity on $\gamma_\alpha \cup Q[0, 1)$ and send κ to γ_α. Let $M_\alpha{}^*$ be the unique structure such that j is an isomorphism from M_α onto $M_\alpha{}^*$. Write $M_\alpha{}^* = (\gamma_\alpha+1, Q[0, 1), E, \leq_{\kappa+1}|, <_{Q[0,1)}, <', X_\alpha)$, where $<_{\kappa+1}|$ is the usual ordering on $\gamma_\alpha+1$. Then $M_\alpha{}^*$ is no longer an elementary substructure of M, but it is elementarily equivalent to M. Therefore $M_\alpha{}^*$ satisfies φ.

Now read the description of φ above as a sentence about $M_\alpha{}^*$. It is now clear that $X_\alpha = GE(E, T[\kappa]) \cap (\gamma_\alpha+1)^k$.

Now set $C = \{\gamma_\alpha : \alpha < \kappa\}$. Let $x \in T[\kappa]^k$. Suppose $ord(x \backslash \kappa) < \gamma_\alpha$. Clearly $j(x) = x(\kappa|\gamma_\alpha)$ since j is the identity below γ_α and sends κ to γ_α. Since j is an isomorphism, we have $x \in GE(E, T[\kappa]) \leftrightarrow j(x) \in X_\alpha \leftrightarrow j(x) \in GE(E, T[\kappa])$, j acting coordinatewise. For the second claim, let $ord(x) \leq \lambda \in C$. Then $ord(x(\lambda|\kappa)\backslash\kappa) < \lambda$, and hence $x(\lambda|\kappa) \in GE(E, T[\kappa]) \leftrightarrow x(\lambda|\kappa)(\kappa|\lambda) = x \in GE(E, T[\kappa])$. □

We fix the closed unbounded $C \subseteq \kappa$ given by Lemma 3.5.2.

Definition 3.5.4 Let $A \subseteq \kappa$, $x \in T[\kappa]^k$. The max coordinates in x are the x_i such that every $x_j \leq_{T[\kappa]} x_i$. The top coordinates/A of x are the max coordinates that lie in $A \times \{0\}$. The high coordinates/A of x are the x_i for which all $x_j \geq_{T[\kappa]} x$ lie in $A \times \{0\}$, and x_i is not max in A. The low coordinates/A of x are the coordinates that are neither top/A nor high/A. It may be that all x_i are low coordinates/A of x. We often write x_i is max in x, x_i is top/A, x_i is high/A, x_i is low/A.

Note that all max coordinates in x have the same first coordinate.

Lemma 3.5.3 *Let $A \subseteq \kappa$, $x \in T[\kappa]^k$.*

i. *All low coordinates/A of x are less than all high coordinates/A of x are less than all top coordinates/A of x, using $<_{T[\kappa]}$.*
ii. *If there is a high coordinate/A of x then the top coordinates/A of x are exactly the max coordinates of x.*
iii. *$x_i \geq_{T[\kappa]} x_j \wedge x_j$ is high/A $\to$ (x_i is high/A $\vee$ x_i is top/A).*

Proof For i, let x_i be low/A, x_j be high/A, $x_i \geq_{T[\kappa]} x_j$. Then all $x_n \geq_{T[\kappa]} x_i$ are in $A \times \{0\}$. Since x_i is not high/A, x_i is max in x. But then x_i is top/A, which is a contradiction. Now let x_i be high/A, x_j be top/A, $x_i \geq_{T[\kappa]} x_j$. Since x_j is max in x, clearly x_i is max in x, which is a contradiction.

For ii, let x_i be high/A. Then all $x_j \geq x_i$ lie in $A \times \{0\}$. If x_j is max in x then $x_j \geq_{T[\kappa]} x_i$, $x_j \in A \times \{0\}$, and so x_j is top/A. If x_j is top/A then x_j is max in x by definition.

For iii, let $x_i \geq_{T[\kappa]} x_j$ and x_j is high/A. Then all $x_n \geq_{T[\kappa]} x_i$ lie in $A \times \{0\}$, which makes x_i high/A or max in x. In the latter case, x_i is top/A. □

Definition 3.5.5 As in Definition 3.4.1, Let $A \subseteq \kappa$. $R_k(A, T[\kappa]) \subseteq T[\kappa]^k \times T[\kappa]^k$ is given by $R_k(A, T[\kappa])(x, y)$ if and only if

i. x, y are order equivalent (using $<_{T[\kappa]}$).
ii. If $x_i \neq y_i$ then all $x_j \geq_{T[\kappa]} x_i$ and $y_j \geq_{T[\kappa]} y_i$ lie in $A \times \{0\}$.

Lemma 3.5.4 *Let $A \subseteq \kappa$ and $R_k(A, T[\kappa])(x, y)$. The following hold.*

i. x, y are order equivalent (using $<_{T[\kappa]}$).
ii. $x_i \neq y_i \rightarrow x_i, y_i \in A \times \{0\}$.
iii. $x_i \in A \times \{0\} \leftrightarrow y_i \in A \times \{0\}$.
iv. $ord(x_i) \in A \leftrightarrow ord(y_i) \in A$.
v. x_i is top/A $\leftrightarrow$ y_i is top/A.
vi. x_i is high/A $\leftrightarrow$ y_i is high/A.
vii. x_i is low/A $\leftrightarrow$ y_i is low/A.

Proof i, ii are immediate from the definition. For iii, let $x_i \in A \times \{0\}$. If $x_i = y_i$ then $y_i \in A \times \{0\}$. If $x_i \neq y_i$ then $x_i, y_i \in A \times \{0\}$. The converse is proved in the same way. For iv, let $ord(x_i) \in A$. If $x_i = y_i$ then $ord(y_i) \in A$. If $x_i \neq y_i$ then $x_i, y_i \in A \times \{0\}$, and so $ord(x_i), ord(y_i) \in A$. The converse is proved in the same way.

For v, let x_i be top/A. Then $x_i \in A \times \{0\}$ is max in x. By order equivalence (using $<_{T[\kappa]}$) and iii, $y_i \in A \times \{0\}$ is max in y. The converse is proved the same way. For vi, let all $x_j \geq x_i$ lie in $A \times \{0\}$, where x_i is not max in x. By order equivalence (using $<_{T[\kappa]}$) and iii, all $y_j \geq y_i$ lie in $A \times \{0\}$ and y_j is not max in x. The converse is proved the same way. For vii, let y_i be not high/A and not top/A. Then y_i is not high/A and not top/A by v, vi. The converse is proved the same way. □

Those readers who wish to continue to stay within ZFC are going to have to now assume $k=2$. For more adventurous readers, assume $k \geq 3$. In the proof of the following Lemma, we will separate the case $k=2$ from the case $k \geq 3$. But after Lemma 3.5.5, there will be no difference.

Lemma 3.5.5 *Let $k=2$, or let $k \geq 3$ and our existing κ be a $(k-2)$-subtle cardinal. There exists infinite $C' \subseteq C$ of order type ω such that $GE(E, T[\kappa])$ is invariant under the relation $R_k(C', T[\kappa]) \subseteq T[\kappa]^k \times T[\kappa]^k$.*

Proof We first assume $k=2$. Let $R_2(C, T[\kappa])(x, y)$. We can almost get $x \in GE(E, T[\kappa]) \leftrightarrow y \in GE(E, T[\kappa])$ without even shrinking C to C′. We will show this under the assumption $x, y \notin (C \times \{0\})^2$, and worry about the general case later. We can obviously assume $x \neq y$. Thus we have $x_1 \neq y_1 \vee x_2 \neq y_2$.

case 1. $x_1 \neq y_1$. Then $x_1, y_1 \in C \times \{0\}$, and therefore $x_2, y_2 \notin C \times \{0\}$. Hence $x_2 = y_2 <_{T[\kappa]} x_1, y_1$. Given this relationship between x, y, we can now apply Lemma 3.5.2, claim 2. Thus raising x_1, y_1 both to $(\kappa, 0)$ does not change the status of membership in $GE(E, T[\kappa])$, and x, y become identical. Therefore $x \in GE(E, T[\kappa]) \leftrightarrow y \in GE(E, T[\kappa])$.

case 2. $x_2 \neq y_2$. Argue as in case 1 with subscripts 1, 2 switched.

In order to handle the general case (still with $k=2$), we must shrink C. Let $C' \subseteq C$ be of order type ω such that membership of ((a, 0), (b, 0)) in GE(E, T[κ]) depends only on the order type of $(a, b) \in C'^2$, using the usual infinite Ramsey theorem from Ramsey (1930). Suppose $x, y \notin (C \times \{0\})^2$ is false. By Lemma 3.5.4, iii, $x, y \in (C \times \{0\})^2$. Let $R_k(C', T[\kappa])(x, y)$. We claim $x \in GE(E, T) \leftrightarrow y \in GE(E, T)$. This is clear since x, y are order equivalent (using $<_{T[\kappa]}$). We have established that GE(E, T) is invariant under $R_2(C', T[\kappa])$.

Now let $k \geq 3$ and assume κ is a (k − 2)-subtle cardinal. At this point, we refer the reader to Appendix A for the relevant definitions. The condition satisfied by κ is as follows (for any closed unbounded $C \subseteq \kappa$, not just the C we are using at this point in the present proof).

(1) Let $f: S_{k-2}(C) \rightarrow S(\kappa)$ be regressive. There is an f-homogenous $C' \subseteq C$ of cardinality k − 1.

This is not quite strong enough for our purposes. Fortunately, in Friedman (2001), Lemma 1.6, we prove the following.

(2) Let $f: S_{k-2}(C) \rightarrow S(\kappa)$ be regressive. There are f-homogenous $C' \subseteq C$ of every cardinality $<\kappa$.

We can obviously assume that our closed unbounded $C \subseteq \kappa$ consists entirely of uncountable cardinals by shrinking if necessary. We will use (2) only to obtain f-homogenous $C^* \subseteq C$ of order type ω.

Let $D \in S_{k-2}(C)$ and $x \in T[\kappa]^k$. We say that x is D-controlled/C if and only if

i. x has a top/C coordinate.
ii. Every high/C coordinate of x lies in $D \times \{0\}$.
iii. Every low/C coordinate of x lies in $\min(D) \times Q[0, 1)$.

Suppose x is D-controlled/C. Define $\gamma(D, x) \in T[\kappa]^k$ to be the result of replacing each high/C coordinate $x_i = (u, 0) \in D \times \{0\}$ of x, with (m, 0), where u is the m-th element of D, counting from 1, and each top coordinate of x with (0, 0). If x is not D-controlled, set $\gamma(D, x) = (0, 0)$.

We claim that every $\gamma(D, x) \in (\min(D) \times Q[0, 1))^k$. This is evident by inspection for D-controlled/C x. For other x, $\gamma(D, x) = (0, 0) \in (\min(D) \times Q[0, 1))^k$.

We use standard tuple coding whereby finite sequences from $T[\kappa] \cup \omega$ are coded by ordinals in a standard one-one way. Thus the standard tuple code of every $\gamma(D, x)$ is $<\min(D)$.

We define regressive $f: S_{k-2}(C) \rightarrow S(\kappa)$ as follows. $f(D) = \{\beta: (\exists x \in GE(E, T[\kappa])(x$ is D-controlled/C $\wedge$ β is the standard code of $\gamma(D, x))\}$.

We now use (2) to obtain an f-homogeneous set $C^* \subseteq C$ of order type ω which is f-homogenous. By the infinite Ramsey theorem from Ramsey (1930), we fix $C' \subseteq C^*$ of order type ω such that for all order equivalent $x, y \in C'^k$ (using $<_{T[\kappa]}$), $x \in GE(E, T[\kappa]) \leftrightarrow y \in GE(E, T[\kappa])$.

We now fix x, y such that $R_k(C', T[\kappa])(x, y)$. We want to prove that $x \in GE(E, T[\kappa]) \leftrightarrow y \in GE(E, T[\kappa])$. Clearly x, y are order equivalent (using $<_{T[\kappa]}$). For the easy case, suppose $x \in (C' \times \{0\})^k \vee y \in (C' \times \{0\})^k$. It is clear that $x, y \in (C' \times$

$\{0\})^k$ using Lemma 3.5.4, iii. Since x, y $\in (C' \times \{0\})^k$ and x, y are order equivalent (using $<_{T[\kappa]}$), we have x $\in$ GE(E, T[κ]) $\leftrightarrow$ y $\in$ GE(E, T[κ]) by the construction of C′.

We can now assume x, y $\notin (C' \times \{0\})^k$, and show x $\in$ GE(E, T[κ]) $\leftrightarrow$ y $\in$ GE(E, T[κ]). By Lemmas 3.5.3 and 3.5.4, if there are no top/C′ coordinates in x or there are no top/C′ coordinates in y, then there are no high/C′ coordinates in x, y, and so all coordinates of x, y are low/C′. In this case, x = y, and we are done.

So we assume that x, y both have top/C′ coordinates.

Let W be the set of all high/C′ coordinates in x and W′ be the set of all high/C′ coordinates in y. Note that x, y both must have the same one or more low/C′ coordinates, because x, y both have coordinates not in $C' \times \{0\}$. Thus W, W′ both omit at least two coordinates, and hence $|W|, |W'| \leq k - 2$.

In fact, since the high/C′ coordinates of x, y appear at the same positions (Lemma 3.5.4, vi), we have $|W| = |W'| \leq k - 2$.

Now extend W, W′ with the same new ordinals from C′ that are greater than max(W ∪ W′), ord(x), ord(y), and all low/C′ coordinates of x, y, arriving at D, $D' \subseteq C'$, each with exactly $k - 2$ elements.

We claim that x is D-controlled/C′ and y is D′-controlled/C′. x, y both have top/C′ coordinates. Every high/C′ coordinate of x lies in $W \times \{0\} \subseteq D \times \{0\}$, and every high/C′ coordinate of y lies in $W' \times \{0\} \subseteq D' \times \{0\}$. Also given the construction of W, W′, D, D′, we see that every low/C′ coordinate of x, y lies in min(D) × Q[0, 1).

We claim that every D-controlled/C′ u is also D-controlled/C. To see this, assume u is D-controlled/C′. Obviously every top/C′ coordinate of u is a top/C coordinate of u. Suppose u_i is high/C′. Then every $u_j \geq_{T[\kappa]} u_i$ lies in $C' \times \{0\}$ and u_i is not max in u. Then every $u_j \geq_{T[\kappa]} u_i$ lies in $C \times \{0\}$, and so u_i is high/C.

We claim that $\gamma(D, x) = \gamma(D', y) \in \min(D \cup D')^k$. To see this, γ(D, x) results from x by replacing each high/C′ coordinate $x_i = (u, 0) \in D \times \{0\}$ of x, with the integer m, where u is the m-th element of D, and each top/C′ coordinate of x with 0. Also γ(D′, y) results from y by replacing each high/C′ coordinate $y_i = (u, 0) \in D' \times \{0\}$ of y, with the integer m, where u is the m-th element of D′, and each top/C′ coordinate of y with 0. By Lemma 3.5.4, the top/C′ and high/C′ coordinates of x, y lie at the same positions, and x, y have the same low/C′ coordinates in the same positions, evidently γ(D, x) = γ(D′, y). The non integer coordinates are the necessarily common low/C′ coordinates in x, y which we have seen are < min(D), min(D′), and hence < min(D ∪ D′).

We claim that if u, v are D-controlled/C and γ(D, u) = γ(D, v), then u $\in$ GE(E, T[κ]) $\leftrightarrow$ v $\in$ GE(E, T[κ]). Let u, v be D-controlled/C and γ(D, u) = γ(D, v). Given this relationship between u, v, we can now apply Lemma 3.5.2, claim 2. Thus raising the top/C coordinates of u, v to $\kappa \times \{0\}$ does not change the status of membership in GE(E, T[κ]), and after this raising, u, v become identical. Therefore x $\in$ GE(E, T[κ]) $\leftrightarrow$ y $\in$ GE(E, T[κ]).

By the definition of f, the standard code β for γ(D, x) lies in f(x) if and only if there exists x* $\in$ GE(E, T[κ]) such that x* is D-controlled/C ∧ β is the standard code of γ(D, x*). We also have that β $\in$ f(y) if and only if there exists y* $\in$ GE(E, T[κ]) such that y* is D′-controlled/C ∧ β is the standard code of γ(D′, y*).

Suppose x ∈ GE(E, T[κ]). Since x is D-controlled/C′, x is D-controlled/C. Hence the standard tuple code of γ(D, x) lies in f(x). Hence the standard tuple code of γ(D′, y) lies in f(y). Let y* ∈ GE(E, T[κ]), y* is D′-controlled/C, and the standard tuple code of γ(D′, y) equals the standard tuple code of γ(D′, y*). Then γ(D′, y) = γ(D′, y*). By the previous claim, y ∈ GE(E, T[κ]) ↔ y* ∈ GE(E, T[κ]). Hence y ∈ GE(E, T[κ]).

Conversely, suppose y ∈ GE(E, T[κ]). Argue as in the previous paragraph, that x ∈ GE(E, T[κ]), by switching x with y. □

We fix $C' \subseteq C$ given by Lemma 3.5.5, and write $C' = \{\lambda_1 < \lambda_2 < \ldots\}$.

We now capture the needed properties of our T[κ], GE(E, T[κ]), and C′, into a single structure with basic internal properties. In so doing, we will be removing all mention of κ and $C' \subseteq C \subseteq \kappa$. Recall the data that we fixed at the outset of this exotic proof. Dimension $k \geq 2$, finite $E \subseteq Q(0, 1)^k$, and $A = \{q_1 < \ldots < q_n = 1\}$, $q_1 > 0$.

We use λ_n for the n-th element of $C \subseteq \lambda$, counting from 1.

Lemma 3.5.6 *There is a structure $M = (D, <_D, P, 0, d_1, \ldots, d_n)$, such that the following holds.*

i. *$<_D$ is a dense linear ordering on D with the left endpoint 0 and the right endpoint d_n.*
ii. *$0 <_D d_1 <_D \ldots <_D d_n$.*
iii. *P is a k-ary relation on D.*
iv. *P is an emulator of E in the following sense. Every pair from P (as a 2 k-tuple) is order equivalent to a pair from E (using $<_D$).*
v. *P is point maximal in the following sense. If any k-tuple is added to P, then we no longer have an emulator of E.*
vi. *P is invariant under the binary relation $R_k(\{d_1, \ldots, d_n\})$ on k-tuples.*

In iv, we do not use E directly, but only a finite list of order types of its elements (k-tuples).

In vi, $R_k(\{d_1, \ldots, d_n\})((x_1, \ldots, x_k), (y_1, \ldots, y_k)) \leftrightarrow (x_1, \ldots, x_k), (y_1, \ldots, y_k)$ are order equivalent (using $<_D$) ∧ if $x_i \neq y_i$ then all $x_j \geq_D x_i$ and all $y_j \geq_D y_i$ are among $d_1, \ldots, d_n$.

Proof Take $D = \{x: x \leq_{T[\kappa]} (\lambda_n, 0)\}$. Take $<_D = <_{T[\kappa]} \cap D^2$. Take $P = GE(E, T[\kappa]) \cap D^k$. Take 0 = (0, 0). Take each $d_i = (\lambda_i, 0)$. Note that D is also $\{x: x <^* (\lambda_n, 0)\}$. In addition, $D^k = T[\kappa]^k \cap \{x: x \leq^{**} (d_n, \ldots, d_n)\}$ because $<^{**}$ orders first according to max, using $<^*$, and then lexicographically using $<^*$. So since P is the initial segment of the greedy T[κ]-emulator of E, GE(E, T[κ]), up through D^k, in the sense of $<^{**}$, it is clear that P is a maximal emulator of E in the sense required by iii, iv, v. Also the relation $R_k(\{d_1, \ldots, d_n\})$ defined here is $R_k(C', T[\kappa]) \cap D^{2k}$. By Lemma 3.5.5, GE(E, T[κ]) is invariant under the relation $R_k(C', T[\kappa]) \subseteq T[\kappa]^k \times T[\kappa]^k$. Hence P is invariant under the relation $R_k(\{d_1, \ldots, d_n\})$. □

In Lemma 3.5.6, we have not yet reached countability, as λ_n is generally uncountable.

Lemma 3.5.7 *The structure M given by Lemma 3.5.6 can be taken to be countable.*

Proof By an obvious sequential construction. The need for a sequential argument arises from density in i, and point maximality in v. The rest of the conditions take care of themselves. Construct finite sets $B_1, B_2, \ldots \subseteq D$ as follows. Take $B_1 = \{0, d_1, \ldots, d_n\}$. Suppose finite B_i has been defined. For each pair of distinct elements of B_i, put an intermediate element of D in B_{i+1}. For each $x \in B_i^k$ such that $P \cup \{x\}$ is not an emulator of E in the sense of iv, choose $y \in P \cup \{x\}$ such that (x, y) is not order equivalent to an element of E^2 (using $<_D$ and numerical <). Then take M restricted, in the obvious sense, to the countable set $\cup_i B_i$. □

We now fix countable $M = (D, <_D, P, 0, d_1, \ldots, d_n)$ as given by Lemma 3.5.7.

Lemma 3.5.8 *M is isomorphic to a system $(Q[0, 1], <, S, 0, q_1, \ldots, q_{n-1}, 1)$, where S is a maximal emulator of $E \subseteq Q[0, 1]^k$ in the usual sense used in MESU/2, and S is invariant under the equivalence relation $R_k(\{q_1, \ldots, q_n\})$ on $Q[0, 1]^k$.*

Proof Let $h: D \to Q[0, 1]$ be any isomorphism from $(D, <_D, 0, d_1, \ldots, d_n)$ onto $(Q[0, 1], <, 0, q_1, \ldots, q_{n-1}, 1)$. This is clear from well known facts about countable dense linear orderings with endpoints, 0, d_n are the left/right endpoints of the first system and 0, 1 are the left/right endpoints of the second system. Then h is an isomorphism from $M = (D, <_D, P, 0, d_1, \ldots, d_n)$ onto $(Q[0, 1], <, h[P], 0, q_1, \ldots, q_{n-1}, 1)$, where h is the image of P acting coordinatewise. It is easy to see that the properties i-vi in Lemma 3.5.8 are preserved under the isomorphism h. Hence h[P] is a maximal emulator of $E \subseteq Q[0, 1]^k$ which is $R_k(\{q_1, \ldots, q_{n-1}, 1\}) = R_k(\{q_1, \ldots, q_n\})$ invariant. □

Theorem 3.5.9 *MESU/2 for dimension k = 2 is provable in ZFC. In fact, Z and even Z_3 suffices.*

Proof The entire proof through Lemma 3.5.8 uses only that κ is an uncountable regular cardinal, except for the second part of the proof of Lemma 3.5.5. But if k = 2 then the second part of the proof of Lemma 3.5.5 is not needed (it uses $k \geq 3$). For k = 2, we can set $\kappa = \omega_1$, as the actual transfinite ordinal ω_1 is available in ZF, and the countable axiom of choice can be used to prove that ω_1 is regular (regularity of κ was used in the proof).

However, we can avoid any use of the axiom of choice, and even stay within Z. We can achieve this using a truncated version of Gödel's constructible hierarchy known to be available in Z. We can build an initial segment of the constructible hierarchy in a well known coded fashion, cut off so that there are exactly three internal infinite cardinalities. This well known kind of construction does not use the axiom of choice or replacement, and the second internal infinite cardinal can serve as the $\kappa = \omega_1$ for the argument. With some additional care, we can in fact stay within Z_3. □

Theorem 3.5.10 *The following hold.*

i. *MESU/1, 2, 3, MED/1, 2, 3 are provable in SRP^+ but not in SRP (assuming SRP is consistent).*

ii. *MESU/1, 2.3, MED/1, 2, 3 are provably equivalent to Con(SRP) over* WKL_0.
iii. *MESU/1, 2, 3, MED/1, 2, 3 are not provable in ZFC (or SRP), assuming ZFC (or SRP) is consistent.*
iv. *MESU/1, 2, 3, MED/1, 2, 3 are independent of ZFC (and even SRP), assuming SRP is 1-consistent.*

Proof We proved Lemma 3.5.8 by using a (k − 2)-subtle cardinal for all $k \geq 3$ (only ZFC for $k=2$ and RCA_0 for $k=1$). So it is clear that we have given a proof of MESU/2 in SRP^+. Now apply Theorem 3.4.8 to see that we have proved all six statements in SRP^+. Now suppose that any of the six are provable in SRP. Then by Theorem 3.4.8, MED/1 is provable in SRP. By Theorem 3.4.11, SRP then proves Con(SRP), and so by Gödel's Second Incompleteness Theorem, SRP is inconsistent.

For ii, we first argue in WKL_0 +Con(SRP). By Corollary 3.1.7, let $\varphi(k)$ be a Π^0_1 formula such that WKL_0 proves $(\forall k)$((MESU/2 for dimension k) $\leftrightarrow \varphi(k)$) is provable in WKL_0, and display such a proof. According to the way Lemma 3.5.8 was proved, we have that for all $k \geq 1$, ZFC+"there exists a k-subtle cardinal" proves MESU/2 for dimension k (actually much better than this). Hence for all $k \geq 1$, ZFC+"there exists a k-subtle cardinal" proves $\varphi(k^*)$, where k^* is the usual closed term for k. Hence, using Con(SRP), we have $(\forall k)(\varphi(k))$. From the displayed proof, we derive $(\forall k)$(MESU/2 for dimension k), which is MESU/2. Thus we have derived MESU/2 in WKL_0 +Con(SRP). Now apply Theorem 3.4.8.

For the other direction of ii, we have RCA_0 +MED/1 proves Con(SRP) by Theorem 3.4.11. Now apply Lemma 3.4.8.

For iii, if any of the six is provable in ZFC (or SRP), then Con(SRP) is provable in ZFC (or SRP), and so by Gödel's Second Incompleteness Theorem, ZFC (or SRP) is inconsistent.

For iv, suppose SRP refutes one of the six statements. Then SRP refutes Con(SRP), and so SRP is not 1-consistent. □

Theorem 3.5.13 *The following are provable in EFA.*

i. *MESU/1, 2, 3, MED/1, 2, 3 for any fixed dimension k is provable in SRP.*
ii. *ZFC proves that for all $k \geq 1$, if there is a max(k − 1, 0)-subtle cardinal then MESU/1 holds for dimension k and MESU/2, 3, MED/1, 2, 3 holds for dimension k+1.*
iii. *For all sufficiently large $k \geq 1$, MESU/1, 2, 3, MED/1, 2, 3 for dimension k are not provable in ZFC, assuming ZFC is consistent.*
iv. *For all sufficiently large $k \geq 1$, MESU/1, 2, 3, MED/1, 2, 3 for dimension k are independent of ZFC, or even any SRP[n] fixed in advance, assuming SRP is consistent.*

Proof For i, let $k \geq 1$. By the way Lemma 3.5.8 was proved, we proved the six statements in dimension k over ZFC using a k-subtle cardinal (and much better), and so we stayed within SRP.

For ii, let $k > 2$. if there is a (k − 1)-subtle cardinal then MESU/2 holds for dimension k+1 (since $k+1 \geq 3$). Now apply Theorem 3.4.8. And for $k=1$, we proved

MESU/2 for dimension 2 in a weak fragment of ZFC (see Theorem 3.5.11). Again apply Theorem 3.4.8.

For iii, by Theorem 3.4.10 and the last claim of Theorem 3.4.11, MED/1 for sufficiently large dimension proves the consistency of ZFC over RCA_0. By Theorem 3.4.8, this is true for MESU/1, 2, 3, MED/1, 2, 3,. Hence we have unprovability in ZFC in sufficiently large dimension, assuming ZFC is consistent.

For iv, let $n \geq 1$. by Theorem 3.4.10 and the last claim of Theorem 3.4.11, MED/1 for sufficiently large dimension proves the consistency of SRP[n] over RCA_0. By Theorem 3.4.8, this is true for MESU/1, 2, 3, MED/1, 2, 3. Hence we have unprovability in SRP[n], in sufficiently large dimension, if SRP is consistent. We also cannot have refutability in any given fixed SRP[n], since we have provability in SRP, assuming SRP is consistent. □

12.3.6 r-Emulation

Recall the general Emulation definitions, ME/DEF/1, 2 of Sect. 12.2. Also the specific versions in Definitions 3.1.2 and 3.1.3 for $M = (Q[0, 1], <)$. Notice the exponent 2 in all of these Definitions, indicating pairs of k-tuples. Here we introduce the very natural parameter $r \geq 1$, which we have spared the reader from thus far.

Definition 3.6.1 S is an r-emulator of $E \subseteq M^k$ if and only if $S \subseteq M^k$ and every element of S^r is M equivalent to an element of E^r. S is a maximal r-emulator of $E \subseteq M^k$ if and only if S is an r-emulator of $E \subseteq M^k$ which is not a proper subset of an r-emulator of $E \subseteq M^k$.

Definition 3.6.2 S is an r-emulator of $E \subseteq Q[0, 1]^k$ if and only if $S \subseteq Q[0, 1]^k$ and every element of S^r is order equivalent to an element of E^r. S is a maximal r-emulator of $E \subseteq Q[0, 1]^k$ if and only if S is an r-emulator of $E \subseteq Q[0, 1]^k$ which is not a proper subset of an r-emulator of $E \subseteq Q[0, 1]^k$.

Note that the elements of S^r, E^r are rk-tuples. Also note that (maximal) emulators are just the (maximal) 2-emulators.

The r-emulators give rise to a strengthened notion of usability. General usability was defined in Sect. 12.2 as MEU/DEF, MEIU/DEF. Usability for $M = (Q[0, 1], <)$ was defined in Definition 3.1.5.

Definition 3.6.3 $R \subseteq M^k \times M^k$ is ME usable* if and only if for all subsets of M^k and $r \geq 1$, some maximal r-emulator contains its R image. $R \subseteq M^k \times M^k$ is ME invariantly usable* if and only if for all subsets of M^k and $r \geq 1$, some maximal r-emulator is R invariant.

Definition 3.6.4 $R \subseteq Q[0, 1]^k \times Q[0, 1]^k$ is ME usable* if and only if for finite subsets of $Q[0, 1]^k$ and $r \geq 1$, some maximal r-emulator contains its R image. $R \subseteq Q[0, 1]^k \times Q[0, 1]^k$ is ME invariantly usable* if and only if for finite subsets of $Q[0, 1]^k$ and $r \geq 1$, some maximal r-emulator is R invariant.

Most of what we have said about emulators, maximal emulators, ME usable, ME invariantly usable, lifts straightforwardly to r-emulators, maximal r-emulators, ME usable*, and ME invariantly usable*. Here we restate the starred form of the previous statements, and indicate their resulting status.

From Sect. 12.2

Maximal Emulation*/1. ME*/1. Every $E \subseteq M^k$ has a maximal r-emulator.

Maximal Emulation*/2. ME*/2. Every $E \subseteq M^k$ has a maximal r-emulator containing any given r-emulator.

Finite Subset Emulation*. Assuming M has finitely many components, every $E \subseteq M^k$ is an r-emulator of some finite $E' \subseteq M^k$ with $E' \subseteq E$. This is provable in RCA_0 for countable M with finitely many components.

Emulation Transitivity* If S is an r-emulator of $E \subseteq M^k$ and E is an r-emulator of $E' \subseteq M^k$, then S is an r-emulator of $E' \subseteq M^k$. Let E be an r-emulator of $E' \subseteq M^k$ and E' be an r-emulator of $E \subseteq M^k$. The r-emulators of $E \subseteq M^k$ are the same as the r-emulators of $E' \subseteq M^k$. The maximal r-emulators of $E \subseteq M^k$ are the same as the maximal r-emulators of $E' \subseteq M^k$. This is provable in RCA_0 for countable M.

Maximal Emulation*/3. ME*/3. (RCA_0) Let M have countable domain and finitely many components. Every subset of M^k has a maximal r-emulator. The following are equivalent.

i. ACA_0.
ii. Every subset of M^k has a maximal r-emulator containing any given r-emulator.
iii. In every equivalence relation M on N, every finite subset of D^2 has a maximal r-emulator containing any given r-emulator.

Theorem 2.2* *(RCA_0) Let M have finitely many components. $R \subseteq M^k \times M^k$ is ME usable* if and only if for all finite subsets of M^k and $r \geq 1$, some maximal r-emulator contains its R image.*

Maximal Emulation*/4. ME*/4. If $R \subseteq M^k \times M^k$ is ME usable* then R is M preserving in the sense that $(\forall x, y)(R(x, y) \rightarrow x, y$ are M equivalent).

Theorem 2.4* *$R \subseteq M^k \times M^k$ is ME invariantly usable* if and only if $R \cup R^{-1}$ is ME usable*. If $R \subseteq M^k \times M^k$ is symmetric then R is ME invariantly usable* if and only if R is ME usable*.*

Theorem 2.5* *(RCA_0) Let $x, y \in M^k$. The following are equivalent.*

i. For finite subsets of M^k, some maximal r-emulator is equivalent at $x, y \in M^k$.
ii. $\{(x, y)\}$ is ME invariantly usable.*
iii. $\{(x, y), (y, x)\}$ is ME usable.*

All of the proofs of the above in Sect. 12.2 go through without modification with the exception of iii → i in ME/3, which we have not thought through.*

From Sect. 12.3.1

Theorem 3.1.2* *(RCA_0) Every $E \subseteq Q[0, 1]^k$ is an r-emulator of a finite subset. E has a recursive maximal r-emulator.*

Lemma 3.1.5* *Same as Lemma 3.1.5 with k, $r \geq 1$ fixed, and emulator replaced by r-emulator.*

Theorem 3.1.6* *(EFA) Consider the statement $\varphi(k, r, E, R)$ = "For finite $E \subseteq Q[0, 1]^k$, some maximal r-emulator S of $E \subseteq Q[0, 1]^k$ has $R[S] \subseteq S$".*

i. *If k, r, E, R are fixed in advance, where $R \subseteq Q[0, 1]^k \times Q[0, 1]^k$ is order theoretic, then $\varphi(k, r, E, R)$ is implicitly Π^0_1 over WKL_0.*
ii. *If k, r, R are fixed in advance, where $R \subseteq Q[0, 1]^k \times Q[0, 1]^k$ is order theoretic, then $(\forall E \subseteq Q[0, 1]^k)(\varphi(k, r, E, R))$ is implicitly Π^0_1 over WKL_0.*
iii. *If k, R are fixed in advance, where $R \subseteq Q[0, 1]^k \times Q[0, 1]^k$ is order theoretic, then $(\forall E \subseteq Q[0, 1]k)(\forall r)(\varphi(k, r, E, R))$ is implicitly Π^0_1 over WKL_0.*

Furthermore, the associated Π^0_1 forms and the equivalence proofs in WKL_0 can be constructed effectively from fixed parameters in a way that RCA_0 can verify.

Corollary 3.1.7* *(EFA) Consider the statement $\varphi(k, R)$ = "$R \subseteq Q[0, 1]^k \times Q[0, 1]^k$ is ME* usable". If k, R are fixed in advance, where $R \subseteq Q[0, 1]^k \times Q[0, 1]^k$ is order theoretic, then $\varphi(k, R)$ is implicitly Π^0_1 over WKL_0*

All of the proofs of the above in Sect. 12.3.1 go through without modification.
From Sect. 12.3.2

Maximal Emulation* Necessary Use. Menu*. If $R \subseteq Q[0, 1]^k \times Q[0, 1]^k$ is ME usable* then R is order preserving.

Maximal Emulation Finite Use*/1. MEFU*/1. Any finite order preserving $R \subseteq Q(0, 1)^k \times Q(0, 1)^k$ is ME usable*.

Maximal Emulation Finite Use*/2. MEFU*/2. Any finite order preserving $R \subseteq Q(0, 1]^k \times Q(0, 1]^k$ is ME usable*.

Maximal Emulation Finite Use*/3. MEFU*/3. Any finite order preserving $R \subseteq Q[0, 1]^k \times Q[0, 1]^k$ not altering both of 0, 1 is ME usable*.

Maximal Emulation Singleton Use*/1. MEOU*/1. Any order preserving $R \subseteq Q[0, 1]^k \times Q[0, 1]^k$ of cardinality 1 is ME usable*.

All of the proofs of the above in Sect. 12.3.1 go through without modification.
MEOU/2 needs to be reconsidered for r-emulators.
From Sect. 12.3.3

Lemma 3.3.1* *(RCA_0) The maximal r-emulators of $E \subseteq Q[0, 1]$, $|E| < r$, are the $S \subseteq Q[0, 1]$, $|S| = |E|$. The maximal r-emulator of $E \subseteq Q[0, 1]$, $|E| \geq r$, is just $Q[0, 1]$.*

Maximal Emulation Large Use*/1. MELU*/1. $R \subseteq Q[0, 1] \times Q[0, 1]$ is ME usable* if and only if for all n there is an n element subset of Q[0, 1] containing its R image.

Proof In RCA_0. Let $R \subseteq Q[0, 1] \times Q[0, 1]$ be ME usable*. Let $|E| = n$ and S be a maximal (n+1)-emulator of E containing its R image. Then $|S| = n$ contains its R image. Conversely, suppose that for all n there is an n element subset of Q[0, 1] containing its R image. Let $E \subseteq Q[0, 1]$ be finite and $r \geq 1$. If $|E| \geq r$ then E has the maximal r-emulator Q[0, 1] containing its R image. If $|E| < r$ then E has the maximal r-emulator S containing its R image, where $|S| = |E|$ and S contains its R image. □

Maximal Emulation Large Use*/2. MELU*/2. $Q(0, 1)^{2<} \times Q(0, 1)^{2<}$ is not ME usable*. It is order preserving, order theoretic, and 0, 1 are not present.

Maximal Emulation Large Use*/4. MELU*/4. For $k \geq 3$, $Q[1/3, 1/2]^{k<} \times Q[1/3, 1/2]^{k<}$ is not ME usable*.

All of the proofs of the above in Sect. 12.3.3 go through without modification.
MELU/3 needs to be reconsidered for r-emulators.
From Sects. 12.3.4 and 12.3.5
Most importantly, we now come to MESU/1, 2, 3 and MED/1, 2, 3.

Theorem 3.4.1* *Same as Theorem 3.4.1 with usable replaced by usable*.*

Maximal Emulation Small Use*/1. MESU*/1. The lower parameterization of any order preserving finite $R \subseteq Q[0, 1]^k \times Q[0, 1]^k$ is ME usable*.

Maximal Emulation Small Use*/2. MESU*/2. For finite $A \subseteq Q(0, 1]$, $R_k(A) \subseteq Q[0, 1]^k \times Q[0, 1]^k$ is ME usable*.

Maximal Emulation Small Use*/3. MESU*/3. Every order preserving $R \subseteq Q[0, 1]^k \times Q[0, 1]^k$ with the Finiteness Condition, not altering 0, is ME usable*.

Maximal Emulation Drop*/1. MED*/1. For finite subsets of $Q[0, 1]^k$, some maximal r-emulator is drop equivalent at (1, 1/2, …, 1/k), (1/2, …, 1/k, 1/k).

Maximal Emulation Drop*/2. MED*/2. Let x, y ∈ $Q[0, 1]^k$. The following are equivalent.

i For finite subsets of $Q[0, 1]^k$, some maximal r-emulator is drop equivalent at x, y.
ii. x, y are droppable or $x_k = y_k = 0$.

Maximal Emulation Drop*/3. MED*/3. Let $x_1, y_1, \ldots, x_n, y_n \in Q[0, 1]^k$. The following are equivalent.

i. For finite subsets of $Q[0, 1]^k$, some maximal r-emulator is drop equivalent at every x_i, y_i.

ii. For finite subsets of $Q[0, 1]^k$ and $1 \leq i \leq k$, some maximal r-emulator is drop equivalent at (x_i, y_i).
iii. For all i, x_i, y_i is droppable or $(x_i)_k = (y_i)_k = 0$.

All of the proofs of the results concerning the above statements in Sects. 12.3.4 and 12.3.5 go through without modification. This includes the Exotic Proof of Sect. 12.3.5.

We conjecture that MESU/2 for $k = 2$ is provable in RCA_0. However, we conjecture that none of MESU*/1, 2, 3, MED*/1, 2, 3 for dimension $k = 2$ is provable in ZFC\P or Z_2. We also conjecture that none of MESU*/1, 2, 3, MED*/1, 2, 3 for dimension $k = 3$ is provable in ZFC. In fact, we conjecture that for each $k \geq 3$, MESU*/1 for dimension $k - 1$ and MESU*/2, 3, MED*/1, 2, 3 for dimension k is provable equivalent to Con(ZFC + "there exists a $(k - 2)$-subtle cardinal").

12.4 General Conjectures

Here we discuss some General Conjectures in Basic Emulation Theory on Q[0,1] which do not specifically pertain to the statements discussed in Sect. 12.3.

General Conjecture 1. GC1. There is an algorithm for determining whether a given order theoretic $R \subseteq Q[0, 1]^k \times Q[0, 1]^k$ is usable. For inputs, use a standardly digitized form of quantifier free formulas over $(Q[0, 1], <)$ with parameters.

GC1 is trivial for fixed dimension $k = 1$ by MELU/1. We have not proved GC1 even for fixed dimension $k = 2$. In fact, we have no significant results about GC1.

The following sharper form of GC1 would not seem to be different than GC1 in any significant way.

General Conjecture 2. GC2. There is a Turing machine with at most $2^{2\hat{}1000}$ states/symbols each, that determines whether a given order theoretic $R \subseteq Q[0, 1]^k \times Q[0, 1]^k$ is ME usable. For inputs, use a standardly digitized form of quantifier free formulas over $(Q[0, 1], <)$ with parameters.

Here $2^{2\hat{}1000}$ is merely a simply described ridiculously large number of states and symbols for any actual algorithm.

We will show below that GC2 is not provable in ZFC, or even in SRP, assuming SRP is consistent. Establishing this unprovability result for GC1 seems to require a new idea, and the unprovability may well be false.

Our result that GC2 is not provable in ZFC (assuming Con(SRP)) does tell us that, in a sense, ZFC (or even SRP) is not sufficient to analyze the ME usability of order theoretic R. There is another sense in which we know that ZFC (or SRP) is not sufficient to analyze this. What is different about the situation with GC2 is that a particular conjecture (GC2) concerning the nature of the ME usability of order theoretic relations is shown to be unprovable in ZFC (or SRP).

This other sense that we know ZFC (or SRP) is not sufficient to analyze the ME usability of order theoretic R is as follows.

Lemma 4.1 *(EFA) Let T be a recursively axiomatized first order system that interprets EFA. Let φ be a (interpreted) Σ_1^0 sentence such that T+φ proves Con(T). Then T refutes φ.*

Proof This is well known. Suppose T+φ proves Con(T). Since T+φ proves "T proves φ", we have that T+φ proves Con(T+φ). So by Gödel's Second Incompleteness Theorem, T+φ is inconsistent, and T refutes φ. □

Theorem 4.2 *(EFA) There is no algorithm α such that ZFC (or even SRP) proves that α correctly decides whether or not a given order theoretic relation is ME usable - assuming SRP is consistent. We do not need the hypothesis that α always returns with an answer.*

Proof Let α be such an algorithm, where α is proved to be correct in SRP. Let n be such that this correctness is proved in SRP[n]. By Theorem 3.4.11, let m be such that MED/1 for dimension m provably implies Con(SRP[n]) over RCA_0. Clearly SRP[n] proves "if α returns that MED/1 holds in dimension m then MED/1 holds in dimension m and so Con(SRP[n])". By Lemma 4.1, SRP[n] proves "it is not the case that α returns that MED/1 holds in dimension m". Hence SRP[n] proves "α returns that MED/1 fails in dimension m". Hence SRP[n] proves that MED/1 fails in dimension m. But SRP proves MED/1. Hence SRP is inconsistent.

Instead of using MED/1 here, we can use the sharper MESU/2. □

General Conjecture 3. GC3. For every order theoretic $R \subseteq Q[0, 1]^k \times Q[0, 1]^k$, the statement "R is ME usable" is either provable in SRP or refutable in RCA_0.

Theorem 4.3 *(EFA)*

i. GC2 → GC1.
ii. Assume Con(SRP). GC3 → GC2.
iii. GC3 is false for any SRP[n], assuming Con(SRP).

Proof i is trivial. For ii, assume Con(SRP). Suppose GC3. We use the algorithm that searches for a proof in SRP or a refutation in RCA_0 of "R is ME usable". We only find one of these since Con(SRP). This algorithm can be given by a small enough Turing machine. Now assume Con(SRP) and GC3 holds for SRP[n]. By Theorem 3.4.11, let m be such that MED/1 (or MESU/2) for dimension m provably implies Con(SRP[n]) over RCA_0. Then SRP[n] proves or refutes Con(SRP[n]). The former case violates Con(SRP). The latter case also violates Con(SRP) since SRP proves Con(SRP[n]). □

We may be very wrong about GC1, 2, 3, and obviously as the dimension k rises we feel less confident. Here is a weak form of Conjecture 3 that we have more confidence in than GC1.

General Conjecture 4. GC4. Let k be least such that there is an order theoretic $R \subseteq Q[0, 1]^k \times Q[0, 1]^k$ for which the statement "R is ME usable" is independent of

ZFC. Then for all order theoretic $R \subseteq Q[0, 1]^k \times Q[0, 1]^k$, the statement "R is ME usable" is provable or refutable in SRP.

What can we say about this least k? Nothing now except $k \geq 2$. But we will still venture a guess.

General Conjecture 5. GC5. Let k (k′) be least such that there is an order theoretic $R \subseteq Q[0, 1]^k \times Q[0, 1]^k$ for which the statement "R is ME usable" is independent of ZFC (ZFC\P). Then $2 < k' < k \leq 8$.

It is natural to modify the conjectures GC1-5 to GC1*-5*, where we replace usable by usable*. Here we greatly strengthen GC5.

General Conjecture 6. GC6. Let k (k′) be least such that there is an order theoretic $R \subseteq Q[0, 1]^k \times Q[0, 1]^k$ for which the statement "R is ME usable*" is independent of ZFC (ZFC\P). Then $k = 3$ and $k' = 2$.

We now show that GC2 is not provable in ZFC, assuming Con(SRP), as promised. The same proof works for GC2*, the version with usable*. We first prove a very general non provability result.

Lemma 4.4 *Let $T+\{\varphi_1, \varphi_2, \ldots\}$ be any consistent theory extending EFA which is not derivable from any $T+\{\varphi_1, \ldots, \varphi_i\}$. Then for no $n \geq 1$ does T prove "there exists a Turing machine with at most n states/symbols each that determines whether any given sentence φ_i is true".*

Proof Let T, $\varphi_1, \varphi_2, \ldots$ and n be as given. For each TM with at most n states/symbols each, let α(TM) be the least i for which α(TM) returns "false" (if i exists). Let r be the maximum of these numbers α(TM). Note that r depends on n. We claim that there exists $s > r$ that the theory $T+\{\varphi_1, \ldots, \varphi_s, \neg\varphi_{s+1}\}$ is consistent. For otherwise, $T+\{\varphi_1, \ldots, \varphi_{r+1}\}$ derives $\varphi_{r+2}, \varphi_{r+3}, \ldots$, contrary to the hypothesis. But obviously $T+\{\varphi_1, \ldots, \varphi_s, \neg\varphi_{s+1}\}$ refutes "TM determines whether any given sentence φ_i is true" for any of these TM's, as they yield a smaller break point or no break point at all. So it is consistent with T that all of these TM's fail to do their task. □

Lemma 4.5 *Assume Con(SRP). For each $k \geq 1$, let φ_k be MED/1 (or MESU/2) for dimension k. For all $n \geq 1$, $SRP[n]+\{\varphi_1, \varphi_2, \ldots\}$ is a consistent theory which is not derivable from any $ZFC+\{\varphi_1, \ldots, \varphi_i\}$.*

Proof These theories are fragments of SRP. Again use Theorem 3.4.11 and Gödel's Second Incompleteness Theorem. □

Theorem 4.6 *Assume Con(SRP). GC2 is not provable in ZFC, or even in SRP. The same holds for GC2*.*

Proof Suppose SRP proves GC2. Let n be such that SRP[n] proves GC2. Then apply Lemmas 4.4 and 4.5, setting $n = 2^{2\hat{}1000}$, to obtain a contradiction. The same argument works for GC2*. □

Appendix A: The Stationary Ramsey Property

Reprinted from Crangle et al. (2014)

All results in this section are taken from Friedman 2001. All of these results, with the exception of Theorem 9.1.1, iv $\leftrightarrow$ v $\rightarrow$ vi, are credited in (Friedman 2001) to James Baumgartner. Below, λ always denotes a limit ordinal.

Definition A.1 We say that $C \subseteq \lambda$ is unbounded if and only for all $\alpha < \lambda$ there exists $\beta \in C$ such that $\beta \geq \alpha$.

Definition A.2 We say that $C \subseteq \lambda$ is closed if and only if for all limit ordinals $x < \lambda$, if the sup of the elements of C below x is x, then $x \in C$.

Definition A.3 We say that $A \subseteq \lambda$ is stationary if and only if it intersects every closed unbounded subset of λ.

Definition A.4 For sets A, let S(A) be the set of all subsets of A. For integers $k \geq 1$, let $S_k(A)$ be the set of all k element subsets of A.

Definition A.5 Let $k \geq 1$. We say that λ has the k-SRP if and only if for every $f{:}S_k(\lambda) \rightarrow \{0, 1\}$, there exists a stationary $E \subseteq \lambda$ such that f is constant on $S_k(E)$. Here SRP stands for "stationary Ramsey property."

The k-SRP is a particularly simple large cardinal property. To put it in perspective, the existence of an ordinal with the 2-SRP is stronger than the existence of higher order indescribable cardinals, which is stronger than the existence of weakly compact cardinals, which is stronger than the existence of cardinals which are, for all k, strongly k-Mahlo (see Theorem A.1 below, and Friedman 2001, Lemma 1.11).

Our main results are stated in terms of the stationary Ramsey property. In particular, we use the following extensions of ZFC based on the SRP.

Definition A.6 $SRP^+ = ZFC +$ "for all k there exists an ordinal with the k SRP". $SRP = ZFC + \{$there exists an ordinal with the k-SRP$\}_k$. We also use SRP[k] for the formal system $ZFC + (\exists\lambda)(\lambda$ has the k-SRP).

For technical reasons, we will need to consider some large cardinal properties that rely on regressive functions.

Definition A.7 We say that $f{:}S_k(\lambda) \rightarrow \lambda$ is regressive if and only if for all $A \in S_k(\lambda)$, if $\min(A) > 0$ then $f(A) < \min(A)$. We say that E is f-homogenous if and only if $E \subseteq \lambda$ and for all B, $C \in S_k(E)$, $f(B) = f(C)$.

Definition A.8 We say that $f{:}S_k(\lambda) \rightarrow S(\lambda)$ is regressive if and only if for all $A \in S_k(\lambda)$, $f(A) \subseteq \min(A)$. (We take $\min(\emptyset) = 0$, and so $f(\emptyset) = \emptyset$). We say that E is f-homogenous if and only if $E \subseteq \lambda$ and for all B, $C \subset S_k(E)$, we have $f(B) \cap \min(B \cup C) = f(C) \cap \min(B \cup C)$.

Definition A.9 Let $k \geq 1$. We say that α is purely k-subtle if and only if

(i) α is an ordinal;
(ii) For all regressive $f:S_k(\alpha) \to \alpha$, there exists $A \in S_{k+1}(\alpha \setminus \{0, 1\})$ such that f is constant on $S_k(A)$.

Definition A.10 We say that λ is k-subtle if and only if for all closed unbounded $C \subseteq \lambda$ and regressive $f:S_k(\lambda) \to S(\lambda)$, there exists an f-homogenous $A \in S_{k+1}(C)$.

Definition A.11 We say that λ is k-almost ineffable if and only if for all regressive $f:S_k(\lambda) \to S(\lambda)$, there exists an f-homogenous $A \subseteq \lambda$ of cardinality λ.

Definition A.12 We say that λ is k-ineffable if and only if for all regressive $f:S_k(\lambda) \to S(\lambda)$, there exists an f-homogenous stationary $A \subseteq \lambda$.

Theorem A.1 *Let $k \geq 2$. Each of the following implies the next, over ZFC.*

i. there exists an ordinal with the k-SRP.
ii. there exists a $(k-1)$-ineffable ordinal.
iii. there exists a $(k-1)$-almost ineffable ordinal.
iv. there exists a $(k-1)$-subtle ordinal.
v. there exists a purely k-subtle ordinal.
vi. there exists an ordinal with the $(k-1)$-SRP.

Furthermore, i, ii are equivalent, and iv, v are equivalent. There are no other equivalences. ZFC proves that the least ordinal with properties i - vi (whichever exist) form a decreasing ($\geq$) sequence of uncountable cardinals, with equality between i, ii, equality between iv, v, and strict inequality for the remaining consecutive pairs.

Proof i $\leftrightarrow$ ii is from Friedman (2001), Theorem 1.28, iv $\leftrightarrow$ v is from Friedman (2001), Corollary 2.17. The strict implications ii $\to$ iii $\to$ iv $\to$ vi are from Friedman (2001), Theorem 1.28. Same references apply for comparing the least ordinals. □

Definition A.13 We follow the convention that for integers $p \leq 0$, a p-subtle, p-almost ineffable, p-ineffable ordinal is a limit ordinal, and that the ordinals that are 0-subtle, 0-almost ineffable, 0-ineffable, or have the 0-SRP, are exactly the limit ordinals. An ordinal is called subtle, almost ineffable, ineffable, if and only if it is 1-subtle, 1-almost ineffable, 1-ineffable.

Appendix B: Formal Systems Used

PFA Polynomial function arithmetic. Based on 0, successor, addition, multiplication, and bounded induction. Same as $I\Sigma_0$ (Hajek and Pudlak 1993, p. 29, 405).

EFA Exponential function arithmetic. Based on 0, successor, addition, multiplication, exponentiation and bounded induction. Same as $I\Sigma_0(\exp)$ (Hajek and Pudlak 1993, p. 37, 405).

RCA_0 Recursive comprehension axiom naught. Our base theory for Reverse Mathematics (Simpson 2009).

WKL_0 Weak Konig's Lemma naught. Our second level theory for Reverse Mathematics (Simpson 2009).

ACA_0 Arithmetic comprehension axiom naught. Our third level theory for Reverse Mathematics (Simpson 2009).

ACA′ Arithmetic comprehension axiom prime. ACA_0 together with "for all $n < \omega$ and $x \subseteq \omega$, the n-th Turing jump of x exists".

Z_2 s order arithmetic as a two sorted first order theory (Simpson 2009).

Z_3 Third order arithmetic as a three sorted first order theory. Extends Z_2 with a new sort for sets of subsets of ω.

Z(C) Zermelo set theory (with the axiom of choice). This is the same as ZF(C) without the axiom scheme of replacement.

ZF(C)\P ZF(C) without the power set axiom (Kanamori 1994).

ZF(C) Zermelo Frankel set theory (with the axiom of choice). ZFC is the official theoretical gold standard for mathematical proofs (Kanamori 1994).

SRP[k] ZFC + $(\exists\lambda)$(λ has the k-SRP), for fixed k. Appendix A.

SRP ZFC + $(\exists\lambda)$(λ has the k-SRP), as a scheme in k. Appendix A.

SRP^+ ZFC + $(\forall k)(\exists\lambda)$($\lambda$ has the k-SRP). Appendix A.

References

Concrete Mathematical Incompleteness. (Book in preparation). Major part is Boolean Relation Theory. https://u.osu.edu/friedman.8/foundational-adventures/boolean-relation-theory-book/.

Cohen, P. J. (1963). The independence of the continuum hypothesis. *Proceedings of the National Academy of Sciences of the United States of America, 50*, 1143–1148; *51*, 105–110.

Crangle, C. E., de la Sienra, A. G., & Longino, H. E. (Eds.). (2014). *Foundations and methods from mathematics to neuroscience: Essays inspired by Patrick Suppes*. Stanford, California: CSLI Publications. (Invariant maximality and incompleteness)

Embedded maximal cliques and incompleteness. Extended abstract. 18 p. Retrieved May 20, 2013, from http://www.math.osu.edu/~friedman.8/manuscripts.html.

Friedman, H. (2001). Subtle cardinals and linear orderings. *Annals of Pure and Applied Logic, 107*(1–3), 1–34.

Friedman, H. (2011). My forty years on his shoulders. In M. Baaz, C. H. Papadimitrou, H. W. Putnam, D. S. Scott, C. L. Harper, Jr. (Eds.) Kurt Gödel and the Foundations of Mathematics, Horizons of Truth (pp. 399–432). Cambridge University Press.

Friedman, H. 108. Tangible mathematical incompleteness of ZFC, August 16, 2018, 68 pages, from https://urldefense.proofpoint.com/v2/url?u=http-3A__www.math.osu.edu_-7Efriedman.8_manuscripts.html&d=DwIFaQ&c=vh6FgFnduejNhPPD0fl_yRaSfZy8CWbWnIf4XJhSqx8&r=MRMGtgr1VLCfD3La5Pl112kEIohcWQXc lm7NEsbRGAhK0P8G3wgu15JF7AhSMnp8&m=bmZRX_fchNpbl0VMQQ-1Q7vWcSjrjh9FF5thevbRl4&s=dzXXJOD-iM70IrP_hyubp4Aq0Hqym-knNpo2ni9lq4I&e=.

Gödel, K. (1931). *On formally undecidable propositions of Principia mathematica and related systems I* (Vol. 1, pp. 145–195).

Gödel, K. The consistency of the axiom of choice and the generalized continuum hypothesis with the axioms of set theory. *Annals of Mathematics Studies, 3* (Princeton University Press)

Gödel, K. (1940). *Collected works* (Vols. I–V, II, pp. 33–101). Oxford: Oxford University Press (S. Feferman et al. eds.).

Gross, O. A. (1962). Preferential arrangements. *American Mathematical Monthly, 69*, 4–8.

Hajek, P., & Pudlak, P. (1993). *Metamathematics of first-order arithmetic, perspectives in mathematical logic*. Springer.
Kanamori, A. (1994). *The higher infinite, perspectives in mathematical logic*. Springer.
Ramsey, F. P. (1930). On a problem of formal logic. *Proceedings of the London Mathematical Society, 39*, 264–286.
Simpson, S. (2009). *Subsystems of second order arithmetic* (2nd ed.). Springer, 1999. ASL.
Sloane, N. J. A. (1964). *The on-line encyclopedia of integer sequences*. https://oeis.org/A000670.

Harvey M. Friedman is Distinguished University Professor of Mathematics, Philosophy, and Computer Science, Emeritus, at The Ohio State University in Columbus, Ohio, and is most well known for his work in the foundations of mathematics. He founded Reverse Mathematics in the 1970's, and more recently, the more ambitious Strict Reverse Mathematics. Friedman has concentrated most heavily on his fifty-year Concrete Mathematical Incompleteness project, which has reached major milestones this past year (2017). These involve uncovering new kinds of discrete and finitary Incompleteness from the usual ZFC axioms for mathematics. Friedman earned his Ph.D. from the Massachusetts Institute of Technology in 1967, was appointed Assistant Professor of Philosophy at Stanford University at the age of 18, and granted tenure three years later. Friedman received the National Science Foundation's Alan T. Waterman Award in 1984, and he delivered the Gödel Lectures (ASL) in 2002 and the Alfred Tarski Lectures in 2007 (UCB).. Friedman is also a classical pianist at the semiprofessional level, and has been the subject of a recent article at Nautilus online (2/17). Friedman retired in July 2012 in order to get more work done.

Chapter 13
Putnam's Constructivization Argument

Akihiro Kanamori

Abstract We revisit Putnam's constructivization argument from his *Models and Reality*, part of his model-theoretic argument against metaphysical realism. We set out how it was initially put, the commentary and criticisms, and how it can be specifically seen and cast, respecting its underlying logic and in light of Putnam's contributions to mathematical logic.

Keywords Constructibility · $V = L$ · Model-theoretic argument · Metaphysical realism

Hilary Putnam's constructivization argument, involving the axiom of constructibility $V = L$ of set theory, is at the cusp of mathematics and philosophy, being the most mathematically pronounced argument that he has put in the service of philosophical advocacy. In his shift in the mid-1970s to his internal realism, the argument appeared in his *Models and Reality* (1980), as a "digression". Nonetheless, with subsequent commentary and criticisms it became considered a substantive piece of what has come to be called Putnam's "model-theoretic argument against metaphysical realism". What follows is a mainly mathematical meditation on the constructivization argument: how it was initially put, the commentary and criticisms, and how it can be specifically seen and cast respecting its underlying logic and in light of Putnam's mathematical work.

Putnam's contributions to mathematical logic—his work in recursion theory, on Hilbert's 10th Problem, on constructible reals and the ramified analytic hierarchy—are mainly from his early years. Whether mathematical results can or should be deployed to support philosophical positions at all, Putnam's subsequent deployment of model-theoretic arguments against an uncompromising realism was a novel and remarkable move.

At the outset, it should be said that we will not directly illuminate how the constructivization argument integrates with Putnam's broad philosophical stance at the time. For one thing, it argues only against a realist concept of set. Rather, we will

A. Kanamori (✉)
Boston University, Boston, MA, USA
e-mail: aki@math.bu.edu

G. Hellman and R. T. Cook (eds.), *Hilary Putnam on Logic and Mathematics*, Outstanding Contributions to Logic 9, https://doi.org/10.1007/978-3-319-96274-0_13

bring out the flow of Putnam's thinking as he put his mathematical experience to work and how in its byways the constructivization argument actually worked.

In what follows, Sect. 13.1 reviews the constructivization argument, as presented in Putnam (1980). Section 13.2 describes the to and fro of commentary and criticisms of it in the literature. Section 13.3 takes a deeper look at the constructivization argument—the mathematical context, the inner logic, and the specific ways in which it can be taken. Section 13.4 coordinates the various criticisms, and in the process, consolidates the mathematical issues about the constructivization argument.

13.1 The Constructivization Argument

Putnam began his (1980) with introductory remarks and then paragraphs headlined "The philosophical problem". He briefly recalled the Skolem-paradox argument about having unintended interpretations of set theory in which nondenumerable sets are "in reality" denumerable. He then specifically recalled the Downward Löwenheim–Skolem Theorem, according to which models can have countable elementary submodels. He pointed out that by the Skolem-paradox argument, "even a *formalization of total science* (if one could construct such a thing), or even a *formalization of all our beliefs* (whether they count as 'science' or not), could not rule out denumerable interpretations." With this showing that "theoretical constraints" "cannot fix the interpretation of the notion *set* in the 'intended' way", he proceeded to argue that even "operational constraints" cannot either. With the Downward Löwenheim–Skolem Theorem, "we can find a countable submodel of the 'standard model' (if there is such a thing)" that also preserves all the information the operational constraints provide. The philosophical problem that then emerges is that if axiomatic set theory does not capture the intuitive notion of set, then "understanding" might; but "understanding" cannot come to more than "*the way we use our language*"; yet even "*the total use of the language* (operational plus theoretical constraints) does not fix a unique interpretation."

In the next paragraphs of Putnam (1980), headlined "An epistemological/logical digression", Putnam presented his constructivization argument, which amplifies the above argument with respect to constructibility. He briefly discussed Gödel's axiom $V = L$, that the set-theoretic universe V coincides with Gödel's universe L of constructible sets, and soon continued:

> [Gödel's] later view was that '$V = L$' is *really* false, even though it is consistent with set theory, if set theory itself is consistent.
>
> Gödel's intuition is widely shared among working set theorists. But does this 'intuition' make sense?
>
> Let *MAG* be a countable set of physical magnitudes which includes all magnitudes that sentient beings in this physical universe can actually measure (it certainly seems plausible that we cannot hope to measure more than a countable number of physical magnitudes). Let *OP* be the 'correct' assignment of values; that is, the assignment which assigns to each member of *MAG* the value that that magnitude actually has at each rational space-time point.

> Then all the information 'operational constraints' might give us (and in fact, infinitely more) is coded into *OP*.
>
> One technical term: an ω-model for a set theory is a model in which the *natural numbers* are ordered as they are 'supposed to be'; that is, the sequence of 'natural' numbers of the model is an ω-sequence.
>
> Now for a small theorem.[2] [[2] Barwise has proved the much stronger theorem that every countable model of ZF has a proper end extension which is a model of ZF + $V = L$ (in *Infinitary methods in the model theory of set theory*, published in ***Logic Colloquium '69***). The theorem in the text was proved by me before 1963.]
>
> THEOREM: *ZF plus* $V = L$ has an ω-model which contains any given countable set of real numbers.

Taking a countable set of reals as routinely coded by a single real, Putnam proceeded to provide an informal proof of his theorem using the Downward Löwenheim–Skolem Theorem to get a countable elementary submodel of L and then applying the Shoenfield Absoluteness Lemma. He noted in passing that "What makes [his] theorem startling" is that while a nonconstructible real cannot be in a β-model of $V = L$, it *can* be in an ω-model.

Putnam continued:

> Now, suppose we formalize *the entire language of science* within the set theory ZF + $V = L$. Any model for ZF which contains an abstract set isomorphic to *OP* can be extended to a model for this formalized language of science which is *standard with respect to OP*—hence, even if *OP* is nonconstructible 'in reality', we can find a model *for the entire language of science* which satisfies *everything is constructible* and which assigns the correct values to all the physical magnitudes in *MAG* at all rational space-time points.
>
> The claim Gödel makes is that '$V = L$' is false 'in reality'. But what on earth can this mean? It must mean, at the very least, that in the case just envisaged, the model we have described in which '$V = L$' holds would not be *the intended model*. But why not? It satisfies all theoretical constraints; and we have gone to great length to make sure it satisfies all operational constraints as well.

Putnam concluded this section:

> What the above argument shows is that if the 'intended interpretation' is fixed only by theoretical plus operational constraints, then if '$V \neq L$' does not follow from those theoretical constraints—if we do not *decide* to make $V = L$ true or to make $V = L$ false—then there will be 'intended models' in which $V = L$ is *true*. If I am right, then 'the relativity of set-theoretic notions' extends to a *relativity of the truth value of* '$V = L$' (and, by similar augments, of the axiom of choice and the continuum hypothesis as well).

13.2 Commentary and Criticisms

Putnam's advocacy of internal realism through articles starting in the later 1970s generated a philosophical literature both extensive and sundry. A focus was on his "model-theoretic argument against metaphysical realism", and eventually the most mathematically pronounced argument, the constructivization argument, itself came under sustained scrutiny in the literature, starting in the later 1990s. What is of particular interest is the extent to which a mathematically-based argument for a philosophical stance has elicited a range of responses about the mathematics and its applicability. In what follows, we review in chronological order the to and fro of commentary and criticisms.

Shapiro (1985), on second-order logic and mathematical practice, briefly attended (p. 724) to the constructivization argument. From a standard fact that he cites, if the set isomorphic to *OP* is nonconstructible then Putnam's final model of ZF + $V = L$ containing the set would not have a (really) well-founded membership relation. But for Shapiro, "one can surely claim that the well-foundedness of the membership relation is a 'theoretical constraint' on (intended) models of set theory."

Levin (1997) mounted a detailed critique of the constructivization argument, in terms of the semantics of first-order logic. On its face a response to Putnam on reference and constructibility, it seems a tissue of conflations about constants and terms and their interpretations model to model. The argument devolves to what *OP* is, its role, its coding as a real number, and whether that real is constructible—all this riddled with confusions and missing the thrust of Putnam's argument.

Velleman (1998) reviewed Levin (1997) and vetted it along the lines above. At the beginning, Velleman pointed out that Putnam's theorem (as stated in the first quoted passage of Sect. 13.1) cannot be provable in ZFC as it implies the consistency of ZFC. (By Gödel's Second Incompleteness Theorem, no theory, unless inconsistent, can establish its own consistency.) "[T]here must be a mistake in Putnam's proof"; the mistake is that "the Löwenheim–Skolem theorem is only applicable to sets, not to proper classes such as L"; and: "For example, the proof can be fixed by adding the hypothesis that there is an inaccessible cardinal κ, and then applying the Löwenheim–Skolem theorem to the set L_κ rather that to L."

Dümont (1999) undertook a "detailed reconstruction" of Putnam's "model-theoretic argument(s)", and ultimately concluded that he "fails to give convincing arguments for rejecting mathematical or metaphysical realism". While mainly concentrating on Putnam's Skolemization argument, Dümont did attend, briefly, to the constructivization argument. Following his overall tack, he took Putnam to have failed to give a convincing answer to the realist who replies (p. 348–9) to "the fact that V = L does not follow from the theoretical and operational constraints": "After all the theoretical and operational constraints have their source in *our* theoretical and empirical investigations and of course our faculties are limited. So our inability to fix one intended model only reflects our restricted access to the independently existing set-theoretical universe."

Bays (2001) mounted a broad critique of the constructivization argument, both its mathematics and its philosophy. He argued first that "a key step in Putnam's argument rests on a mathematical mistake", discussing its philosophical ramifications; second, that "even if Putnam could get his mathematics to work, his argument would still fail on purely philosophical grounds", and third, that "Putnam's mathematical mistakes and his philosophical mistakes are surprisingly closely related".

As Velleman (1998) had done, Bays indicated (p. 366f) that Putnam's proof of his theorem (as stated in the first quoted passage of Sect. 13.1) is mistaken, as the Downward Löwenheim–Skolem Theorem cannot be applied to L, a proper class, and that the proof can be patched up e.g. by assuming that there is an inaccessible cardinal. Bays, however, argued (p. 339f) that such patch-ups involving additional assumptions "do very little toward salvaging [Putnam's] overall philosophical argument". If in ZFC + XYZ one establishes that there is a model of ZFC + $V = L$, then XYZ would be part of the theoretical constraints yet would not hold in the model. The problem is "intrinsic" (because of Gödel's Second Incompleteness Theorem).

After criticizing Putnam's argument on philosophical grounds, Bays at the end made a connection between the mathematical and the philosophical. Putnam is not being fair to the realist, as (p. 349):

> When the realist tries to 'stand back' from his set theory to talk about that theory's interpretation—to specify, for instance, that this interpretation must be transitive, or well founded, or satisfy second-order ZFC—Putnam accuses him of 'begging the question.' Although Putnam's own model-theoretic talk should be viewed as talk *about* set theory, the realist's talk must be viewed as talk *within* set theory.

Gaifman (2004), on non-standard models in a broader perspective, brought up the constructivization argument, pointing out that "Putnam's proof contains a mathematical error" and that one needs an "additional assumption" to be believed by the realist. With this granted, Gaifman went on, favorably: "if s [coding OP] is not in L, the [final] model is not well-founded, but this makes no difference; we can carry out all our physical measurements, while assuming that $V = L$".

Gaifman proceeded to point out how a realist can appreciate the investigation of various structures, e.g. in which "false" propositions hold. He objected to Putnam's approach of treating "the problem as one that should be decided by appeal to general pragmatic criteria [operational constraints] and some blurry ideal of rationality [theoretical constraints]".

Bellotti (2005) examined the constructivization argument and critiques thus far. Getting to Bays (2001), Bellotti opined (p. 404–5) that his charge that Putnam made a mathematical mistake "seems unfair, since Putnam is not clear about the theory in which he is working". *Contra* Bays, Bellotti argued that effecting Putnam's argument with an additional assumption, e.g. having an inaccessible cardinal, does not weaken Putnam's philosophical point. Such an additional assumption can be taken to be part of "our best theory of the world", and "Putnam *can* obtain a final model which satisfies the necessary assumption". On the other hand, Bellotti agreed with what Bays (2001) had at the end (quoted above), that Putnam is not fair to his opponents, in that "he does not allow them what he allows himself", e.g. arbitrating what is an intended model. Following Shapiro (1985), Bellotti focused on the ill-foundedness of the final model (p. 408):

> ...Putnam's models for nonconstructible reals are so definitely unintended (they are not well founded, although they 'believe' themselves to be such) to lose much of their disquieting character for any philosophical reflection on unintended models of set theory.

In a reply to Bellotti (2005), Bays (2007) mainly reaffirmed his (2001) position. Going on at considerable length, he nuanced and finessed, one specific point *contra* Bellotti: Putnam is working and needs to work in a fixed theory. Bays newly opined (p. 133f), taking account of recent criticism, that "the issues involved in the other parts of [Putnam's] argument are more fundamental", with "the big conceptual questions " being "intended" versus "unintended" models, "standard" versus "non-standard" models, and the role of second-order logic. At the end, Bays concluded:

> ...I think it's still important to focus some of our attention on the purely mathematical problems in Putnam's argument. It's not that they are the *only* problems in this argument or even that they're the *deepest* problems in this argument; it's that they're the problems which are most closely connected to the things which make this argument philosophically interesting."

Finally, Button (2011), in an account framed by Bays' criticisms, set out the details and imperatives of the "metamathematics of Putnam's model-theoretic arguments". Button discussed at length what he took to be some related mathematics, e.g. the Completeness Theorem and weak set theories, anticipating concerns and reactions of the metaphysical realist. He did forward a simple overall line of argument, that although "Bays' challenge poses considerable problems for the constructivization argument", "it has no impact at all on the Skolemization or the permutation arguments".[1] For these two arguments, only a conditional is needed: 'if there is a model at all, there is an unintended one".

13.3 Constructivization Revisited

Spurred by the commentary and criticisms, we here take a deeper look at Putnam's constructivization argument—the text, the mathematical context, the underlying logic, and the specific ways, in the end, in which it can be taken. Putnam's footnote just before his theorem (cf. in the first quoted passage of Sect. 13.1), though not discussed by any of the commentators, provides textual evidence, with its two items each serving as points of beginning in what follows.

How did Putnam actually conceive of and render his argument? In the footnote, he wrote that he had proved the theorem to be deployed before 1963. We look at the historical context here to get an appropriate construal of his theorem.

Putnam's (1963) was a short yet seminal paper on constructible sets of integers.[2] In it, Putnam established, with ω_1^L being the least uncountable ordinal in the sense of L:

(∗) There is an ordinal $\alpha < \omega_1^L$ such that

there is no set of integers in $L_{\alpha+1} - L_\alpha$.

Gödel, of course, had established that $L_{\omega_1^L} = \bigcup_{\alpha<\omega_1^L} L_\alpha$ contains every constructible set of integers, thereby establishing the relative consistency of the Continuum

[1]The permutation argument is another model-theoretic argument from Putnam (1980); any theory with a model has multiple distinct yet isomorphic models given by permuting elements, and so there is a fundamental semantic indeterminacy.

[2]Sets of integers are routinely identifiable with, and called, reals, but we stick with the thematic trajectory here for a while.

Hypothesis. Putnam's theorem revealed that sets of integers are not steadily constructed up the L_α hierarchy, with his proof of $(*)$ actually showing that there are arbitrarily large $\alpha < \omega_1^L$ such that there is no set of integers in $L_{\alpha+1} - L_\alpha$. Putnam (1963) also showed that by the Shoenfield Absoluteness Lemma, which had just appeared in Shoenfield (1961), for any Δ_2^1 ordinal γ, there is an $\alpha < \omega_1^L$ such that there is no set of integers in $L_{\alpha+\gamma} - L_\alpha$.[3] This early and astute use of the Lemma is consonant with its use in Putnam's proof of his theorem for his constructivization argument.

Putnam's $(*)$ stimulated the dissertation work of his student Boolos on the recursion-theoretic analysis of the constructible sets of integers, this leading to their Boolos and Putnam (1968). According to Jensen in his classic (1972, p. 230): "To my knowledge, the first to study the fine structure of L for its own sake was Hilary Put[nam] who, together with his pupil George Bool[o]s first proved some of the results in Sect. 13.3."

How Putnam (1963) proceeded with the proof of $(*)$ anticipated his later constructivization argument. At the outset, he credited Cohen with the method; on his way to forcing, Cohen (1963) had shown that there is a minimal $\in$-model of set theory, and to do this he closed off $\{0, 1, 2, \ldots, \omega\}$ under set-theoretic operations and most crucially, under the instances of the Replacement Schema, assuming that this is possible and appealing to the Löwenheim–Skolem Theorem to get the countability of the resulting model.

Recasting this, Putnam argued for $(*)$ by initially appealing to the Downward Löwenheim–Skolem Theorem to get a countable elementary submodel of L, and proceeding to its transitive isomorph $\langle M, \in\rangle$, so that $M = L_\gamma$ for some countable γ. He then pointed out that there is no set of integers in $L_{\omega_1+1} - L_{\omega_1}$ by Gödel, and hence by elementarity that there is an ordinal $\alpha \in M$ such that there is no set of integers in $L_{\alpha+1} - L_\alpha$.

With VB (von Neumann–Bernays) being his working set theory, Putnam next pointed out: "...essentially the preceding argument can be formalized in VB. Of course, we cannot construct a model of *all* of VB in VB and also prove that it is a model." He then described formalizing argument in $\langle L_{\omega_1+2}, \in\rangle$ instead.

Finally, all this relativizes to L, and so there is an $\alpha < \omega_1^L$ such that there is no set of integers in $L_{\alpha+1} - L_\alpha$.

Now to Putnam's theorem (as stated in the first quoted passage of Sect. 13.1) for the constructivization argument, essentially:

$(**)$ For any real s, there is an ω-model of ZF + $V = L$ containing s.

The first point is that this cannot be a theorem of ZF, simply because of its asserting the existence of a model of ZF and hence the consistency of ZF. This would have been clear to anyone versed in mathematical logic as Putnam certainly was, and his (1963) remarks on VB above bears this out. *Putnam did not make a mathematical mistake in stating the theorem*, for surely he did not intend to state a theorem of ZF.

[3]In fact, this holds, by a straightforward modification of his argument, for *any* $\gamma < \omega_1^L$.

Proceeding to Putnam's proof of (∗∗) as given in Putnam (1980), one next sees the connection to his proof of (∗), described above. It is quite so, as commentators have observed, that using the Downward Löwenheim–Skolem Theorem on L requires additional resources beyond ZFC. This could be said also of Cohen's (1963) proof and of Putnam's (1963) argument. However, one sees in these argumentations from the early 1960s that they were proceeding informally to get at the fact of the matter. Putnam (1963) understood that there is the Cohen minimal model conditionally "if there is any well-founded model" (p. 269), and noted that his argument for (∗) with the Downward Löwenheim–Skolem Theorem can be carried out in $\langle L_{\omega_1+2}, \in\rangle$, as a model of full set theory is not required.

Similarly, *Putnam's* (∗∗) *is a theorem of informal mathematics*, stating a fact of the matter to be accepted by the metaphysical realist. His proof, getting quickly to the crucial use of Shoenfield Absoluteness, was meant, it would seem, to provide sufficient deductive ballast to usher the realist to the truth of (∗∗). If one does insist on a ZFC theorem, then the following is appropriate for an appeal to Shoenfield Absoluteness:

(1) If for any real s there is an $\in$-model of ZF containing s, then
for any real s there is an ω-model of ZF + $V = L$ containing s.

Given a constructible real s, there is by the hypothesis an $\in$-model M of ZF containing s and hence (by Cohen's argument!) there is such a model of form $\langle L_\gamma, \in\rangle$. Hence, the Π^1_2 statement formalizing "$\forall s \exists\, \omega$-model of ZF + $V = L$ containing s" is satisfied in L, and the result follows by Shoenfield Absoluteness.

That (1) is a *conditional* assertion in ZFC leads to a pivotal point about Putnam's model-theoretic arguments. Both his Skolemization and constructivization arguments are rhetorically in the form of a *reductio*, and the underlying logic can be carried by the conditional: if there is a model at all, then there is an unintended one. This being said, one can see what Putnam would have had in mind for the mathematics to be invoked by looking again at his footnote just before his theorem (cf. the first quoted passage of Sect. 13.1).

Putnam began the footnote with "Barwise has proved the much stronger theorem", that:

(2) Every countable model of ZF has a proper end extension
which is a model of ZF + $V = L$.

If $\langle M, E\rangle$ and $\langle N, E'\rangle$ are models of ZF, then the second is an *extension* of the first if $M \subseteq N$ and the membership relation E' extends the membership relation E; moreover, it is an *end extension* if for any $a \in M$ and $b \in N$, $b\,E'\,a$ implies that $b \in M$, i.e. elements of M have no new members in N. Barwise's theorem was a culmination of both the investigation of end extensions of models of set theory and the application of infinitary logic to the construction of models of set theory. His (1971) proof can be described as a complex application of the Barwise Compactness Theorem and the Shoenfield Absoluteness Lemma; the proof, rendered in the elegant terms of "admissible covers", appears as the last in his book (Barwise 1975). Barwise's theorem is evidently a strong "upward" Löweinheim–Skolem Theorem, in that one gets an end extension that also satisfies $V = L$. The analogy extends to having a sort of Skolem paradox for models of set theory, with any countable model of ZF being extendible to a canonically slimmest kind of model. This thematically suggests a role in the constructivization argument.

With Putnam in the footnote writing of Barwise's theorem as "much stronger" than his, we take the tack of deploying Barwise's theorem itself, rather than Putnam's, to effect a specification of the constructivization argument:

Both the Downward Löweinheim–Skolem Theorem and Barwise's theorem are conditional theorems of ZFC. With the former, having a (set) model of ZF that contains an abstract set isomorphic to *OP* amounts to having a countable such model. With the latter, having a countable model of ZF having an abstract set isomorphic to *OP* amounts to having such a model that also satisfies $V = L$. Thus, we have a ZFC rendition of Putnam's "Any model of ZF which contains an abstract set isomorphic to *OP* can be extended to a model for this formalized language of science which is *standard with respect to OP*" (cf. the second quoted passage of Sect. 13.1). Of course, the theorems used provide a close relationship between the resulting models and the initial one.

Putnam's Skolemization argument really turns on assuming that there is some model of set theory compatible with theoretical and operational constraints, and then showing that there is a countable one. Its logical structure analogous, his constructivization argument turns on assuming that there is some model of set theory compatible with theoretical and operational constraints, and then showing that there is one satisfying $V = L$. The first implication deployed the Downward Löweinheim–Skolem Theorem, and the second can be effected with Barwise's "upward" theorem.

Putnam (1980) had deployed ω-models in order to preserve the sense of *OP* coded as a real. For a specification of his argument just turning on ω-models, one can argue as above with the following immediate corollary of Barwise's theorem:

(3) If there is a countable $\in$-model of ZF containing a real s, then there is an ω-model of ZF + $V = L$ containing s.

With (1) schematized as $\forall s\varphi \longrightarrow \forall s\psi$, (3) is seen as the stronger version $\forall x(\varphi \longrightarrow \psi)$.

In summary, Putnam's constructivization argument was directed against a realist concept of set. An "epistemological/logical digression" as he put it, it has the rhetorical form of his Skolemization argument, that if there is a model at all, then there is an unintended one. Putnam simply pointed to a mathematical fact of the matter (∗∗) for his argument, but if the realist insists, one can present a conditional ZFC theorem (1) to him. In fact, there are stronger ZFC results, e.g. (2) and (3), that can be invoked. The constructivization argument has various aspects, various ways of putting it and of taking it. However, its overall philosophical thrust and import would not seem to depend on its underlying mathematics. Several results and theorems can be cited or invoked, each perhaps toning the argument in different directions, but not affecting its overall philosophical arc.

13.4 Critical Coordination

In this final section, we coordinate various criticisms (Sect. 13.2) that have been made in the literature of the constructivization argument, and in the process, consolidate mathematical issues about the argument beyond what was brought out in Sect. 13.3. In the broad, Putnam famously attempted to use model theory, i.e. mathematics, to draw metaphysical conclusions. The particular, constructivization argument, depending on a mathematical contingency new at the time, became surprisingly pivotal in the philosophical literature decades later. Mathematics having a precision, there were specifics that could be aired and argued, and with some more confident than others about the mathematics, commentators generated a fine-grained mesh of interpretation and assessment. While adjudication is often not be the order of the day for philosophical arguments, those involving mathematical results can arguably be illuminated by seeing how they turn on or can be taken according to the mathematics. In what follows, we initially follow a simplified dialectical arc.

Bays (2001, 2007) has been the most persistent and uncompromising in his criticism of Putnam's constructivization argument. Caught up in the mathematics, Bays urged repeatedly that Putnam's proof of his theorem is "mistaken" and maintained that there is an "intrinsic" problem here because of the Second Incompleteness Theorem, and then that the overall argument, in rhetorically pursuing such paths, is compromised.

Taken to an extreme, if no theorems asserting the existence of models are to be allowed at all, then Putnam's argument would collapse through vacuity. This simple *reductio* could not be what Bays was pursuing; he did acknowledge, for both Putnam's Skolemization and constructivization arguments, that one is assuming first that there is a model and then getting an unintended one. However, there is an ostensible asymmetry in how Bays proceeded:

Putnam had buttressed his Skolemization argument with the Downward Löwenheim–Skolem Theorem applied to "the standard model (if there is one)" which could formally be the proper class V of all sets, to get at a fact of the matter for the realist. If one insists on a ZFC theorem, then one can appeal to the Downward Löwenheim–Skolem Theorem for sets, starting from a set model and getting a countable one. Bays acknowledged the Skolemization argument in passing, this conditional avenue to getting an unintended model from a given model presumably being operative in Putnam's argument.

For the constructivization argument, Putnam had used the Downward Löwenheim–Skolem Theorem on L to get at a fact of the matter. Bays objected to this as such and did not pursue the underlying conditionality. However, if one insists on a ZFC theorem, (1) or (2), explicit in Putnam's footnote, could have been invoked, as discussed in Sect. 13.3.

Putnam's constructivization argument, it would seem, has a certain sense and an overall thrust. Its components can be addressed and debated variously, with its mathematical underpinnings renderable, e.g. with (2). Bays, in focusing on Putnam's "mathematical mistake"—and moreover treating it as symptomatic of Putnam's philosophical mistakes in general—seems to have fastened onto a relatively inconsequential eddy and distorted the overall flow of argumentation.

Bellotti (2005), in arguing against Bays' (2001) contention that Putnam had made a mathematical mistake in ZFC, first pointed out that Putnam had not specified his working theory. Belotti then became focused on possible extensions that could serve as that theory, to be part of "our best theory of the world". One additional assumption beyond ZFC sufficient for Putnam's theorem that Bellotti mentioned is akin to the antecedent of (1) (cf. Sect. 13.3). Taking the conditional (1), one can stay in ZFC as the working theory while advancing the constructivization argument rhetorically against the realist.

At the end, Bellotti (2005) affirmed (p. 407) his "most serious objection to Putnam", that the final models for nonconstructible reals are "definitely unintended", having an ill-founded membership relation. On this Bellotti followed Shapiro (1985), who had approached the issue from the perspective of second-order logic. The set-theoretic Axiom of Foundation asserts that the membership relation is well-founded, and if one is working in second-order logic, the axiom would indeed require any model to have a (really) well-founded membership relation. *Contra* Shapiro, one sees, however, that Putnam was working in first-order logic. Indeed, his Skolemization argument would not even get off the ground in second-order logic, as the Löwenheim–Skolem Theorem would not hold. One can pursue this sort of *reductio* to vacuity of course, but it would be by changing the very ground of the argument. *Contra* Bellotti, if one stays in first-order logic but requires intended models to be well-founded, this imposition of a second-order condition still goes against the very tenor of the constructivization argument. Theoretical and operational constraints are to be seen at work inside the final model, and (real) well-foundedness is something one only sees from outside the model.

Button (2011) did point out, *contra* Bays, that Putnam's Skolemization argument turned on the conditional: if there is a model at all, then there is an unintended one. Button also pointed out how the Completeness Theorem, provable in a weak set theory, can carry this conditional, while Putnam had appealed to the Downward Löwenheim–Skolem Theorem. As part of his extended analysis, he could have seen Putnam's footnote in his constructivization argument, from which it becomes evident how it too turns on the conditional, which can be carried by Barwise's theorem (2).

Separate from this to and fro, Gaifman (2004) interestingly waded through the mathematics of the constructivization argument, landing on a different shore. After acknowledging that Putnam's theorem can be based on some additional assumption(s) to be granted by the realist, Gaifman also pointed out that whether the final model is well-founded or not makes no difference—only what holds in the model, like $V - L$, is substantive to the argument. After this however, Gaifman took basic issue with ever launching such an argument, in view of the viability of non-standard models for the realist.

Stepping back, one sees that the mathematics of Putnam's constructivization argument has been chewed over variously, with the Sect. 13.3 articulation very much possible to hold up as a conditional challenge to the realist. Looking past the mathematics, the commentators, including Bays, went on to address substantive issues about how further to put and take the constructivization argument and determine the extent to which it is philosophically effective. These turn mainly on possible skeptical responses and where the realist stands dialectically in relation to the argument's components and what and how its moves are to be accepted. Be that as it may, however, the mathematics does stand as an interesting and robust part of the argument that Putnam put into play.

References

Barwise, J. (1971). Infinitary methods in the model theory of set theory. In R. O. Gandy & C. M. E. Yates (Eds.), *Studies in logic and the foundations of mathematics* (Vol. 61, pp. 53–66). Logic Colloquium '69. Amsterdam.

Barwise, J. (1975). *Admissible sets and structures. an approach to definability theory*. Berlin: Springer.

Bays, T. (2001). On Putnam and his models. *The Journal of Philosophy*, *98*, 331–350.

Bays, T. (2007). More on Putnam's models: A reply to Bellotti. *Erkenntnis*, *67*, 119–135.

Bellotti, L. (2005). Putnam and constructibility. *Erkenntnis*, *62*, 395–409.

Boolos, G. S., & Putnam, H. (1968). Degrees of unsolvability of constructible sets of integers. *The Journal of Symbolic Logic*, *33*, 497–513.

Button, T. (2011). The metamathematics of Putnam's model-theoretic arguments. *Erkenntnis*, *74*, 321–349.

Cohen, P. J. (1963). A minimal model for set theory. *Bulletin of the American Mathematical Society*, *69*, 537–540.

Dümont, J. (1999). Putnam's model-theoretic argument(s). A detailed reconstruction. *Journal for General Philosophy of Science*, *30*, 341–364.

Gaifman, H. (2004). Non-standard models in a broader perspective. In A. Enayat & R. Kossak (Eds.), *Non-standard models of arithmetic and set theory* (Vol. 361, pp. 1–22). Contemporary mathematics. Providence: American Mathematical Society.

Jensen, R. B. (1972). The fine structure of the constructible hierarchy. *Annals of Mathematical Logic*, *4*, 229–308.

Levin, M. (1997). Putnam on reference and constructible sets. *British Journal for the Philosophy of Science*, *48*, 55–67.

Putnam, H. (1963). A note on constructible sets of integers. *Notre Dame Journal of Formal Logic*, *4*, 270–273.

Putnam, H. (1980). Models and reality. *The Journal of Symbolic Logic*, 45, 464–482. Delivered as a presidential address to the Association of Symbolic Logic in 1977.

Shapiro, S. (1985). Second-order languages and mathematical practice. *The Journal of Symbolic Logic*, *50*, 714–742.

Shoenfield, J. R. (1961). The problem of predicativity. In Y. Bar-Hillel, E. I. J. Poznanski, & A. Robinson (Eds.), *Essays on the foundations of mathematics* (pp. 132–139). Jerusalem: Magnes Press.

Velleman, D. (1998). MR1439801, Review of Levin, Putnam on reference and constructible sets. *Mathematical Reviews 98c*, 1364.

Akihiro Kanamori is a Professor of Mathematics at Boston University. A set theorist, he has worked in infinite combinatorics and on consistency results and large cardinal hypotheses. On this last subject, his book The Higher Infinite (Springer, 1994) has become the standard text. He has co-edited, with Matthew Foreman, the three-volume The Handbook of Set Theory (Springer, 2010). His latest research interests lie in the history and philosophy of mathematics.

Chapter 14
Putnam on Mathematics as Modal Logic

Øystein Linnebo

Abstract Two uses of modal logic to explicate mathematics—due primarily to Hilary Putnam and Charles Parsons—are compared and contrasted. The approaches differ both technically and concerning ontology. Some reasons to push the former approach in the direction of the latter are articulated and discussed.

14.1 Introduction

One of the many privileges of being a graduate student at Harvard in 1996–2002 was the opportunity to attend several of Hilary Putnam's lecture series and seminars. As anyone who has shared this privilege knows, these sessions were enormously wide ranging and full of insights and often surprising connections. A recurring theme, however, was that of different but 'equivalent' descriptions of one and the same aspect of reality. An early example of this central theme in Putnam's thinking is found in his philosophy of mathematics, especially (Putnam 1967), which defends the possibility of two complementary philosophical 'pictures' of mathematic: a broadly platonistic 'object' picture and a non-platonistic 'modal' picture.

On a traditional platonist conception, mathematics is concerned with a fixed and determinate universe of abstract objects. Indeed, this is part of the analogy between mathematics and the empirical sciences on which mathematical platonism to a large extent is based. Just as astronomy, say, is concerned with a fixed and determinate universe of stars, galaxies, and gas clouds, so mathematics is concerned with its own fixed and determinate universe of numbers, sets, and spaces. Other than being non-spatiotemporal and causally inefficacious, the objects of mathematics exist in the same way as those of astronomy and are all 'available' to be talked about and quantified over in the same straightforward and unproblematic way as stars and galaxies.

As Putnam (1967) made clear, there are alternative conceptions of mathematics which reject the static conception of its subject matter in favor of what I like to

Ø. Linnebo (✉)
University of Oslo, Oslo, Norway
e-mail: linnebo@gmail.com

G. Hellman and R. T. Cook (eds.), *Hilary Putnam on Logic and Mathematics*, Outstanding Contributions to Logic 9, https://doi.org/10.1007/978-3-319-96274-0_14

think of as a more dynamic conception. On such alternative views, it is impossible to ascribe to mathematics a fixed and determinate universe of mathematical objects. Any such universe can be used to define an even larger such universe, which is no less legitimate or mathematically interesting than the previous one. So on such views, we must come to terms with an inherently dynamic character of the subject matter of mathematics. This will most likely require supplementing ordinary quantification theory (of first or higher order) with resources that are better suited for representing the dynamic process of considering ever larger universes.

One way to break with the traditional platonist's static point of view is suggested by Putnam's picture of mathematics as modal logic. Mathematics need not be understood as concerned with a distinct universe of its own, populated by abstract objects. It can equally well be seen as concerned with possible realizations of mathematical structures—including, of course, *concrete* realizations—and generalizations about what necessarily holds in all these structures. There is no maximal or all-encompassing structure with which mathematics is concerned. Rather, for any possible structure, there could be an even larger or richer structure. Central aspects of this view can be traced back to Zermelo (1930). Its most systematic development is found in the work of Geoffrey Hellman, in particular (Hellman 1989).

Another way to break with the traditional platonist's static point of view was pioneered by Charles Parsons. Here, the idea is not to 'trade in' one's mathematical objects in favor of modal claims about possible realizations of structures but rather to locate some modally characterized features in the mathematical objects themselves. The mathematical universe is not 'flat'. Rather, some of its objects stand in relations of *ontological dependence*, and the existence of some of its objects is *merely potential* relative to that of others.

> A multiplicity of objects that exist together *can* constitute a set, but it is not necessary that they *do*. Given the elements of a set, it is not necessary that the set exists together with them. [...] However, the converse does hold and is expressed by the principle that the existence of a set implies that of all its elements. (Parsons 1977, pp. 293–4)

As Parsons emphasizes, this approach can also be used to explicate the influential iterative conception of sets, which tends to be explained by suggestive but loose talk about a 'process' of 'set formation'. It would be better, Parsons claims, to replace this talk of time and construction with 'the more bloodless language of potentiality and actuality' (Parsons 1977, p. 293). This use of modal notions to characterize the universe of sets harks back to Cantor, whose famous letters to Hilbert (in 1897) and Dedekind (in 1899) seek to explain why certain 'multiplicities' do not form sets in terms of their members' inability to 'exist together' or cannot be regarded as 'a unity, as "one finished thing"' (Ewald 1996, pp. 931–932). In recent years, this modal approach to set theory has been further developed and explored.[1]

In fact, there is a third way too to break with the platonist's traditional static conception, represented by Fine (2005, 2006). This third way is based on an imperatival notion of postulation. This makes it natural to regard Aristotle, with his deep

[1]See Linnebo (2010, 2013), Studd (2013, 2016).

and influential notion of potential infinity, as an Abraham whom all three traditions should recognize as a distant but common forefather.[2]

My aim in this paper is to compare and contrast the first two ways of articulating a more dynamic conception of the subject matter of mathematics. As we will see, there are substantial differences in technical development and concerning ontology. The idea of mathematics as modal logic invites a nominalist view (and has been defended as such by Hellman), while the tradition inspired by Parsons is firmly committed to the existence of mathematical objects. But underneath these differences, there are some important similarities, especially as concerns Putnam's own version of the tradition that *he* inspired. Moreover, I shall argue that there are reasons to push Putnam's approach even further in the direction of Parsons'.

14.2 Putnam's Approach to Modal Set Theory

Set theory provides a particularly nice arena in which to compare the two traditions with which we are concerned. Both traditions are motivated in large part by the problem of the 'open-endedness' of the hierarchy of sets. However many sets have been formed, it is possible to form even more. There can be no end to the process of set formation. It is clear that this idea played a major role in Putnam's thinking. For instance, in the concluding paragraph, Putnam claims that

> [t]he real significance of the Russell paradox, from the standpoint of the modal-logic picture, is this: it shows that *no* concrete structure can be a standard model for the naive conception of the totality of all sets; for any concrete structure has a possible extension that contains more "sets."[3]

A closely related idea is found in Zermelo's groundbreaking 1930 article.

> But [the set-theoretic paradoxes] are only apparent 'contradictions', and depend solely on confusing set theory itself, which is not categorically determined by its axioms, with individual models representing it. What appears as an 'ultrafinite non- or super-set' in one model is, in the succeeding model, a perfectly good, valid set with both a cardinal number and an ordinal type, and is itself a foundation stone for the construction of a new domain. (Zermelo 1930, p. 1233)

As we have seen, the idea of 'open-endedness' plays a prominent role in the tradition inspired by Parsons as well. It is impossible for all of the possible sets to coexist. For whatever objects coexist, it is possible for there to be a set with precisely these objects as its members. And on pain of paradox, this set cannot be one of the sets with which we started.

---

[2]Though as Stewart Shapiro reminds me, the descendants of Aristotle get along much better than those of Abraham. I believe this says something about the broader value of the form of reasoned debate that philosophy often illustrates.

[3]See Putnam (1967, p. 22). Cf. also p. 21, where Putnam locates 'any philosophical significance' of his modal approach in its ability to avoid the need for maximal models of Zermelo set theory.

In slogan form, the idea of open-endedness is that the process of set formation can always be extended. How can this slogan be developed into a more sharply characterized view? Two questions stand out as particularly important.

1. What, exactly, are the extensions that are said to be possible?
2. What is the modality with respect to which the extensions are said to be possible?

Until the final section, our concern will be exclusively with the first question.

In order to understand Putnam's answer to this question, we need the notion of a *standard model* of second-order set theory. First, as Hellman (1996) has emphasized, it is natural to understand talk about such models as just plural or second-order talk about many things that are arranged appropriately. (For ease of communication, we shall nevertheless use set theoretic notation.) Next, such a model is said to be *standard* if it is well-founded and (loosely speaking) contains no gaps when we look downwards along the membership relation (to members, or their members, or so on) or sideways. A precise definition can be given in our second-order language. Consider a model based on a domain M and a membership relation $R \subseteq M \times M$, in terms of which the membership predicate $\in$ is interpreted. The model is said to be standard if (i) the membership relation R is well founded, and (ii) the model has the following maximality property:

> Consider any a in M. Let X be the collection of objects that bear R to a. Then, for any subcollection $Y \subseteq X$, there must be some b in M such that Y is the collection of objects that bear R to b.

Let ZFC2 be second-order Zermelo–Fraenkel set theory with the axiom of choice. Second-order Zermelo set theory (Z2) is ZFC2 minus Foundation and Replacement.

Putnam's answer to the first question is that *any standard model of Z2 can be extended to a larger standard model.* This answer is completely neutral with regard to the elements of the standard models. But Putnam clearly envisages the standard models as populated by concrete objects, since he talks about them as being based on 'pencil points' connected by 'arrows' (presumably also drawn in pencil) (Putnam 1967, p. 20).

Our next task is to provide an interpretation of ordinary first-order set theoretic discourse. To do so, we need to provide a translation from the language of ordinary set theory into the language that talks about possible models and their extensions. A simple example suffices to convey the idea, which is quite intuitive. Consider the claim that for every ordinal there is a greater ordinal: $\forall\alpha\exists\beta(\alpha < \beta)$. This claim is translated as the follows:

> Necessarily, for every standard model and every object α that plays the role of an ordinal in this model, possibly there is an extended standard model containing an object β that also plays the role of an ordinal, and according to which α is smaller than β.

What are the standard models in question? As we have seen, Putnam takes them to be standard models of Z2. We will instead follow Hellman and take them to be standard models of ZFC2 (to which we will henceforth refer simply as standard models).

Careful readers will want a proper definition of the translation. This requires us to work with translations relativized to a model. Write $[\phi]_M$ for the translation of ϕ

relative to M. Let $[t_1 \in t_2]_M$ be Et_1t_2, where E is this model's interpretation of $\in$. The only other atomic formulas are identities, which are translated as themselves. The connectives are translated compositionally; for instance, $[\neg\phi]_M$ is $\neg[\phi]_M$. Next, a universal quantification over sets, relative to a model M, is translated as the necessitated universal quantification over standard models M^+ that extend M, followed by universal quantification over the members of M^+; and the reference model is now reset to M^+. In symbols:

$$[\forall x\phi]_M \quad = \quad \Box(\forall M^+ \sqsupseteq M)(\forall x \in M^+)[\phi]_{M^+}$$

where the variables M and M^+ are restricted to standard models, and $M^+ \sqsupseteq M$ means that M^+ is a (not necessarily proper) extension of M. Finally, we define a non-relativized *Putnam translation* of sentences by letting the translation of ϕ, which we write ϕ^{PT}, be $[\phi]_\varnothing$.[4]

How plausible is this as a translation of ordinary set theoretic discourse? As Putnam (1967) realized and Hellman (1989) articulated in more detail, there is no problem as far as logic or the theorems of set theory are concerned. (Some more demanding requirements are considered in Sect. 14.6.) However, it is only very recently that this insight has been given what I regard as a fully adequate formulation and proof. Sam Roberts (forthcoming, Sect. 2.6) shows that there are natural and plausible theories of ordinary non-modal set theory and of mathematics as modal logic such that a sentence ϕ is a theorem of the former just in case its Putnam translation ϕ^{PT} is a theorem of the latter. In technical parlance, the former theory is *faithfully interpreted* in the latter. The non-modal set theory in Roberts' result is the system Z* + In of (first-order) Zermelo set theory plus the claim that every set is contained in some strongly inaccessible rank. Another theory, MSST (for modal structural set theory) seeks to capture Putnam's and Hellman's own ideas concerning mathematics as modal logic. This theory, which is based on a combination of free logic and the modal logic S5, has as its most distinctive axiom *the extendability principle*, which says that necessarily, every standard model possibly is extended by a larger standard model:

(EP) $$\Box\forall M\Diamond\exists M^+(M \sqsubset M^+)$$

where the variables range over standard models and '$\sqsubset$' means 'is properly extended by'. (This principle is of course closely related to Zermelo's claim that '[w]hat appears as an 'ultrafinite non- or super set' in one model is, in the succeeding model, a perfectly good, valid set'.) Roberts proves that the translation PT provides a faithfully interpretation of Z* + In in MSST; that is:

$$\mathrm{Z}^* + \mathrm{In} \vdash \phi \quad \text{iff} \quad \mathrm{MSST} \vdash \phi^{\mathrm{PT}}$$

[4]Strictly speaking, the empty set is not a model. But this causes no problem in practice.

Notice that Z^* + In does not include the axiom of Replacement, although the axiom holds in each of the standard models that MSST quantifies over. Indeed, while Z^* + In clearly proves the existence of n inaccessibles for each natural number n, Roberts (forthcoming, fn. 55) demonstrates that it does not prove the existence of the ω'th inaccessible.

14.3 A Version of Parsons' Approach to Modal Set Theory

I turn now to the approach inspired by Parsons. My focus will be on a streamlined version that I developed in Linnebo (2013).[5] Recall the two questions prompted by our slogan formulation of open-endedness. The first question—about the nature of the extensions that are said to be possible—now receives a very simple answer. *Any objects potentially form a set.*

In order to formalize this answer, I rely on plural logic, understood as in the tradition deriving from Boolos (1984). This logic allows plural variables, such as xx and yy, each of which is allowed to have many values from the domain, rather than just a single value, as in the case of ordinary singular variables. The plural variables are also allowed to be bound by quantifiers. Finally, we introduce a new logical constant $\prec$ for plural membership. Thus, $x \prec yy$ means that x is one of yy. From a technical point of view, plural logic is clearly just a version of (extensional, monadic) second-order logic, gently modified so as to require that all pluralities be non-empty.

The mentioned answer—that necessarily any given things possibly form a set—can now be formalized as

$$(\text{Collapse}^{\Diamond}) \qquad \Box\forall xx\, \Diamond\exists y\, \text{Set}(y, xx)$$

where $\text{Set}(y, xx)$ says that y is the set whose elements are precisely xx. I call this principle *potential plural collapse*. It can be thought of as dynamic version of naive set comprehension. As Stephen Yablo nicely puts it, the principle is our 'main engine of set production' (Yablo 2006, p. 150).

Next, we need principles of extensionality for sets. But extensionality now needs to be formulated so as to hold not just within each possible world but also across possible worlds. That is, we must require that a set have precisely the same members at every world at which it exists. We begin with the necessitation of the ordinary principle of extensionality. In addition, we adopt two principles stating that membership is stable as we go from world to world:

[5]This approach differs from Parsons's in making more explicit use of plural logic and in offering a much simpler translation from the language of ordinary set theory into the language of modal set theory. (Parsons' translation is a combination of the double negation translation of an intuitionistic language, composed with Gödel's translation of this language into the modal language.)

$$(\text{STB}^+ \in) \qquad u \in x \to \Box(u \in x)$$

$$(\text{STB}^- \in) \qquad u \notin x \to \Box(u \notin x)$$

(We adopt analogous stability principles for plural membership $\prec$.) Earlier investigations confirm that these principles, plus one more, adequately capture the transworld extensionality of sets.[6] Our occasional talk about possible worlds is thus merely heuristic.

As on Putnam's approach, we can translate from the language of ordinary set theory into the language of modal set theory. The translation is pleasingly simple. A universal quantification over sets is translated as a necessary universal quantification, and an existential quantification, as a possible existential.[7] That is, $\forall x$ and $\exists x$ are translated as $\Box\forall$ and $\Diamond\exists$, respectively. The connectives are translated compositionally. Let $\phi^\Diamond$ be the result of translating ϕ in this way.

As on Putnam's approach, we would like to show that this translation preserves facts about logical entailment and theoremhood in the relevant theories. In order to do so, we need more details. Let us begin with the modal logic. As we have seen, Putnam and Hellman use S5. For my own purposes, however, I have elsewhere[8] recommended a slightly weaker modal logic known as S4.2, which is the result of adding to the familiar system S4 the axiom

$$(\text{G}) \qquad \Diamond\Box\phi \to \Box\Diamond\phi.$$

This modal logic is chosen in order to represent some key structural aspects of the generation of mathematical objects. The stages along which this generation can unfold are partially ordered. This justifies the axioms of S4, which hold in all reflexive and transitive frames. The justification for adding the axiom (G) lies in a cumulativity requirement on the process of set formation: the licence to form a set never disappears but can, if need be, be exercised at a later stage. This means that our partial order has to be directed—that is, any two worlds w_1 and w_2 can be extended to a common world w_3—which in turn can be seen to justify (G).[9]

We can now state our main results. Let $\mathcal{L}$ be a non-modal language, which may have plural resources. Let $\vdash$ be the relation of deducibility in $\mathcal{L}$ based on classical

[6] See Linnebo (2013, Sect. 6.1).

[7] Parsons' own translation is substantially more complicated, as explained in footnote 5.

[8] Again, see Linnebo (2010, 2013).

[9] Although the modal logic of my approach is strictly weaker than the one used on the Putnam approach, the opposite holds for their theories of quantification: Roberts' is free, while mine is not. Consequently, my logic entails the Converse Barcan Formula, which requires that the domains be non-decreasing along the accessibility relation. This means that necessity on my approach corresponds, on the Putnam approach, to necessity assuming the continued existence of all the objects there are. (More precisely, '$\Box\ldots$' on my approach corresponds to '$\forall xx\big(\forall y(y \prec xx) \to \Box(Exx \to \ldots)\big)$' on Putnam's, where E is a plural existence predicate.) In light of this, (G), in my setting, makes a claim that goes beyond anything ensured by Roberts' logic in his setting, namely that two possible extensions of the ontology can be 'merged' into a single common extension. This observation will be important as we go along.

sentential logic, the standard axioms of identity, and the standard introduction and elimination rules for the quantifiers of all orders, *but without any plural comprehension axioms*, which we set aside for separate treatment.[10] Let $\vdash^{\Diamond}$ be provability by $\vdash$, S4.2, and axioms stating that every atomic predicate is stable, but with no higher-order comprehension. Then we have:

$$\phi_1, \ldots, \phi_n \vdash \psi \quad \text{iff} \quad \phi_1^{\Diamond}, \ldots, \phi_n^{\Diamond} \vdash^{\Diamond} \psi^{\Diamond}$$

Moreover, we can formulate a natural and plausible modal set theory, MS, such that the 'potentialist' translation of ϕ as $\phi^{\Diamond}$ provides a faithful interpretation of first-order Zermelo-Fraenkel set theory in MS. The modal set theory MS has as its heart the principle of potential plural collapse, (COLLAPSE$^{\Diamond}$), much as the modal structuralist theory from the previous section has the Extendability Principle. It also relies on a kind of reflection principle, to the effect that any purely first-order claim whose potentialist translation holds is also possible; that is[11]:

(Refl) $$\phi^{\Diamond} \rightarrow \Diamond\phi$$

14.4 Metaphysical Entanglement

We are now ready to compare the two approaches to modal set theory. The comparison will be structured around their respective answers to our guiding questions from Sect. 14.2.

The first question concerns the extensions that are said to be possible. What are these? The answers given by our two approaches differ in two respects. One difference concerns a purely structural matter, namely the size of the jumps involved in the extensions. Putnam proceeds in big strides. Any standard model can be extended to a larger standard model. We know from Zermelo (1930) that each of these standard models is isomorphic to a rank V_κ for some strongly inaccessible κ.[12] So Putnam's extensions always take us straight to the next strongly inaccessible rank. By contrast, on the Parsons approach, we can proceed in steps that are tiny and involve as little as a single set being added. The significance of this difference will be discussed in the next section. Another difference concerns the metaphysical nature of the extensions. On the Putnam approach, the extensions can be based on any objects whatsoever, provided that these are arranged so as to play the structural role of sets. On the Parsons approach, by contrast, the extensions are populated by sets, not just objects arranged so as to look like sets.

[10] For instance, the deducibility relation can be given by the logic PFO of Linnebo (2012a) minus the plural comprehension scheme.

[11] See Linnebo (2013) for details. As we will see below, it is problematic to add an analogous reflection principle to Putnam's approach.

[12] Thus, (EP) is closely related to the claim that for every inaccessible there is a larger inaccessible.

The extra freedom obtained by permitting the standard models to be populated by non-mathematical objects seems appealing but turns out to be treacherous. As a result of this freedom, the Putnam approach incurs strong and potentially problematic commitments about the metaphysics of concrete objects. One example concerns the extendability principle (EP), which states that necessarily any such standard model can be extended by a larger standard model. But is this principle really defensible, given the complete freedom concerning the members of these standard models? Do we really know that there cannot be 'metaphysically shy' objects, which can live comfortably in universes of small infinite cardinalities, but which would rather go out of existence than to cohabit with a larger infinite number of objects?[13] Another variant of this problem arises when we consider applied set theory. In order to apply set theory to some given objects, the approach in question presumably needs it to be possible for these objects to coexist with strongly inaccessibly many objects, arranged so as to make up a standard model built on top of these urelemente. But this claim will be false if there can be objects that necessarily cannot coexist with inaccessibly many other objects.

Might the problem be solved by modifying the extendability principle? It will not do simply to restrict the principle to models based on non-shy objects, as this would threaten the application of mathematics to shy objects. A more promising option, suggested to me by Hellman, is to relax the extendability principle such that it only makes demands 'up to isomorphism': 'Necessarily, for any model M, possibly there is a model M' which is isomorphic to M and which possibly has a proper extension.' While this is promising, we need to be shown how the modal structuralist has the resources to formulate the transworld isomorphism claim.

Another problem was pointed out to me by Sam Roberts. The most natural way to connect up ordinary set theory and Putnam's modal alternative involves an appeal to Zermelo's quasi-categoricity theorem, which says that any two standard models are either isomorphic or such that one is isomorphic to an initial segment of the other.[14] However, the mentioned connection requires us to be able to compare models not only within a possible world but across possible worlds.[15] But the ability to do so comes under pressure from the phenomenon of incompossible objects. Borrowing an example from Timothy Williamson, suppose you have one shaft and two blades. Then it is possible for you to make one knife by attaching blade number one to the shaft, and another, by instead attaching blade number two. But it is impossible for you to make both knives simultaneously.[16] Now, suppose you have a standard model M. Then it is possible for there to be an extension of M involving the knife based on

[13]A possible example would be an ontology of (first-order) facts, according to which there is the fact that there are so-and-so many things just in case there are that many things. (Thanks here to Peter Fritz.).

[14]See e.g. Hellman (1989, pp. 68ff).

[15]As Hellman (1989, pp. 42–3) is well aware.

[16]Of course, it is a consequence of Williamson (2013)'s *necessitism*—the view that necessarily everything necessarily exists—that there can be no incompossibles. Williamson's preferred analysis of the example is that the two coexisting possible knives cannot simultaneously be 'chunky', which in this case comes to being realized in spacetime. However, necessitism provides no solace in our

blade number one, and another extension involving the knife based on blade number two. However, because of their incompossible members, it is impossible for the two extensions to coexist, as would be required in order to apply the quasi-categoricity theorem.

Might the problem be avoided by circumventing this direct appeal to quasi-categoricity? Hellman (1989, p. 43) discusses an analogous problem in the case of arithmetic and proposes a solution, which is easily adapted to the present setting. The idea is to adopt an 'accumulation principle' to ensure that two structural possibilities can always be realized in one and the same possible world:

$$\Diamond\exists M(M \models \phi) \wedge \Diamond\exists N(N \models \psi) \rightarrow \Diamond\exists M\exists N(M \models \phi \wedge N \models \psi)$$

Roberts (forthcoming) makes an improved proposal, which he shows to overcome a shortcoming of Hellman's proposal and which plays an essential role in his proof of the result about faithful interpretability. This proposal is based on a substantive assumption. Loosely speaking, we must assume that if a claim about models and some pattern of possible model extensions is possible, then *necessarily* this claim is possible.[17] Or—indulging in talk about possible worlds—we must assume that if some pattern of model extensions is available at a world w, then the analogous pattern is available at any other world to which w has access. While Roberts' proposed solution is promising, we are still owed an account of why modal structuralists are entitled to accept the assumption on which it rests.

Incompossibles give rise to another problem as well, which emerges when we try to add a set theoretic reflection principle to a Putnam-style approach. Hellman attempts to do so in his (forthcoming), where the chief candidate is[18]:

$$\phi^{\mathrm{PT}} \rightarrow \Diamond\exists M(M \models \phi)$$

To see this, let K_1 and K_2 be descriptions of the mentioned possible knives. Let ϕ be $\exists x\ K_1(x) \wedge \exists x\ K_2(x)$. Thus, ϕ^{PT} is

$$\Diamond\exists M(\exists x \in M)K_1^{\mathrm{PT}}(x) \ \wedge \ \Diamond\exists M(\exists x \in M)K_2^{\mathrm{PT}}(x)$$

present context, as its fixed maximal domain of objects would clash with our emphasis on set theoretic open-endedness.

[17] I have in mind Roberts' 'stability' axiom, (S), which he shows goes significantly beyond (the adaption to our present setting of) Hellman's accumulation principle. The axiom is the Putnam translation of the logical truth $\forall\bar{x}(\exists\bar{y}\phi \rightarrow \forall z\exists\bar{y}(z = z \wedge \phi))$.

[18] As Hellman (2015) observes, this principle needs to be restricted so as to prevent reflection on Extendability, which (as I had pointed out) would result in an inconsistency. While this problem concerns the interaction of reflection and higher-order quantification, the problem described in what follows is qualitatively different, as it arises for a purely first-order sentence. It should be noted that Hellman has now given up this attempt to do reflection in a modal-structural setting (prompted in large part by Roberts (forthcoming)); see Hellman (2015, fn. 22) and Hellman (forthcoming).

Now, in order to account for applied set theory, it is plausible that all the standard models in play interpret non-mathematical vocabulary standardly. This ensures that ϕ^{PT} is equivalent to the truth $\Diamond\exists x\, K_1(x) \wedge \Diamond\exists x\, K_2(x)$.[19] However, by reflecting on this truth, we obtain $\Diamond\exists M(M \models \phi)$, which is equivalent to $\Diamond\phi$. But this falsely states that the two knives are compossible.

Let me be very clear about my complaints in this section. I am not asserting that metaphysically shy objects are in fact possible or that there might not be some clever way to circumvent the problems generated by the phenomenon of incompossibles. My point is only that the extra freedom of Putnam's approach, which initially seemed purely advantageous, has the unintended side effect of incurring potentially problematic metaphysical commitments, which are avoided on the Parsons approach.

14.5 The Size and Shape of the Jumps

Does it matter that the Putnam approach proceeds in big strides, while the Parsons approach allows for tiny steps?

One might think that the difference favors the former approach, as it justifies a stronger system of non-modal set theory, involving an unbounded sequence of strong inaccessibles. I do not think this is a genuine advantage. It is easy to modify my approach so as to make it more similar in this respect to Putnam's. All we need to do is require that the relevant possible worlds be closed under Powerset and a plural version of Replacement.[20] Bracketing the question of nominalism (to which we shall return), I contend that this closure property is just as plausible as the Extendability Principle.

My own view is the opposite, namely that the structural difference between the two kinds of jump favors my approach over Putnam's. By always jumping to the next standard model, Putnam's approach builds in a commitment to strong set theoretic principles such as Powerset and (a second-order version of) Replacement. But the open-endedness of set theory is a phenomenon that is recognized by many constructivists and predicativists as well, despite their rejection of a determinate totality of subsets of the natural numbers. It would be advantageous if our analysis of open-endedness was acceptable to all major defenders of the phenomenon.

Can a version of Putnam's approach be developed which is structurally similar to my own and thus avoids the strong but unnecessary commitments? A model of set theory N is said to be an *end extension* of another such model M just in case N is an extension of M which adds no new elements of any sets present already in N. Instead of considering an increasing sequence of standard models, we can consider models of set theory ordered by the relation $\sqsubseteq$ of being an end extension. Clearly, this is a

---

[19] We are here relying on the plausible assumption that the knives in question aren't 'metaphysically shy'.

[20] The latter states that, if xx are no more numerous than some things yy which form a set, then xx too form a set.

partial order. In order to imitate an essential feature of my approach which ensures that the translations of the quantifiers really behave logically like quantifiers, we additionally need $\sqsubseteq$ to be directed, in the sense that necessarily any two extensions M_1 and M_2 of some given model possibly share a *common* extension N (that is an N such that $M_1 \sqsubseteq N \wedge M_2 \sqsubseteq N$). But this directedness property is false, as can be seen by reflecting on the phenomenon of incompossible objects.

It therefore has to be investigated whether the need for directedness can be circumvented. This takes us back to the considerations of the previous section, where we discussed how the need for quasi-categoricity might be circumvented.[21]

14.6 Equivalent Descriptions or Levels of Explicitness?

While Parsons insists that all the extensions of set theoretic structures be populated by genuine sets, the Putnam approach offers greater flexibility. Hellman uses this extra flexibility to defend a nominalistic interpretation of the approach, which seeks to use modal resources to eliminate all ontological commitment to abstract mathematical objects. Putnam's own view is more complex. He is explicit that his 'purpose is not to start a new school in the foundations of mathematics (say, "modalism")' (Putnam 1967, p. 19). Instead, he regards the two pictures—the 'object' picture and the 'modal' picture (p. 11)—as equally legitimate but 'dual', in something like the sense in which there is a wave-particle duality in quantum mechanics (pp. 11 and 19). This duality means that Putnam is not a nominalist. Unlike Hellman, he is not claiming that there are no mathematical objects, only that it is possible to give a theory of the world that is based on modality rather than abstract mathematical objects, and which is not derivative from or inferior to its more platonistic rival.

I share Putnam's general view that there is much to be learnt from considering different pictures of one and the same subject matter, and that this is a salutary antidote to foundationalism. Even so, I find this particular alleged duality hard to make sense of. It is of course true that the theories associated with the two pictures are mutually interpretable. But mutual interpretability is not a strong form of equivalence. There are many philosophically interesting properties that are not preserved under interpretability. Richard Heck (2000) provides a nice example. Given his training in mathematics, Frege obviously knew that Euclidean geometry is interpretable in real analysis. And he famously regarded the latter as analytic. Nevertheless, Frege held that Euclidean geometry is not analytic but rather synthetic a priori.

The question is thus what Putnam takes the equivalence between the two pictures to consist in, beyond just mutual interpretability. He admits that the theories are not synonymous. (Although he claims that for mathematical purposes, they might as well have been so regarded.) Rather, the theories are said to have the same mathematical content or describe the same mathematical facts. Yet Putnam admits that on their own,

[21] An obvious adaptation of Roberts' Stability axiom would work from a technical point of view. So again, modal structuralists' entitlement to this axiom needs to be assessed.

these remarks add little of substance (p. 8). Perhaps more promising, the theories are also said to be 'equivalent descriptions' in Reichenbach's sense (p. 8). This means in particular that

> there is no particular advantage to taking one of the two theories as fundamental and regarding the other one as derived. The two theories are, so to speak, on the same explanatory level. (p. 8)

This rules out the view that one of the theories is merely an interesting and perhaps expedient way of reformulating the other and thus is nothing but a *façon de parler*. Yet at the end of the day, the desired equivalence remains elusive. We are given little more than the metaphorical idea of two pictures that provide equally permissible ways of 'carving up' reality and which therefore make the same demands on reality. While suggestive, these metaphors require unpacking.

Although the Parsons approach too operates with two different languages and theories, it contains no analogue of Putnam's claim about duality. The sets that are described by the non-modal theory have certain features, such as ontological dependence and potential existence, which are made explicit only by the modal set theory. The relation between the two theories is therefore not one of duality but is simply a matter of levels of explicitness.

In fact, there are reasons to doubt the tenability of Putnam's duality between a platonistic 'object' picture and a nominalistic 'modal' picture. I believe each of the pictures needs to be softened in a way that will push the resulting view in the direction of Parsons'. Let us begin with the modal nominalism. As Parsons observes:

> there is no reason to believe that structures of the required kinds are possible where the objects involved have any of the characteristic marks of concreteness. In cardinality, they will outstrip anything that can be represented in the physical world. (Parsons 1990, p. 330)

We may perhaps be able to make sense of there possibly being an omega sequence of stars. But with the enormous structures that are required by contemporary set theory, it is hard to see in what sense it is possible for there to be *concrete* realizations. The possibilia to which Putnam's 'modal' picture appeals must be radically different from ordinary physical objects and thus not much different from abstract objects.

Next, consider the platonistic side of Putnam's duality. As we have seen, Putnam takes very seriously the open-endedness that is made explicit in 'modal' picture. He insists that no model of set theory—whether concrete or abstract—can be maximal:

> Even God could not make a model for Zermelo set theory that it would be *mathematically* impossible to extend, and no matter what "stuff" He might use. (Putnam 1967, p. 21)

Putnam is therefore committed to a completely unrestricted version of the Extendability Principle. By simultaneously accepting the 'object' picture, however, he is in danger of making a mistake that Hermann Weyl warned us against, namely to 'attempt to turn the field of possibilities opening to infinity into a closed realm of absolute existence' (Weyl 1930, pp. 19–20). As Weyl realized, to make this mistake is to court paradox. In our present context, the paradox takes the following form. Consider the totality of sets on Putnam's 'object' picture. By applying the Extendability Principle, it follows that there might be an even larger standard model. And by the equivalence of Putnam's two pictures, this has consequences back in the 'object' picture: for it follows that the universe of sets continues beyond the sets with which we started. But this contradicts our assumption that these were all the sets.

In order to block this argument, we need to explain why the Extendability Principle cannot be applied to the hierarchy of sets associated with the 'object' picture. A more potentialist conception of the sets—which doesn't regard them as 'a closed realm of absolute existence'—promises to do just that. The Extendability Principle allows us to extend any set theoretic structure whose members coexist and in this sense form a 'closed realm'. On a potentialist conception, however, all the sets cannot coexist, which means that the set theoretic hierarchy cannot be regarded as a 'closed realm'. This blocks the application of the Extendability Principle and thus also the paradoxical argument.

Summing up, I have argued that both of the pictures that Putnam regards as dual to one another need to be modified. The 'modal' picture fails to deliver a genuine form of nominalism. And the 'object' picture needs to be given a potentialist construal, which will set it apart from a more traditional platonistic conception. Both adjustments represent moves in the direction of the approach inspired by Parsons.

14.7 Modality

Our last topic will be the long deferred question about the sense in which extensions of set theoretic structures are said to be possible.

It is not entirely clear how Putnam understands this modality. At times, he speaks about relying on 'necessity in Quine's narrower sense of logical validity' (p. 11), and at other times, about 'mathematical possibility' (pp. 21 and 22). Hellman is similarly unspecific and writes that 'we shall make limited use of a logic-mathematical modality—a notion of logical possibility—as part of the structuralist language' (p. 15). One thing is clear, however. Putnam and Hellman's notion of necessity is meant to be stricter than the usual post-Kripkean notion of metaphysical necessity. Or equivalently, more situations are meant to be possible with respect to the former modality than with respect to the latter. Or again—helping ourselves to talk about possible worlds, if only for heuristic purposes—Putnam and Hellman's modality is meant to be tied to a larger sphere of possible worlds than the notion of metaphysical modality. Let us accordingly refer to this as the 'more worlds' conception of the modality at work.

Some of Parsons' writings suggest that he too shares the 'more worlds' conception of the modality to be used in mathematics. For instance:

> Among nonepistemic modalities four should be distinguished (roughly in order of stringency of necessity):
>
> (i) Physical or natural
> (ii) Metaphysical or broadly logical
> (iii) Mathematical modality

(iv) Logical modality in the strict sense.

(Parsons 2008, pp. 84 and 86)

This passage suggests a stepwise lifting of constraints imposed on the possibilities that we are willing to countenance. What is naturally possible is what is permitted by the laws of nature and is not otherwise contradictory or incoherent. When we move from natural to metaphysical modality, we lift the requirement that the situations in question be compatible with the laws of nature and instead require only that the situations be permitted by the natures or essences of objects and their properties and relations.[22] Next, when we proceed from metaphysical to mathematical modality, we suspend constraints of a metaphysical nature and count a situation as possible provided that it is compatible with all the laws of mathematics. What about the final notion of 'logical modality in the strict sense'? This is fairly quickly set aside by Parsons, who finds it to be 'either [...] an awkward notion generally or not in the end to differ from mathematical modality' (Parsons 2008, p. 91).

This conception of mathematical modality may well work for the purposes of Parsons' discussion of structuralism (which is the context in which the quoted passages occur). But it is unclear whether it will deliver what he needs for reasoning about the modal aspects of sets. One problem concerns the possible non-existence of pure sets. For example, Parsons needs the possibility of the empty set existing without its singleton; after all, the former is 'formed before' the latter. But the existence of $\{\varnothing\}$ appears to be necessary not only metaphysically but also mathematically. For presumably the axioms of set theory count as laws of mathematics. And it follows from these laws that the mentioned set exists. Its existence should therefore be a matter of mathematical necessity. Parsons is aware of the problem and writes that

> it is not evident that mathematical modality is a unitary notion. [...] I attributed to Putnam a notion of mathematical possibility that allows it to be mathematically possible that there should be no sets of uncountable rank [...]. I consider that the natural notion for many purposes. (Parsons 2008, p. 92)

There is a second problem as well, which threatens not only Parsons' approach but also Putnam's. Both approaches make modal claims about the possible existence of sets (or of other objects playing the structural role of sets). What if these modal claims are true but all the possibilities that witness their truth have undesirable properties? For instance, what if it is possible for there to be the set $\{\varnothing\}$ (or an object playing its structural role)—but only if some substantive non-mathematical assumption is met, such as the existence of Socrates? Then $\{\varnothing\}$ (or any object playing its structural role) would ontologically depend not just on its sole element—as intended—but also on Socrates! There is an analogous argument from applied impure set theory. The use of modality to analyze impure set theory requires that we not have this kind of perverse interaction between the possibility of adding mathematical objects and the non-mathematical circumstances. For instance, assume we wanted to add uncountable sets, say in order to do real analysis à la Dedekind, but that this is possible only

[22] Cf. Fine (1994).

by changing the non-mathematical circumstances. Then we would be prevented from using the new sets introduced to reason about the non-mathematical circumstances as they in fact are. To solve this cluster of problems, we need the possibility of changing the mathematical ontology (or of adding realizations of relevant mathematical structures) while keeping all the (other) non-mathematical facts fixed. But at least as it stands, the 'more worlds' conception of the mathematical modality makes no provision for such carefully controlled changes.[23]

The two problems give us reason to explore a different understanding of the modality that is used in the explication of set theory. On the approach pursued so far, we have considered expanding the sphere of possible worlds by lifting constraints on what counts as possible. This yields different *circumstantial* modalities, which keep the interpretation of the language fixed but allow the circumstances to vary. Another option is to consider *interpretational* modalities, which keep the circumstances fixed but allow the interpretation of the language (and indirectly also the mathematical ontology) to vary. The most well-known example of an interpretational modality is the one used in the standard Tarskian definition of logical consequence. Of course, this particular modality cannot serve our purposes, as the modal operators would otherwise be allowed to change the interpretation of $\in$, which would wreak havoc to modal set theory. But perhaps we can do better.

At times, Parsons too seems to have in mind an interpretational understanding of his modal operators. Recall his metaphysical claims that a multiplicity of objects that exist together *can*—but *need not*—constitute a set, and that the existence of a set implies that of all its elements. As soon as he has made these claims, he provides the following gloss.

> The same idea [as the modal claims in question] would be expressed in semantic terms by the supposition that we can use quantifiers and predicates in such a way that the range of the quantifiers and the objects satisfying any one of the predicates can constitute single objects, but these objects are not already captured by our discourse. (Parsons 1977, p. 294)

This suggests an interpretational understanding of the mathematical modality.[24] If so, this prompts the big question of how an interpretational modality can have the metaphysical consequences that are here attributed to it. As already indicated, the familiar Tarskian brand of interpretational modality certainly has no such consequence.

Putnam makes a suggestion that I believe points in the right direction, namely that mathematical objects are a kind of reified possibilities.[25] He suggests, echoing John Stuart Mill, that sets be regarded as 'permanent possibilities of selection' (Putnam 1967, p. 12). The set of some things is just the reified possibility of collecting precisely

---

[23]Might this conception nevertheless be *compatible* with holding fixed the non-mathematical circumstances? I cannot rule this out. If modal resources could be added that enable such carefully controlled changes, this would yield a version of the 'two separate modal dimensions', to be discussed below, but where both dimensions concern the circumstances, not the interpretation of the language.

[24]The same impression is given by Parsons (1974), where it is clear that what shifts are interpretations, not the circumstances.

[25]Similar ideas are found in Kitcher (1983, ch. 6).

these things. Importantly, the collecting in question need not be a physical process but can be the semantic act of singling out the relevant things and allowing ourselves to talk about their set.[26] This conception sits well with another idea, which is central to the Parsons tradition, namely that the existence of a set is potential relative to that of its members. *First* we must have the relevant things. *Then* we can start to talk about their set. This conception also yields a tolerably clear sense in which a set demands nothing more of reality than the existence of its members. If the members of the would-be set exist, then the possibility of collecting them is already there. Nothing additional is required.[27]

I would like to end by observing how an interpretational understanding of the modality used in the explication of set theory solves the two problems noted earlier in this section. The key is to observe that on this understanding, we have two separate 'modal dimensions' that can be varied more or less independently of one another: the non-mathematical circumstances and the interpretation of the quantifiers and mathematical vocabulary.[28] A geometrical analogy may help. Think of the circumstances as laid out along a horizontal axis, and the interpretations, along a vertical one. Each dimension can be varied while keeping the other one fixed. We thus have a 'horizontal' move, which holds fixed the interpretation (and thus also the purely mathematical ontology), while varying the circumstances. We also have a 'vertical' move, which holds fixed the circumstances, while varying the interpretation (and thus also the mathematical ontology). Let $\Box$ and $\blacksquare$ represent the notions of necessity associated with these two moves, respectively. (Until now, we have represented both of these notions by $\Box$.) Then, for any non-mathematical formula ϕ, we have $\phi \rightarrow \blacksquare\phi$. And for any purely mathematical formula ϕ, we have $\phi \rightarrow \Box\phi$.

On this conception, there is no problem about the non-existence of pure mathematical objects. All such objects can be taken to exist of metaphysical necessity. When their non-existence is said to be possible, the relevant modality is the interpretational one, not the metaphysical modality or any of its circumstantial cousins. This solves the first problem. The second problem too is solved. For the two-dimensional modal structure permits precisely the kind of 'vertical' move that our analysis of mathematics requires, which keeps the non mathematical circumstances fixed but varies the interpretation and with that, also the mathematical ontology.

I wish to close by returning to the relation between mathematical matters and non-mathematical ones, which has been a main theme of this article. We have seen that the approaches to modal set theory inspired by Putnam and Parsons have deep

---

[26]Cf. Hellman (1989, Ch. 2), which also suggests semantic or linguistic means of selecting to 'form sets'. Allowing this kind of 'set formation' represents no danger of paradox provided we allow the domain to expand. Consider for instance the objects rr that are all and only the non-self-membered objects. When we introduce their set $r = \{rr\}$, the domain expands. Paradox would follow only if we insisted—misguidedly—that r be in the original domain. See Linnebo (2010) for a detailed analysis.

[27]Of course, more needs to be said. I say some of it in Linnebo (2012b) and more in Linnebo (2018).

[28]In fact, there will be some connections between the 'dimensions', as an interpretation will depend on the objects in terms of which it is defined. But this complication does not matter for present purposes.

and interesting features in common. However, I have argued that the latter approach consistently has the upper hand because of its willingness to accept an ontology of mathematical objects, which can be cleanly separated from the non-mathematical objects. Our discussion in this section suggests that a similar separation will be beneficial when it comes to the modalities that are invoked. Such a separation is effected by our broadly two-dimensional conception of the relevant modalities.[29]

References

Benacerraf, P., & Putnam, H. (Eds.). (1983). *Philosophy of mathematics: Selected readings* (2nd ed.). Cambridge: Cambridge University Press.

Boolos, G. (1984). To be is to be a value of a variable (or to be some values of some variables). *Journal of Philosophy*, *81*(8), 430–449. Reprinted in (Boolos, 1998).

Boolos, G. (1998). *Logic, logic, and logic*. Cambridge: Harvard University Press.

Ewald, W. (1996). *From Kant to Hilbert: A source book in the foundations of mathematics* (Vol. 2). Oxford: Oxford University Press.

Fine, K. (1994). Essence and modality. In J. Tomberlin (Ed.), *Language and logic* (Vol. 8, pp. 1–16)., Philosophical perspectives Ridgeview: Atascadero.

Fine, K. (2005). Our knowledge of mathematical objects. In T. S. Gendler & J. Hawthorne (Eds.), *Oxford studies in epistemology* (Vol. 1, pp. 89–109). Oxford: Oxford University Press.

Fine, K. (2006). Relatively unrestricted quantification. In A. Rayo & G. Uzquiano (Eds.), *Absolute generality* (pp. 20–44). Oxford: Oxford University Press.

Heck, R. G, Jr. (2000). Cardinality, counting, and equinumerosity. *Notre Dame Journal of Formal Logic*, *41*(3), 187–209.

Hellman, G. (1989). *Mathematics without numbers*. Oxford: Clarendon.

Hellman, G. (1996). Structuralism without structures. *Philosophia Mathematica*, *4*(2), 100–123.

Hellman, G. (2015). Infinite possibilities and possibilities of infinity. In R. E. Auxier, D. R. Anderson, & L. E. Hahn (Eds.), *The philosophy of Hilary Putnam*. La Salle: Open Court.

Hellman, G. (forthcoming). Reflections on reflection in a multiverse. In E. Reck (Ed.), *Logic, philosophy of mathematics, and their history: Essays in honor of W.W. Tait*. London: College Publications.

Kitcher, P. (1983). *The nature of mathematical knowledge*. Oxford: Oxford University Press.

Linnebo, Ø. (2010). Pluralities and sets. *Journal of Philosophy*, *107*(3), 144–164.

Linnebo, Ø. (2012a). Plural quantification. *Stanford Encyclopedia of philosophy*. http://plato.stanford.edu/archives/fall2012/entries/plural-quant/.

Linnebo, Ø. (2012b). Reference by abstraction. *Proceedings of the Aristotelian Society*, *112*(1pt1), 45–71.

Linnebo, Ø. (2013). The potential hierarchy of sets. *Review of Symbolic Logic*, *6*(2), 205–228.

Linnebo, Ø. (2018). *Thin objects: An abstractionist account*. Oxford: Oxford University Press.

Parsons, C. (1974). The liar paradox. *Journal of Philosophical Logic*, *3*(4), 381–412. Reprinted in Parsons (1983).

[29]I am grateful to Peter Fritz, Jon Litland, Stewart Shapiro, and the participants in a workshop on Modality in Mathematics and the Oslo Mathematical Logic Seminar for helpful discussion of the material presented in this article. Special thanks to Geoffrey Hellman and my former doctoral student Sam Roberts, whose astute comments on earlier versions of the article have prompted to substantial improvements. This article was completed during a period of research leave enabled by the Oslo Center for Advanced Study, whose support I gratefully acknowledge.

Parsons, C. (1977). What is the iterative conception of set? In R. Butts & J. Hintikka (Eds.), *Logic, foundations of mathematics, and computability theory* (pp. 335–367). Dordrecht: Reidel. Reprinted in Benacerraf and Putnam (1983) and Parsons (1983).
Parsons, C. (1983). *Mathematics in philosophy*. Ithaca: Cornell University Press.
Parsons, C. (1990). The structuralist view of mathematical objects. *Synthese*, *84*, 303–346.
Parsons, C. (2008). *Mathematical thought and its objects*. Cambridge: Cambridge University Press.
Putnam, H. (1967). Mathematics without foundations. *Journal of Philosophy*, *LXIV*(1), 5–22. Reprinted in Benacerraf and Putnam (1983) and Putnam (1975).
Putnam, H. (1975). *Mathematics, matter and method*. Cambridge: Cambridge University Press.
Roberts, S. (forthcoming). Modal structuralism and reflection. *Review of Symbolic Logic*. https://doi.org/10.1017/S1755020318000138.
Studd, J. (2013). The iterative conception of set: A (bi-)modal axiomatisation. *Journal of Philosophical Logic*, *42*(5), 697–725.
Studd, J. (2016). Abstraction reconceived. *British Journal of Philosophy of Science*, *67*(2), 579–615.
Weyl, H. (1930). Levels of infinity. In P. Pesic (Ed.), *Levels of infinity: Selected writings on mathematics and philosophy* (pp. 17–31). Mineola: Dover.
Williamson, T. (2013). *Modal logic as metaphysics*. Oxford: Oxford University Press.
Yablo, S. (2006). Circularity and paradox. In T. Bolander, V. Hendricks, & S. A. Pedersen (Eds.), *Self-reference*. Chicago: CSLI Press.
Zermelo, E. (1930). Über Grenzzahlen und Mengenbereiche. *Fundamenta Mathematicae*, *16*, 29–47. Translated in Ewald (1996).

Øystein Linnebo is Professor of Philosophy at the University of Oslo, having previously been a Professor at Birkbeck, University of London, and held positions at Bristol and Oxford. He obtained his PhD in Philosophy from Harvard in 2002 and an MA in Mathematics from Oslo in 1995. Linnebo's main research interests lie in the philosophies of logic and mathematics, metaphysics, and early analytic philosophy (especially Frege). He has published more than 50 scientific articles and is the author of two books, *Philosophy of Mathematics* (Princeton University Press, 2017) and *Thin Objects: An Abstractionist Account* (Oxford University Press, 2018).

Index

G. Hellman and R. T. Cook (eds.), *Hilary Putnam on Logic and Mathematics*,
Outstanding Contributions to Logic 9, https://doi.org/10.1007/978-3-319-96274-0

GPSR Compliance
The European Union's (EU) General Product Safety Regulation (GPSR) is a set of rules that requires consumer products to be safe and our obligations to ensure this.

If you have any concerns about our products, you can contact us on

ProductSafety@springernature.com

In case Publisher is established outside the EU, the EU authorized representative is:

Springer Nature Customer Service Center GmbH
Europaplatz 3
69115 Heidelberg, Germany

www.ingramcontent.com/pod-product-compliance
Ingram Content Group UK Ltd.
Pitfield, Milton Keynes, MK11 3LW, UK
UKHW020248290926
15470UKWH00002B/799
* 9 7 8 3 3 1 9 9 6 2 7 3 3 *